TEACHER'S MANUAL

Introduction to General Science

A CAMBRIDGE WORK-A-TEXT

PATRICIA GRASSO, **Author**

ALFRED E. CHANT, **Consultant**

TABLE OF CONTENTS

ISBN: 0-87065-969-3

Printed in the United States of America 3 4 5 6 7 8 9 0

GLOBE BOOK COMPANY
Englewood Cliffs, New Jersey

INTRODUCTION

As we approach the twenty-first century, humans have already explored the moon and are looking to the planets and stars. Nuclear power operates ships, electrical power plants and rockets. Solar and geothermal energies may one day supply much of the earth's energy needs. Studies of basic life processes have resulted in the development and improvement of preventive medicine and, in some cases, the discovery of cures for certain diseases. These and other accomplishments of modern science are a part of our lives now, and even more discoveries are in progress as you read these pages. An understanding of science and its processes and concepts can no longer be considered necessary only for scientists. People in general need at least a basic understanding of scientific principles if they are to live effective and productive lives.

Introduction to General Science is designed to help prepare students for the everchanging and more complex scientific developments of the future. Perhaps these very same students will play an important role in these developments; perhaps they will be asked simply to have a basic understanding of them. Whatever the future brings, science and scientific achievements will play an ever-increasing role in the life of every individual.

PHILOSOPHY OF THE BOOK

A science course that is merely an encyclopedic collection of facts that students must memorize is meaningless. Students must be given the opportunity to develop skills such as observing, identifying, describing, comparing and classifying. Throughout this book students are encouraged to develop a scientific curiosity about their environment. They are urged to ask what, how, and why—questions that lead to and reinforce critical thinking and problem-solving skills. With the added encouragement of the teacher, students will be further motivated to explore those scientific areas of special interest to them. In so doing, they will have the opportunity to discover that the study of science can be an enriching and rewarding experience.

The importance of reading skills and their relation to problem-solving skills in science is evidenced throughout the book. Important concepts are fully developed, yet the text is written in a clear, straightforward style without the complexities that are so often confusing to the average student. Readability was carefully controlled using the Dale/Chall formula.

The activities in this book were selected based not only on their relationship to the concepts being taught, but according to how appropriate they were for students at this grade level. The activities, therefore, are well within the students' abilities and range of experiences. The objective of these and other features of the book is to give students every possible opportunity to experience success in science.

ORGANIZATION OF THE BOOK

The text is divided into units, and the units are divided into chapters. Each chapter develops several major concepts which, in turn, correspond to instructional objectives.

A SUGGESTED TEACHING APPROACH

The following sections of this guide contain a general description of the various features of the book and offer suggestions for their use. Of course, every teacher has his or her own particular method of instruction depending on educational philosophy, student ability, class size, and so on. In the end, therefore, it is up to the teacher to decide which suggestions are appropriate and workable.

Using the text. The content is based on a review of science curricula and is composed of those concepts or major ideas appropriate to this grade level. The decision to include the instructional objectives as part of the pupil's text is based on the belief that students need a focus or purpose for reading and doing. Knowing what is expected of them makes it easier for students to approach a given task.

The material in the pupil's text and the information in the teacher's annotations provide the necessary background needed to perform the "Discover for Yourself" activities that appear in the text.

Using the "Get Ready To Read" sections. Preceding each unit is a section entitled "Get Ready To Read." This section consists of three pages of a series of student-directed readings and questions that focus on specific comprehension skills.

The purpose of "Get Ready To Read" is to help students develop the kinds of comprehension skills needed to read and understand science materials. It is important for students to realize that once they learn a skill and are able to use it, they should then be able to apply it to a variety of situations. Often students have the necessary comprehension skills but find it difficult to apply them to new and different problems.

The comprehension skills developed in reading correspond to and reinforce the problem-solving skills needed to succeed in science; they are mutually beneficial. The teacher, therefore, may want to expand on the reading activity as student need demands and time permits. You may find, for example, that some students are having difficulty using reading skills such as *comparing and contrasting*, and *ordering and sequencing*. As this will affect their ability to classify, an important problem-solving skill in science, you may want to devote additional time to these comprehension skills. Also, a visit to the library, during which the librarian explains and shows the library's system of classification, would reinforce the concept of grouping. To further enhance the students' reading experiences you may want to refer to the bibliography of children's science books on page TG7 of this guide.

Using "Discover for Yourself." The authors are aware of the need to teach students to read with comprehension. This is evidenced in part by the inclusion of the "Get Ready To Read" sections. However, it is necessary and possible to combine both content *and* process. As stated earlier, process skill development helps to develop and reinforce comprehension skills while motivating students to be doers. (See the table on page TG5 for a list of the activities, skills developed and materials needed.)

"Discover for Yourself" are simple, hands-on activities. They are designed to give students the opportunity to develop science skills such as observing, measuring, comparing, classifying, and inferring. At the same time, they are meant to help reinforce and enrich the concepts developed in the chapter or lesson.

All the activities require simple, easy-to-find materials available locally. If at all possible, begin to collect materials in advance using the Materials List table as a guide. Whenever possible, you may want to involve your students in the process. This will give them a sense of responsibility and participation and will motivate them to do the activity.

Be sure that students read the discovery activities in advance so that they can gather materials needed, and become aware of safety precautions where applicable. **Although the activities are designed for maximum safety, students at this level should never be allowed to perform activities with heat or open flame without the direct supervision of a teacher or other qualified adult.**

Additional activities are in Appendix A of the Teacher's Annotated Edition and can be used to introduce or supplement a unit of study. You may make copies of the activities for your students to do in class or at home. (The answers to these activities are given at the back of this section in the Teacher's Annotated Edition.)

Using the evaluation sections. There are many check points throughout the text and each corresponds to the objective covered in the lesson or chapter.

A brief review of objectives follows each lesson. These review questions may be completed by the student as a self-check and serve to give both student and teacher continuous feedback on the student's progress. In this way, if a student is having difficulty mastering a certain objective, he or she can ask for help immediately.

"Review" at the end of each chapter may be administered as an end-of-chapter test, or may be used informally as a review of the chapter material. The question-types vary so as to give students experience in dealing with different test formats. Students' answers may differ from those provided in the Teacher's Annotated Edition, and any response by a student should be considered on its own merits.

Unit tests are provided in Appendix B of the Teacher's Annotated Edition. The items on each test are keyed to specific objectives in the unit. These unit tests can be copied and distributed to the class or individual students for formal testing. Answers to the test items appear in the appendix also.

Using illustrations and photographs. The numerous illustrations and photographs throughout this series not only help to clarify facts and concepts and thereby aid student comprehension, but are enjoyable to look at also. Each unit is illustrated with drawings and photographs of objects (familiar and unfamiliar) and living things, some of which may not be possible to bring into the classroom. There are charts, tables, and diagrams that are used to supplement and clarify the text. Photographs of students actively engaged in scientific investigations are used to encourage student participation.

Using the Glossary/Index. One of the chief problems students experience in reading science material is reading and understanding the rather difficult technical vocabulary. In this book every attempt was made to keep the reading level at or below the grade level for which the book is intended. At the same time, however, it is important that students become familiar with and understand certain scientific terms.

When a term is introduced for the first time in a lesson, that term appears in boldface type, is defined (usually in context), and is repeated three or more times. The boldface words are defined again in the Glossary/Index and spelled phonetically according to *The American Heritage School Dictionary*. (A pronunciation key can be found at the beginning of the Glossary/Index.)

Using a glossary and an index are important skills that should be developed as soon as a student is ready. Students usually enjoy keeping their own personal "word banks" using information they have looked up in their Glossary/Index.

Using the teacher's annotations. Teachers have selected the annotations as the most valuable part of a teacher's guide. The annotations serve several purposes. They provide teachers with

a. background information and definitions of science terms;
b. unusual or interesting facts and thought-provoking questions to introduce a lesson and motivate class discussion;
c. additional ideas for developing, reinforcing, and supplementing concepts contained in the pupil's text;
d. observations, results, cautions and suggestions related to the "Discover for Yourself" activities;
e. suggestions for activities in addition to those in the pupil's text;
f. the answers to all questions and possible student responses and/or behavior

all of which are printed directly on a full-sized copy of the student's page. This eliminates the need for two books or for flipping pages back and forth.

DISCOVER FOR YOURSELF ACTIVITIES
Introduction to General Science

UNIT/ CHAPTER	ACTIVITY	PROCESS SKILLS DEVELOPED	MATERIALS NEEDED FOR ONE SETUP	PAGE
I/1	Observing microscopic organisms	Comparing, classifying	stagnant pond water (or aquarium water), microscope slide, eyedropper, microscope, coverslip	10
2	Classifying some common animals	Observing, comparing, classifying	reference books on animals	17
	Classifying some common plants	Observing, comparing, classifying	leaves, paper, glue or tape, reference books on plants	20
3	Comparing the seeds of 2 kinds of flowering plants	Observing, comparing	lima bean seeds, corn seeds, dish, warm water	27
4	Demonstrating fossil information	Observing, comparing, stating a conclusion	plaster of paris, water, spoon, coffee can, shoe box, petroleum jelly, leaf, shell or bone	32
II/2	Observing the effects of the enzyme, ptyalin, on starch	Observing, comparing	unsalted crackers, 3 test tubes, Benedict's solution, Bunsen burner, matches, labels	54
3	Examining the blood cells of different animals	Observing, comparing	prepared slides of human blood and frog blood, two microscopes	62
	Demonstrating the effects of exercise on the rate of heartbeat	Measuring, comparing, inferring	paper, pencil, watch or clock with second hand	62
4	Analyzing exhaled air for carbon dioxide content	Observing, inferring	limewater, beaker, drinking straw	68
5	Demonstrating how the iris works	Observing, comparing, inferring	flashlight, clock (or watch)	79
7	Observing bone marrow	Observing, describing, inferring	animal leg bone from a market or butcher shop	89
III/1	Demonstrating a decomposition reaction	Observing, inferring, stating a conclusion	2 test tubes, test tube rack, water, 3.0% solution of hydrogen peroxide, wood splints, matches	113

UNIT/ CHAPTER	ACTIVITY	PROCESS SKILLS DEVELOPED	MATERIALS NEEDED FOR ONE SETUP	PAGE
2	Demonstrating static electricity	Observing, comparing, inferring	a few newspaper sheets, pieces of fur, rubber, wool, cotton, silk, plastic, coins	122
	Constructing a series circuit	Observing, demonstrating, stating a conclusion, constructing a model	insulated copper wire, switch, 2 dry cells, 3 light bulbs in porcelain sockets	127
	Constructing a parallel circuit	Demonstrating, observing, stating a conclusion, constructing a model	same materials as preceding activity	128
3	Demonstrating how light travels	Observing, describing, stating a conclusion	4 pieces of cardboard, hole puncher, 4 blocks of wood, clay, candle, matches	137
	Investigating the effects of transparent, translucent and opaque materials on light	Observing, comparing classifying	flashlight, clear plastic wrap, tissue paper, wax paper, wood, aluminum foil, glass of water, cardboard, pane of glass, glass of milk, frosted glass	137
4	Demonstrating how heat affects contraction and expansion	Observing, comparing, stating a conclusion	steel ring with a long handle, steel ball with a long handle, Bunsen burner, matches	148
	Demonstrating the rates at which different metals conduct heat	Observing, comparing, inferring, stating a conclusion	ring stand, Bunsen burner, matches, iron, aluminum, brass and copper rods (all the same size), wax, thin wire, scissors	151
5	Demonstrating how the pitch of a wind instrument depends on the length of a vibrating column of air	Comparing, inferring, stating a conclusion	3 empty soda bottles (same size), water	159
IV/1	Demonstrating the movement of the earth's crust during an earthquake (see Teacher's annotation)	Constructing a model, observing, inferring	large balloon, cardboard box, soil, modeling clay	182
2	Identifying the physical properties of some common minerals	Observing, comparing, describing	mineral samples (e.g. quartz, mica, talc, pyrite, calcite, feldspar), shiny copper penny, knife, streak plate or unglazed tile	185
	Classifying some common rocks	Observing, comparing, classifying	rocks, labels, guide-book to rocks in your area	191
6	Demonstrating Newton's Third Law of Motion (the principle of rocketry)	Constructing a model, observing, describing, inferring	Heat resistant glass bottle with rubber stopper, thin wire, scissors, Bunsen burner, matches, water, stick	234

BIBLIOGRAPHY OF STUDENTS' BOOKS

Unit I: Introduction to Biology

Amon, Aline. *Roadrunners and Other Cuckoos*. New York: Atheneum, 1978.

Anderson, Lucia. *The Smallest Life Around Us*. New York: Crown Publishers, 1978.

Batten, Mary. *The Tropical Forest: Ants, Ants, Animals & Plants*. New York: Thomas Y. Crowell, 1973.

Callahan, Philip. *Bird Behavior*. New York: Four Winds Press, 1975.

Laycock, George. *Wingspread: A World of Birds*. New York: Four Winds Press, 1972.

———. *Death Valley*. New York: Four Winds Press, 1976.

Leen, Nina. *Snakes*. New York: Holt, Rinehart & Winston, 1978.

Limburg, Peter. *What's in the Name of Wild Animals*, illustrated by Murray Tinkelman. New York: Coward, McCann & Geoghegan, 1977.

Malatesta, Anne and Ronald Friedland. *The White Kikuyu: Louis S. B. Leakey*. New York: McGraw-Hill, 1978.

Patent, Dorothy Hinshaw. *Sizes and Shapes in Nature—What They Mean*. New York: Holiday House, 1979.

Selsam, Millicent. *Play With Plants*. New York: William Morrow & Company, 1978.

Soucie, Anita Holmes. *Plant Fun: Ten Easy Plants to Grow Indoors*, illustrated by Grambs Miller. New York: Four Winds Press, 1974.

Strong, Arline. *Veterinarian, Doctor for Your Pet*. New York: Atheneum, 1977.

Zappler, Lisbeth. *Nature's Oddballs*. New York: Doubleday, 1978.

Unit II: Introduction to Human Biology

Marr, John S. *The Food You Eat*. New York: M. Evans & Co., 1973.

Unit III: Introduction to Physical Science

Bendick, Jeanne. *Why Things Work: A Book about Energy*. New York: Parents' Magazine Press, 1972.

Branley, Franklyn M. *Energy for the Twenty-first Century*. New York: Thomas Y. Crowell, 1975.

Kettelkamp, Larry. *Lasers: The Miracle Light*. New York: William Morrow & Company, 1979.

Lauber, Patricia. *Tapping Earth's Heat*. Champaign: Garrard Publishing Company, 1978.

Lefkowitz, R. J. *Matter All Around You*. New York: Parents' Magazine Press, 1972.

Scott, John M. *What Is Sound?* New York: Parents' Magazine Press, 1973.

Unit IV: Introduction to Earth and Space Science

Cazeau, Charles J. *Earthquakes*. Chicago: Follett, 1975.

Cohen, David. *What's Happening to our Climate?* New York: M. Evans & Company, 1979.

Helfman, Elizabeth. *Our Fragile Earth*. New York: Lothrop, Lee & Shepard, 1972.

Lambert, David. *The Earth and Space*. Warwick Press, 1979.

MacDonald Educational. *Weather*, First Library Series. New York: MacDonald Educational, 1975.

Martin, Alice Fitch and Parker, Bertha Morris. *Rocks and Minerals*. Racine, Wisconsin: Golden Press, 1974.

Moche, Dinah. *Search for Life Beyond Earth*. New York: Franklin Watts, 1978.

Moore, Patrick. *The New Challenge of the Stars*. Skokie: Rand McNally, 1978.

Simon, Seymour. *Danger from Below—Earthquakes: Past, Present, and Future*. New York: Four Winds Press, 1979.

TEACHER'S ANNOTATED EDITION

Introduction to General Science

A CAMBRIDGE WORK-A-TEXT

PATRICIA GRASSO, **Author**

ALFRED E. CHANT, **Consultant**

GLOBE BOOK COMPANY
Englewood Cliffs, New Jersey

AUTHOR BIOGRAPHIES

Sister Patricia Grasso is currently a post-doctoral research fellow in the Department of Biochemistry of the Albany Medical College of Union University in Albany, New York. She recently received a doctorate in anatomy from the Schools of Medicine and Dentistry of Georgetown University in Washington, D.C. She was formerly Chairperson of the Science Department and a teacher of biology at Catholic Central High School in Troy, New York. Sister Grasso has published articles in *The American Zoologist, Endocrinology, Biology of Reproduction, The American Biology Teacher* and *Today's Catholic Teacher.*

Alfred E. Chant is Curriculum Supervisor at Fairfield School in Massapequa, New York and Reading Clinician at Hofstra University. He has given several in-service workshops for elementary science teachers and is the author or co-author of books in the areas of science, reading development and mathematics.

ISBN: 0-87065-968-5 Student's Edition

ISBN: 0-87065-969-3 Teacher's Annotated Edition

Printed in the United States of America 2 3 4 5 6 7 8 9 0

Acknowledgments

Administrative Editor: Joanne E. Kane

Coordinating Editor: Kathleen Mayer

Art and Production Manager: Arthur Michalez

Editors: Eileen Mitchell, Claire Wetter

Cover Photograph: Wilson Goodrich of Tom Stack and Associates

Text photographs:

Unit 1: p. 4 – U.S. Department of Agriculture; p. 5 – YMCA; pp. 8, 23 – Carolina Biological Supply; p. 9 – LaPine Scientific Inc.; pp. 13, 15 *middle left,* 19 – U.S. Department of Agriculture; pp. 15 *top*, 31, 35, 37 *top left, top right, middle right* – American Museum of Natural History; p. 15 *middle right* – Australian Information Service; p. 16 *top left* – Chicago Natural History Museum; p. 16 *middle left* – Miami Seaquarium; p. 16 *right* – New Holland/Sperry; pp. 17, 28 – Walter Schwarz; p. 32 – Les Mahon; p. 37 – National Geographic

Unit 2: p. 44 – Comstock, Inc.; p. 45 – Girl Scouts of America; p. 47 *top left, bottom left,* 54 *top* – Carolina Biological Supply; p. 47 *top center, top right, bottom right,* 61 – Ward's Natural Science Establishment; p. 51 – Les Mahon; pp. 54 *bottom,* 62, 68 – Walter Schwarz; p. 59 – American Red Cross; p. 67 – YMCA; p. 70 – American Cancer Society; p. 73 – Bethlehem Steel Co.; p. 78 – Barton Silverman of *The New York Times;* p. 83 – Mimi Forsythe of Monkmeyer Press Photo Service; p. 85 – Bausch & Lomb; p. 89 – Schwinn Bicycle Co.; p. 97 – National League for Nursing

Unit 3: p. 108 – New York Convention and Visitors Bureau; p. 109 – The Bettman Archive; p. 117 – U.K. Atomic Energy Commission; pp. 121, 165 – Honeywell; pp. 122, 138 – Walter Schwarz; p. 135 – Chicago Association of Commerce & Industry; p. 148 – N.Y. Convention & Visitors' Bureau; p. 152 – U.S. Air Force; p. 155 – *The New York Times;* p. 159 – Les Mahon; p. 161 – Bell Telephone; p. 167 – Department of Environmental Conservation; p. 168 – Research Cottrell; p. 170 – Westinghouse

Get Ready to Read Unit 4: p. 173 – United Nations: p. 174 – Leila Violano

Unit 4: p. 176 – Hal H. Harrison of Monkmeyer Press Photo Service; p. 177 – Norwegian National Tourist Bureau; pp. 183, 185 *top*, 188 *bottom,* 201 *middle right, bottom* – U.S. Geological Survey; p. 185 *bottom* – Walter Schwarz; p. 188 *top* – Santa Fe Railway; p. 189 – Ward's Natural Science Establishment; p. 190 – American Museum of Natural History; p. 194 – National Geographic; p. 197 – United Press International; p. 200 – Swiss National Tourist Office; p. 201 *middle left* – U.S. Geological Survey/EROS Data Center; p. 207 – U.S. Navy Applied Physics Laboratory; pp. 219, 222, 223, 233, 237 – NASA: pp. 225, 226 – Mt. Wilson/Palomar Observatory

TABLE OF CONTENTS

Get Ready To Read UNIT 1

COMPARING AND CONTRASTING CHARACTERISTICS

In this book you will read and learn about many topics relating to you and the world around you. To help you better understand the information in this biology unit of *Introduction To General Science,* you will need to consider the likenesses and differences in living things. You can do this by comparing and contrasting the information that is presented to you in words, pictures, diagrams, and other visual aids in this book.

Read the following story about two sisters. As you read, ask yourself: how are these two girls alike? In what ways are they different? Then do the exercises that follow the story.

Jackie and Judy are identical twins. That is, they look almost exactly alike. Both girls are about 5 feet tall and weigh about 80 pounds. Some people cannot tell them apart at first glance because their curly red hair, blue eyes, and freckles seem to match. But they have differences that you can't see by just looking.

"Step on it," said Jackie. "I'm in a real hurry and I don't want to be late. Can't you drive any faster?"

"I can go faster," said Judy, "but it wouldn't be safe. The speed limit is 55 mph, and that's as fast as I'll go."

"Well," Jackie replied, "this is a very important party. I'm going to be the best dancer there. In fact, I'm planning on winning the dance contest."

"Oh, dances, dances," Judy remarked. "Doesn't music mean anything more to you than just dancing? To me, music should be appreciated all by itself—like the music I play on my violin."

"Well, thank goodness, we're almost there," Jackie said under her breath.

Now, answer these questions.

1. Which girl do you think is a safe driver? Judy
2. List three physical characteristics that are the same about the girls.
   height ______ hair/eye color ______ weight/freckles ______
3. Which sister does the word *funloving* best describe? Jackie
4. The words in the story that refer to two sisters who were born at the same time and look exactly alike are identical twins ______.
5. Which girl probably is a music student? Judy

Point out that some of the characteristics may be descriptive of just one girl, some of both girls, and some you can only guess at from the story. (e.g. energetic, self-confident)

6. Match each word below to the girl it describes. If the word describes both girls, draw a line to both faces.

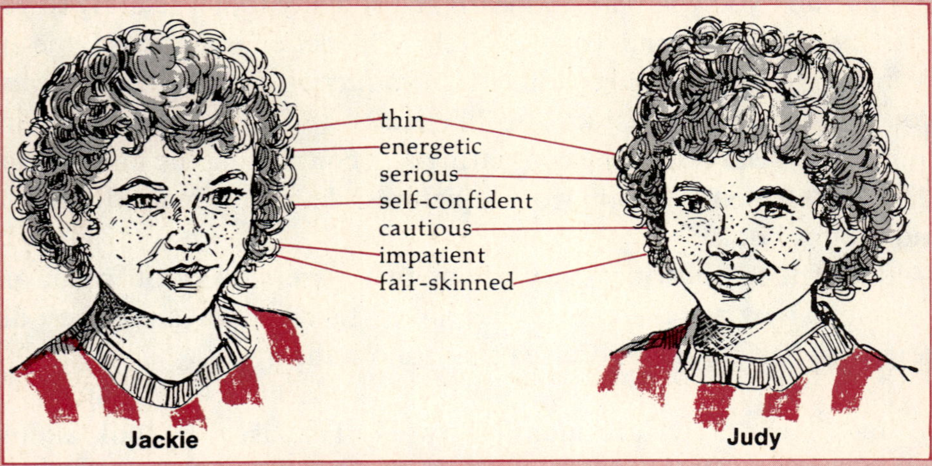

HOW ARE THINGS BOTH ALIKE AND DIFFERENT?

Study the information from the labels of these two milk products. The labels show the amount of nutrition in each serving.

Yummy Yogurt A

Percentage of U.S. recommended daily allowance (USRDA) *contains less than 2% of USRDA of these nutrients

Nutrient	%
Protein	30
Vitamin A	2
Vitamin C	*
Thiamin	4
Riboflavin	30
Niacin	*
Calcium	40
Iron	*
Vitamin B_{12}	20
Phosphorus	35

Creamy Cream Cheese B

Percentage of U.S. recommended daily allowance (USRDA) *contains less than 2% of the USRDA of these nutrients

Nutrient	%
Protein	4
Vitamin A	4
Vitamin C	*
Thiamin	*
Riboflavin	2
Niacin	*
Calcium	2
Iron	*

Both foods contain less than 2% of Vitamin C, Niacin, and Iron.

Which food contains more protein per serving? yummy yogurt

Vitamin B_{12} is *not* contained in creamy cream cheese.

Which food is higher in nutrients? yummy yogurt

As you read about plant and animal life in Unit 1, try to use the skills of comparing and contrasting characteristics. You will find that you have a broader understanding of the subject.

Ask students to bring in labels from other food products, particularly those they eat themselves. Have them make comparisons.

CLASSIFYING FROM GENERAL TO SPECIFIC

Just as characteristics of things can be compared and contrasted, characteristics of things can be used as a reason for putting them in groups. Grouping things according to a plan is called classifying. Classifying helps us to organize information so that it can be understood more easily.

HOW CAN THINGS BE GROUPED?

Here are some grouping plans in which things can be classified:
1. by size 2. in alphabetical order 3. by year 4. by shape
5. by purpose (what things are used for) 6. by color

Some things can be grouped, or classified, in more than one way. Put the numeral or numerals of the grouping plan(s) next to the things that need to be grouped.

coins in a collection __1, 3__ record albums __2,__

words in a dictionary __2__ sporting goods __1, 4, 5__

paints in a paint store __6__ dishes in a china cabinet __5__

a child's building blocks __1__ shoes in a shoe store __1__

You may want students to group other things to give them practice. Some examples would include their books, school supplies (paper, pens, pencils, compass, etc.)

Every time you go shopping you make use of a system of classifying things. To find what you need, you look under the proper heading in the store's directory.

Look at this directory from Reid's Department Store. Then answer the questions that follow.

1. Jean wants to buy her little brother a toy train. On which floor would she find it? __2nd floor__
2. Where would Bill find paper plates, napkins, and tablecloths for his party? __main floor__
3. Dad needs a new pair of shoes. On which floor should he look for them? __main floor__
4. Name some things you would expect to find in the Furniture and rug department. In household goods. Answers will vary.

Main Floor
Cosmetics
Party Goods
Men's Clothing
Men's Footwear

2nd Floor
Women's Sportswear
Women's Dresses
Children's World
Women's Shoes

3rd Floor
Furniture, Rugs
Household Goods
China & Glassware

As you read *Introduction To General Science* you will see that the book is organized by units that are the main topics of the book. The units are then divided into *chapters* that are an organization of the facts and concepts studied in the unit. Within each chapter, the text is divided again into *subheadings*. The subheadings are printed in boldface type and tell you just what you will be learning about in the text that follows.

UNIT 1 Introduction to Biology

What life activities are these basketball players carrying on?

They are carrying on the life activities of movement, respiration and sensitivity.

New science words: nucleus, organic, sensitivity

Skills developed: comparing, classifying

Chapter 1

WHAT ARE LIVING THINGS MADE OF?

Take a moment to think about where you live. Is it in the middle of a big city, in a quiet suburb, or on a farm? Wherever your home is, you will find many living and nonliving things. Everything on planet Earth can be placed into one of these two groups. All living things carry on certain activities that separate them from nonliving things.

A. Living and Nonliving Things

After doing this lesson you will be able:

A1 • to list five activities that make living things different from nonliving things

A2 • to tell the difference between organic and inorganic substances

A3 • to find out what protoplasm is and describe it

All living things can carry on certain activities. These activities, called **life activities,**
A-1 include (1) movement, (2) nutrition, (3) respiration, (4) sensitivity and (5) reproduction. Nonliving things cannot carry on these life activities.

1. Movement. *All living things are able to carry on some kind of movement.* Animals can move from place to place. Plants bend their stems and turn their leaves to receive the sunlight. A nonliving thing cannot move by itself.

2. Nutrition. *All living things use food for growth and for release of energy.* A nonliving thing does not use food. If it becomes larger, it is not because it has used food. A snowball becomes larger if it is made larger.

3. Respiration. *Most living things need oxygen.* Oxygen is used to burn food and release energy. The activity by which living things take in and use oxygen to let out energy is called respiration.

Anaerobic bacteria do not use oxygen.

4. Sensitivity. *All living things react to changes in their environment.* Light, sound, heat and chemicals are just some of the things that can change in the environment. Living things can react to these changes. Things that are not living do not react to changes.

Changes in the environment of an organism, whether external or internal, are called *stimuli.*

5. Reproduction. *All living things reproduce.* Animals produce (make) other animals like themselves. Plants produce other plants like themselves. If living things could not reproduce, life would disappear from Earth. Nonliving things do not produce more of their own kind.

Only living things carry on life activities. Things that are not living do not carry on any life activities.

Scientists classify all matter as **organic** or **inorganic.** All living things are made up of

Use a growing seedling, such as a bean plant, to demonstrate that plants, as well as animals move. Also, many time-lapse motion pictures are available that show plant movement.

A2 organic substances. Organic substances have carbon. Substances that do not have carbon are inorganic. For example, coal and oil are organic, while metals are inorganic.

All living things are made up of a sub-
A3 stance called **protoplasm.** What is protoplasm? *Protoplasm is the basic living material.* It is always made of carbon, oxygen, hydrogen, nitrogen and very often sulfur and phosphorus.

Protoplasm also has a very large amount of water in it. Protoplasm is usually colorless.
A3 It feels and looks like jelly. Only living things can make new protoplasm or repair damaged protoplasm.

Since you are a living thing, you must be made of protoplasm. Yet, you don't look or feel like jelly. You have a definite shape. Most plants and animals aren't colorless and jellylike. They, too, have a shape. Where, then, is the protoplasm?

Check Up A

A1. Name five life activities.

(1) Movement

(2) Nutrition

(3) Respiration

(4) Sensitivity

(5) Reproduction

A2. What is the difference between organic and inorganic substances? Organic substances contain carbon; inorganic substances do not.

A3. What is protoplasm? What does it look and feel like? Protoplasm is the basic living material. It is colorless and looks and feels like jelly.

B. Cells

After doing this lesson you will be able:

- B1 • to tell what the building blocks of all living things are called
- B2 • to name the person who invented one of the first microscopes
- B3 • to name the person who first saw part of a cell
- B4 • to tell the functions of the
  - (1) nucleus
  - (2) cytoplasm
  - (3) cell membrane
  - (4) vacuole
  - (5) cell wall
  - (6) chloroplast
- B5 • to list four parts found in both plant and animal cells
- B6 • to list two things found only in plant cells

The protoplasm of all living things is found
in **cells.** Cells are *the building blocks of all* B1
living things. Some living things are made of only one cell. Others are made up of millions of cells. Most cells are so small that they can be seen only with the aid of a microscope. Others can be seen by the unaided eye. It is the cell that carries on all the life activities of a living thing. What does a cell look like?

A Dutch lens maker, Anton van Leeu- B2
wenhoek (1632–1723), invented one of the first microscopes. The microscope made it possible to see very small things. Later
Robert Hooke (1635–1703), an English sci- B3
entist, improved the microscope. He thought he saw the first cell. Actually, he saw only

How does the structure of a green plant cell differ from that of an animal cell?

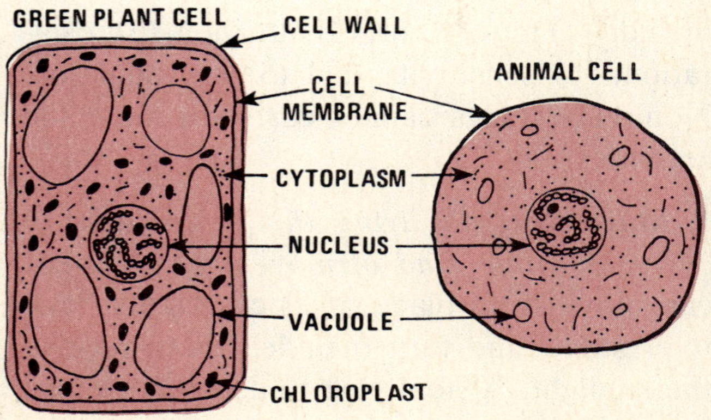

Plant cells can be distinguished from animal cells by the presence of cell walls and chloroplasts.

A few kinds of cells, such as very simple algae and bacteria, do not have a well-defined nucleus. Human red blood cells are also enucleate.

one part of the cell—the cell wall. (You will learn more about the cell wall later in this chapter.)

Scientists using powerful microscopes are able to observe what makes up cells. They have discovered that most cells are made up of two kinds of protoplasm: (1) the nucleus and (2) cytoplasm.

1. The **nucleus** is usually found in the center of the cell. It is thick and usually has a round
B4(1) shape. The nucleus controls most of the activities of the cell. It also plays a big part in the reproduction of the cell.

2. **Cytoplasm** is the protoplasm that is outside of the nucleus. The cytoplasm carries
B4(2) out all the life activities of the cell *except* cell reproduction.

All living things are made up of cells. All cells are made up of protoplasm. Most cells also have cytoplasm and a nucleus. If you looked at a plant cell and an animal cell under a microscope, you would see how these cells are alike and how they are different. Look at the pictures of plant and animal cells on page 6.

All cells have a **cell membrane.** This is the outermost living layer of all cells. The cell membrane is made up of living proto-
B4(3) plasm. It is very thin and gives a cell its shape. It allows oxygen, carbon dioxide, water, food materials and other substances to enter the cell. It also allows wastes to leave the cell. It also keeps out harmful materials.

Both plant and animal cells have **vacuoles.** Vacuoles are fluid-filled spaces in a cell.
B4(4) They serve many purposes in the cell. Some vacuoles hold food. In some simple forms of living things, the food is broken down in special vacuoles so it can be used by the cell. Some vacuoles hold the waste materials and water that are made by the cell.

B5 Both plant and animal cells have a nucleus, cytoplasm, a cell membrane and vacuoles. How do they differ? Plant cells have parts that animal cells do not have. One part
is the **cell wall** which is outside the cell B6
membrane of a plant cell. The cell wall is not made up of living protoplasm. It is made of **cellulose.** Cellulose is a nonliving part of the plant cell. The cell wall increases in size
as a plant gets older. The wall gives the plant B4(5)
extra support and strength.

Other parts found in green plant cells, B6
but not in animal cells, are **chloroplasts.**
Chloroplasts are very small and contain green B4(6)
coloring matter called **chlorophyll.** Chlorophyll is needed by the green plant so it can make its own food. Animals cannot make their own food. Most animal food comes from green plants.

Chlorophyll is the most common matter in nature that can trap and store energy from the sun.

Check Up B

B1. What are the building blocks of all living things? Cells

B2, 3. Who invented one of the first microscopes? Who improved it and first saw a part of a cell?

Anton van Leeuwenhoek

Robert Hooke

B4. Write a function of each of the following cell parts:

(1) Nucleus Controls most cell activities; plays a part in reproduction.

(2) Cytoplasm Carries out all the life activities of the cell except reproduction.

(3) Cell membrane Allows oxygen, carbon dioxide, water, food materials to enter and wastes to leave the cell.

The cell wall of the plant cell is composed of cellulose. Chemists use cellulose to make paper, cellophane and explosives. (Raw cotton is 91% cellulose.)

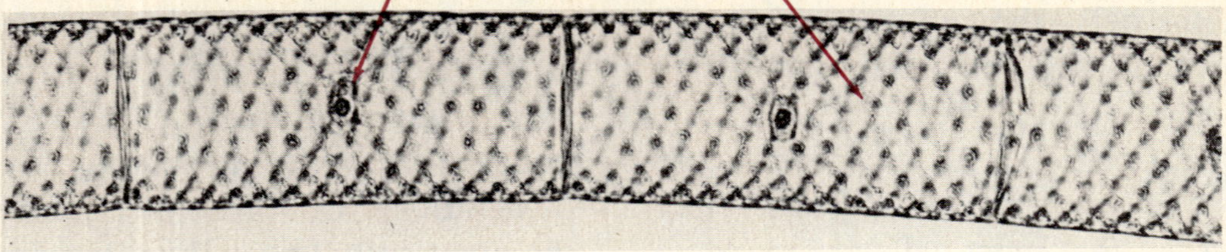

The spirogyra is usually found in long strands of many spirogyra cells living together end to end. Can you identify the nucleus and the spiral chloroplasts?

Photo courtesy of Carolina Biological Supply Co.

Spirogyra are often seen as green stringy scum on ponds and streams. Use a microscope to examine some pond scum. (Magnification of about 200X is best.) Students should be able to see the spiral chloroplasts in the cells.

(4) Vacuole Some hold and break down food; others hold waste and water.

(5) Cell wall Provides support and strength for plant cells.

(6) Chloroplast Contains chlorophyll needed by plants to manufacture food.

B5. List four parts found in both plant and animal cells.

(1) Nucleus

(2) Cytoplasm

(3) Cell membrane

(4) Vacuoles

B6. What two parts are found only in plant cells?

(1) Cell wall

(2) Chloroplasts

C. One-Celled Organisms

After doing this lesson you will be able:

- C1 • to name a one-celled, plantlike organism and the process by which it makes its food
- C2 • to name two one-celled animal-like organisms
- C3 • to find out how the amoeba gets food, digests food, gets rid of wastes and reproduces
- C4 • to find out how the paramecium gets food, digests food, gets rid of wastes and reproduces

The **spirogyra** is a kind of green, fresh- C1
water algae. Algae are one-celled, plantlike organisms. Each cell of the spirogyra has one or more spiral-shaped chloroplasts. The chloroplasts hold chlorophyll which allows a green plant to carry on **photosynthesis.**
Photosynthesis is the process by which green C1
plants make their own food. The spirogyra can make its own food so it does not depend on any living thing.

The cytoplasm of the spirogyra is always moving within the cell. This movement helps carry food, oxygen and other substances to all parts of the cell. The cell membrane controls the entrance of carbon dioxide and water into the cell. It also controls the passage of oxygen and other wastes out of the cell.

When the one-celled spirogyra reaches a certain size, the nucleus and the cytoplasm divide equally to produce two new one-celled spirogyra. This division is called **reproduction.** Each new cell formed by this process has a nucleus, cytoplasm, a cell wall, chloroplasts and a cell membrane. This is one way the spirogyra produces more of its own kind.

Two simple one-celled animallike organisms that can be seen under a microscope C2
are the ameba and the paramecium.

The **ameba** lives in water. It cannot make its own food. It gets its food by feeding on the many small plants and animals that live in the water.

The cell membrane is very important to the ameba. It allows certain substances to enter and leave the cell. Oxygen enters the cell for respiration. **Respiration** is the process

The amoeba is often called the "ghost" of the one-celled world because of the way it changes its shape. White blood cells are very similar to amoebas. They surround and destroy bacteria and other foreign substances in the body.

The amoeba moves very slowly by means of streaming cytoplasm. This streaming movement is called cyclosis.

by which oxygen is taken in and combined with food for the release of energy. The waste products, carbon dioxide and water, are given off. The amoeba gets oxygen it needs from the water in which it lives.

C3 The amoeba gets its food by surrounding it. Once the food is inside the amoeba, part of the cell membrane becomes the lining of a
C3 **food vacuole.** Food is digested here and put into a form the amoeba can use. It then passes into the cytoplasm where it is carried to all parts of the cell. The digested food is then used in the release of energy or in the making of new protoplasm.

Absorption is the process by which digested food can pass into different parts of an animal or plant. After digestion, the solid waste materials leave the cell at any point in the cell membrane.

Liquid waste materials are collected in a
C3 **contractile vacuole.** When the contractile vacuole is filled, it squeezes shut and pushes the liquid wastes out of the amoeba through the cell membrane.

When the amoeba reaches a certain size, the nucleus and cytoplasm divide in half and
C3 form two new amoeba. Each new amoeba has a nucleus, cytoplasm and a cell membrane. After a short while, these two new amoeba reach a certain size and divide again. Each will produce two more amoeba and so on.

The **paramecium** has a definite shape. It has tiny hairlike parts called **cilia** that line the outside of the cell membrane. The paramecium uses cilia to move around. To get food, the cilia beat the water like the oars
C4 of a boat. This movement creates a current which causes food to be swept into a mouth-like structure called the **oral groove.**

The paramecium, like the amoeba, must get its food from the water in which it lives. The food passes from the oral groove through a tubelike structure called the **gullet** and then
C4 into a food vacuole. Here, it is digested by chemicals from the cytoplasm.

The projections of the cytoplasm by which the amoeba surrounds its food are called "pseudopods" (false feet).

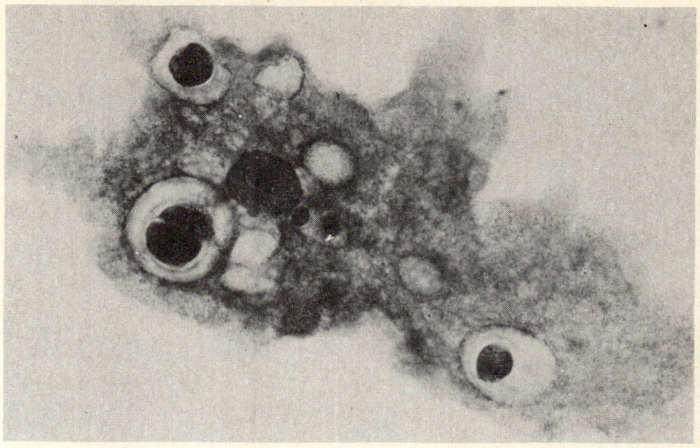

Note the food vacuoles in this amoeba proteus. Food is digested here and made into a form the amoeba can use.

A paramecium has two contractile vacuoles to hold liquid wastes. The wastes are pushed out of the paramecium in the same way as the amoeba. However, solid wastes do C4
not leave directly through the cell membrane. They leave through an opening in the membrane called the **anal spot.**

A paramecium has two nuclei (plural for nucleus). Very few cells have two nuclei. The smaller nucleus is the **micronucleus** (**micro** means "small"). It controls reproduction. The larger nucleus is the **macronucleus** (**macro** means "large"). It controls all the other activities of the paramecium.

A paramecium reproduces in the same way as the amoeba. The nuclei and the cyto- C4
plasm split and form two new paramecia (plural for paramecium).

The paramecium.

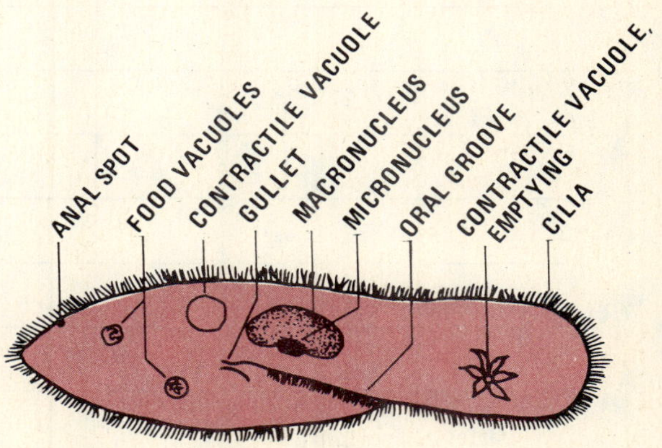

Students should realize that these simple one-celled plantlike and animallike organisms perform the same life activities that the complex higher plants and animals do. (Movement, nutrition, respiration, sensitivity and reproduction.)

Check Up C

C1. Name a one-celled plantlike organism. How does it make its own food?

Spirogyra. It makes its own food by the process of photosynthesis.

C2. Name two one-celled animal-like organisms.

(1) Amoeba

(2) Paramecium

C3. How does the amoeba (1) get food, (2) digest food, (3) get rid of waste products and (4) reproduce?

(1) Surrounds it.

(2) In a food vacuole.

(3) Collected in a contractile vacuole and forced out through the cell membrane.

(4) By dividing in half.

C4. How does the paramecium (1) get food, (2) digest food, (3) get rid of wastes and (4) reproduce?

(1) Movement of cilia creates a current which causes food to be swept into the oral groove.

(2) In a food vacuole.

(3) Liquid wastes – contractile vacuole and forced out through cell membrane. Solid wastes – anal spot.

(4) By dividing in half.

Discover for yourself
how you could observe organisms living in a drop of water.

PURPOSE: To identify microorganisms • To develop the skills of comparison and classification

You need: stagnant pond water (or aquarium water), eyedropper, microscope slide, microscope, coverslip

a. Place a drop of pond water on a microscope slide. Put a coverslip over the drop of water. Place the slide under a microscope and look at the drop of water under low power.

1. Describe any plant cells you see. Answers will vary.

2. Describe any animal cells you see. Answers will vary.

3. What cell parts can you see? Answers will vary.

b. Look at the same drop of water under the high power of the microscope.

What cell parts can you see now that you could not see before? Answers will vary.

Students will probably see organisms other than amoeba, paramecium and spirogyra. Use reference books from the library or a high school biology textbook to help with identification.

Unit 1, Chapter 1
REVIEW

Matching Exercise Match each term in **Column A** with its description in **Column B**. Write the letter of the description in the space beside the correct term.

	Column A	*Column B*
f	**1.** cell	**a.** does not have carbon
a	**2.** inorganic	**b.** holds chlorophyll
j	**3.** nucleus	**c.** fluid-filled space in a cell
c	**4.** vacuole	**d.** lets oxygen and other substances enter a cell and waste products pass out of the cell
d	**5.** cell membrane	**e.** carries out all life activities except cell reproduction
h	**6.** protoplasm	**f.** building block of all living things
e	**7.** cytoplasm	**g.** reacting to changes in surroundings
b	**8.** chloroplast	**h.** colorless, jellylike substance that is the basic living material
i	**9.** nutrition	**i.** using food for growth and energy
g	**10.** sensitivity	**j.** thick, round structure usually found at the center of a cell

True/False In the space provided, write **T** if a statement is true. If it is false, change the underlined word or words to make the statement true.

membrane **1.** Both plant cells and animal cells have a cell wall.

cytoplasm **2.** The nucleus of a cell carries out all of the life activities of the cell except for reproduction.

T **3.** Vacuoles hold the waste materials and water that are made by a cell.

plant **4.** Chloroplasts are found only in animal cells.

ameba **5.** Both the spyrogyra and the paramecium reproduce by dividing equally to produce two new cells.

paramecium **6.** The ameba has cilia that line the outside of the cell membrane.

T **7.** All living things are made up of organic substances.

inorganic **8.** Substances that do not have carbon are organic.

cell membrane **9.** The cell wall is the outermost living layer of all cells.

cellulose **10.** The cell wall is made up of living protoplasm.

T **11.** The response of living things to changes in their environment is sensitivity.

green plants **12.** Photosynthesis is the process by which plants and animals make their own food.

Cells ______ **13.** <u>Protoplasms</u> are the building blocks of all living things.

Absorption ______ **14.** <u>Nutrition</u> is the process by which digested food can pass into different parts of a plant or animal.

micronucleus ______ **15.** The <u>macronucleus</u> of the paramecium controls reproduction.

Multiple Choice Exercise Circle the letter of the word or phrase that <u>best</u> completes each statement.

1. The nonliving material found in the cell wall of a plant is **(a)** cytoplasm **(b)** protoplasm **(c)** cellulose **(d)** photosynthesis.
2. The process by which digested food passes to the different parts of a plant or animal is **(a)** respiration **(b)** digestion **(c)** oxidation **(d)** absorption.
3. All organic substances contain **(a)** protoplasm **(b)** carbon **(c)** oxygen **(d)** cytoplasm.
4. The building blocks of all living things are **(a)** photosynthesis **(b)** membranes **(c)** cells **(d)** nuclei.
5. Cell reproduction is controlled by the **(a)** cell membrane **(b)** nucleus **(c)** vacuole **(d)** cellulose.
6. All of the following are found in both plant and animal cells, except **(a)** chloroplasts **(b)** cytoplasm **(c)** cell membranes **(d)** vacuoles.
7. The process by which green plants make their own food is **(a)** photosynthesis **(b)** absorption **(c)** oxidation **(d)** digestion.
8. The outermost living layer of all cells is the **(a)** nucleus **(b)** cell wall **(c)** vacuole **(d)** cell membrane.
9. In the paramecium, reproduction is controlled by the **(a)** macronucleus **(b)** micronucleus **(c)** oral groove **(d)** cilia.
10. The process by which oxygen is taken into a cell and combined with food is **(a)** photosynthesis **(b)** absorption **(c)** oxidation **(d)** digestion.

Application Exercise Next to each arrow, write the name of the cell part that is shown. Then, on the line below each diagram, label the cell as a plant cell or an animal cell.

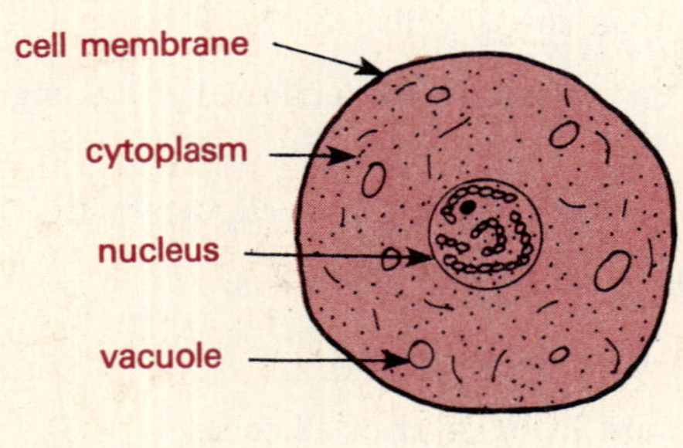

Animal Cell

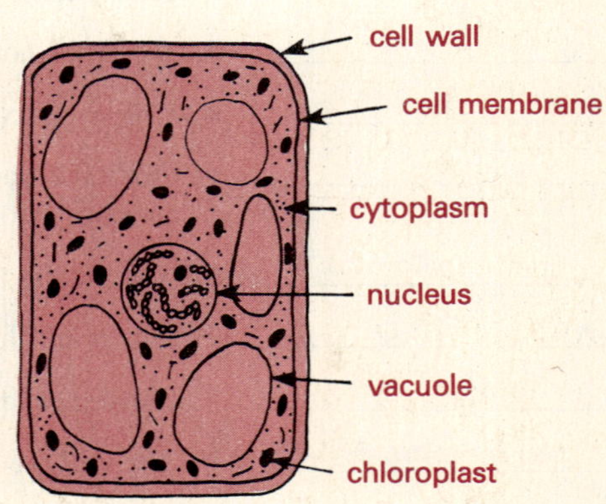

Plant Cell

Mushrooms belong to the fungus family. Fungi are non-green plants. The main portion of the mushroom grows below the ground. The umbrella-shaped parts you see here are the reproductive parts of the plant. Wild mushrooms should not be eaten because they may be poisonous.

New science words: phylum, vertebrate, parasite

Skills developed: comparing, classifying

Chapter 2
HOW LIVING THINGS ARE CLASSIFIED

You already know about one way things on earth are grouped: living and nonliving things. As you look around, you see different kinds of living things. Some living things look alike and some do not look alike. Scientists realized long ago that to make it easy to study living things, they needed a system for grouping them.

Living things are placed into four major groups called **kingdoms:** the **protist kingdom,** the **moneran kingdom,** the **plant kingdom** and the **animal kingdom.** Each kingdom is divided into smaller groups called **phyla** (singular, phylum). The phyla are divided still further.

How do **biologists** (scientists who study living things) decide to which of the divisions a living thing belongs? They are grouped according to their likenesses. This method of grouping makes it easier for scientists to study the different forms of life. This grouping of living things by likenesses is called **classification.**

A. The Protist Kingdom

After doing this lesson you will be able:

A • to list the phyla that make up the protist kingdom

In the last chapter, you were introduced to a group of living things that are not like common plants or animals. The members of

Karl von Linne (1707–1778), a Swedish botanist, is the father of modern taxonomy, the science of classification. He classified more than 4,000 species of animals and an even larger number of plants. He is better known by his Latin name, Carolus Linnaeus.

A this group belong to the protist kingdom. This kingdom includes three phyla: (1) protozoa, (2) algae and (3) fungi.

1. Protozoa include amebas and paramecia. The word protozoa means "first, or simple, animals." Protists may be either single-celled or multi-celled, but all are so small that they can only be seen with a microscope.

2. Algae are sometimes called "one-celled green plants." All algae have chlorophyll and carry on photosynthesis. Algae live in the ocean, in fresh water, and wet places on land. The spirogyra is a kind of algae that lives in freshwater ponds.

3. Fungi include molds, mildews and mushrooms. Some fungi get their food at the expense of other living things. This type of organism is called a parasite.

Lichens are found on rocks and trees. They look like a thin, gray crust. Lichens are made up of algae and fungi. The algae make food for both organisms. The fungi take in water and protect the algae.

Check Up A

What are the three phyla that make up the protist kingdom?

(1) Protozoa

(2) Algae

(3) Fungi

B. The Moneran Kingdom

After doing this lesson you will be able:

B • to name the two major phyla in the moneran kingdom

Until recently, the organisms in the moneran kingdom were classified as part of the protist kingdom. The moneran kingdom is made up of living things that have only one cell and do not have a well-defined nucleus.
B There are two phyla in this kingdom: (1) bacteria and (2) viruses.

1. Bacteria are found everywhere on Earth. They are very small. In fact, they can barely be seen with an ordinary light microscope. Some bacteria cause sickness. Others are used to make foods, such as cheese and sauerkraut. Some bacteria, by the process of decay, recycle the chemicals in dead organisms.

2. Viruses are the smallest living things. They only live inside other cells where they upset the life activities of these cells. Viruses are the cause of many illnesses, such as polio, the flu (influenza), colds, smallpox, mumps, and some forms of cancer.

Check Up B

What are the two phyla that make up the moneran kingdom?

(1) Bacteria

(2) Viruses

C. The Animal Kingdom

After doing this lesson you will be able:

C1 • to name the two major groups of animals: those that have a backbone and those that do not

C2 • to name nine animal phyla and give an example of an animal in each phylum

C3 • to tell the difference between a cold-blooded animal and a warm-blooded animal

C4 • to name the five classes of chordates and give an example of an animal in each class

C5 • to classify human beings by kingdom, phylum and class

Scientists found that some animals have a backbone while others do not. Thus, they

The three animals are similar: they all have a head, eyes, mouth, legs and wings. They are carnivorous; they eat the flesh of other animals. They are different: the bird and bat have a backbone and ears; the bird has feathers, a beak, clawed feet, and it can lay eggs; the bat has fur, teeth and toes; the wasp has antennae and a poisonous stinger.

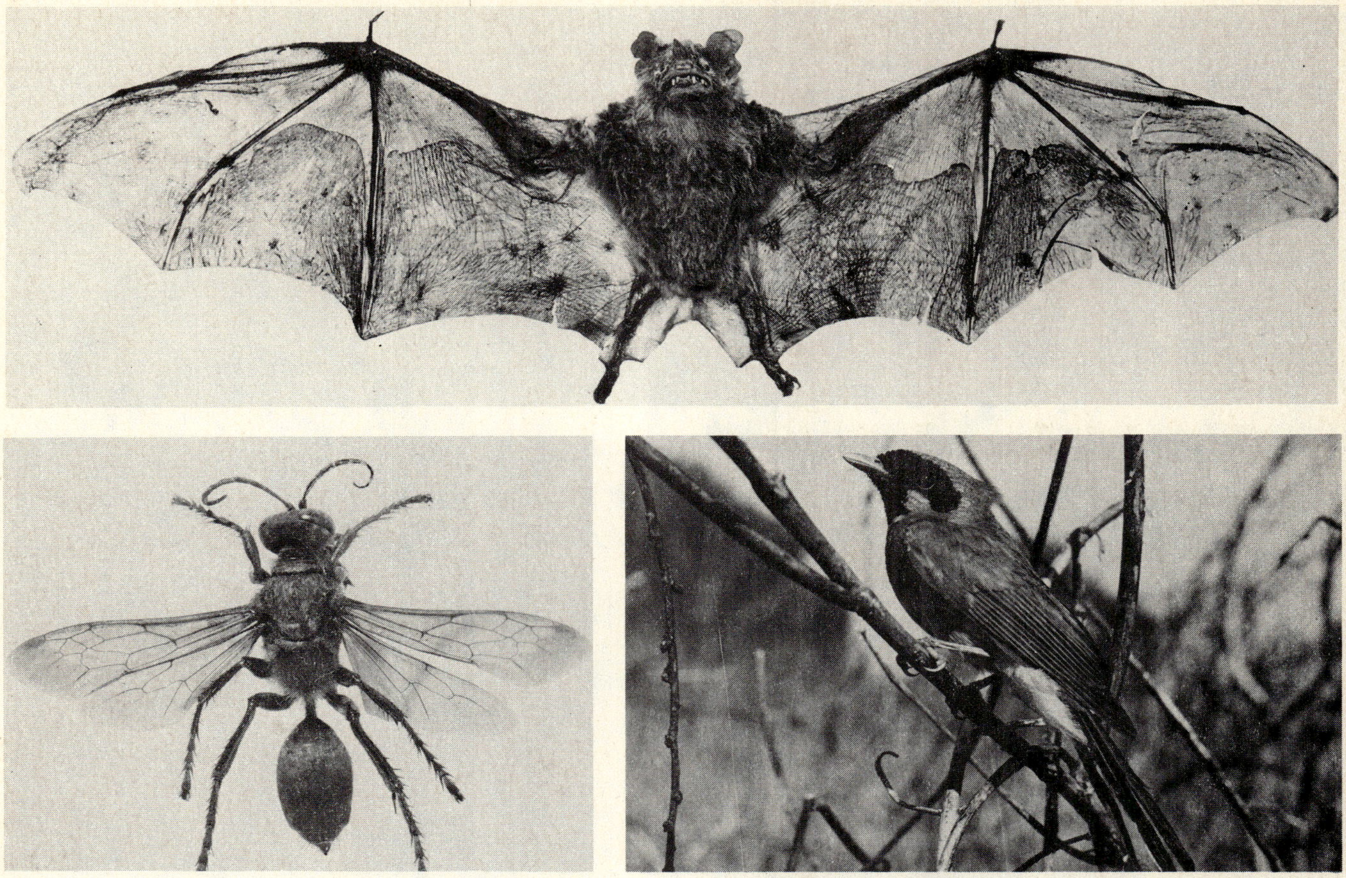

A bat, a bee and a bird can fly. How else are they alike? How are these animals different? How would you classify them?

made two different groups of animals: the vertebrates and the invertebrates. Animals
C1 with a backbone are **vertebrates.** Those without a backbone are **invertebrates.** There are nine phyla in the animal kingdom. Animals in the first eight phyla have no backbone.

C2 **1. Porifera.** Animals with pores, or tiny openings in their bodies, are in the phylum **porifera,** which means "pore-bearing." It includes the sponges that are found in both freshwater and saltwater. Because these animals are made up of more than one cell, they are called **multicellular** animals. They take in water, which carries food, through the pores.

C2 **2. Coelenterates.** Jellyfish, hydra and coral are some examples of the phylum **coelenterata,** or "hollow-bodied" animals. The animals of this phylum are multicellular. Their cells are arranged in two layers surrounding a cavity (space). They have saclike bodies.

3. Flatworms. The flatworms, phylum C2
platyhelminthes, are the simplest worms. Their bodies look like long ribbons. Planaria, liver flukes and tapeworms are examples of flatworms. Some flatworms, like the tapeworm and liver fluke, live off some other living thing. Thus, they are parasites.

4. Roundworms. A second group of worms C2
is the roundworms, phylum **nematoda.** The members of this phylum have long, thin, smooth bodies. Most roundworms are parasites. The trichina, or "pork worm," is found in the muscles of hogs and people who eat uncooked, infected meat. Hookworms are sometimes found in a person's intestines where they can cause much damage.

5. Segmented worms. The segmented C2
worms, phylum **annelida,** are the most com-

All three are warm-blooded animals. The elephant and the horse are plant eaters. The seal is a meat eater. The elephant shown is found on the plains of Africa. A marine animal, the seal lives along coastlines of North America, South America and Australia.

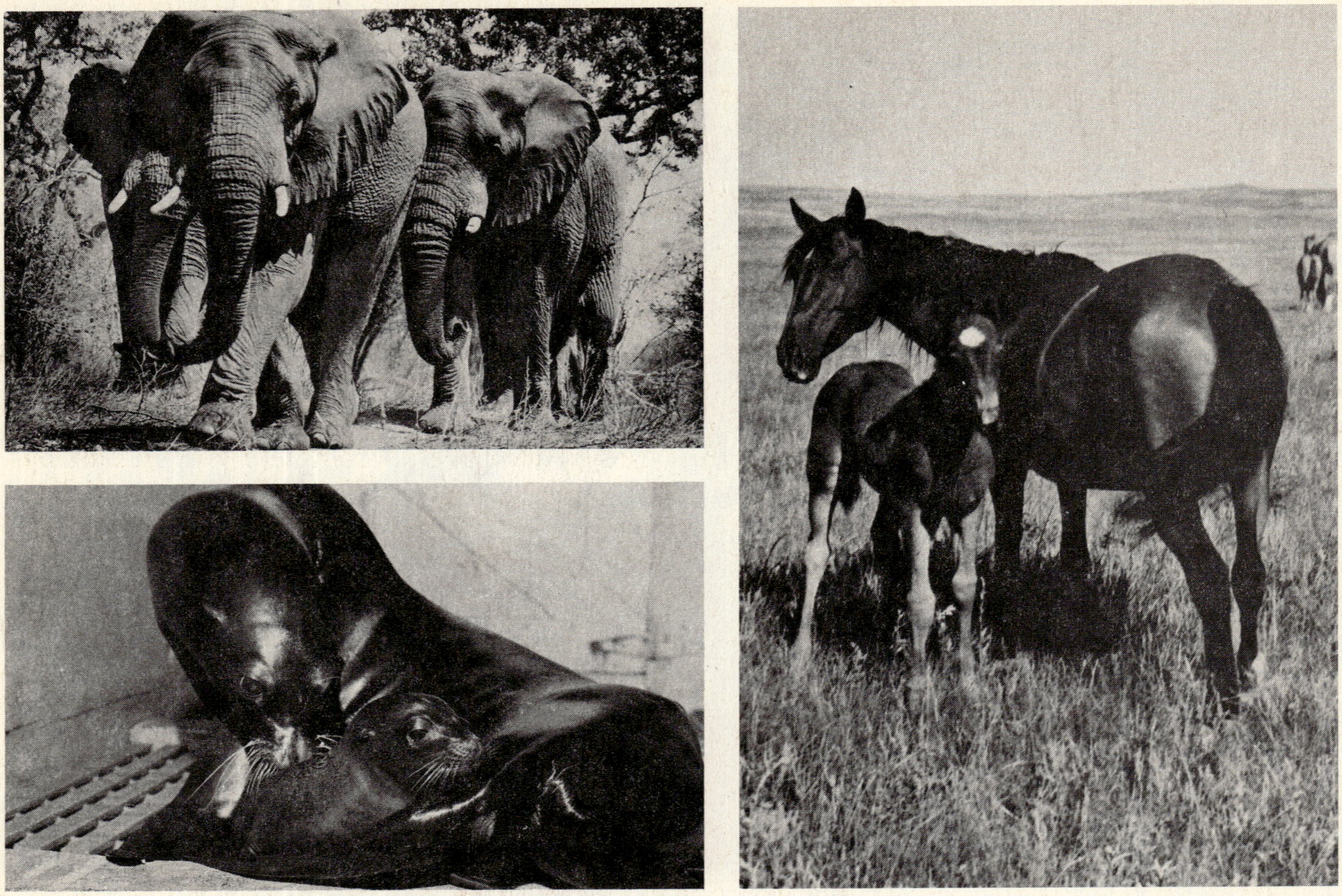

These animals are mammals. What characteristics do they share? How are they different? Compare their environments. C2

plex. Their bodies consist of many sections called segments. These worms have a body cavity and well-developed digestive system. Some animals in this group are the sandworm, earthworm and leech.

C2 **6. Echinoderms.** The sea urchin, starfish and sea cucumber are some animals that are members of the phylum **echinodermata.** Animals of this group live only in saltwater. They have a hard outer covering with needle-like spikes which give the name "echinoderm" to the phylum. Some have rows of tiny tube-shaped feet for movement.

C2 **7. Mollusks.** This phylum, **mollusca,** makes up a large group of animals. They live on land and in freshwater or saltwater. They have soft bodies which in some cases are protected by a hard outer shell. Most mollusks have a muscular foot that they use for movement. Some of the best-known examples of this phylum are the clam, squid, snail, oyster and octopus.

8. Arthropods. The phylum **arthropoda** has C2
more different kinds of animals than any other phylum of the animal kingdom. They live everywhere on the earth – in the air, on land and in the sea. All arthropods, or "jointed-leg animals," have a hard body covering which is really an outside skeleton. They have segmented bodies, legs and antennae (feelers). Some arthropods have wings. Shrimps, centipedes, spiders and flies belong to this phylum. Animals like the shrimp and crab have five pairs of legs. Flies, mosquitoes, grasshoppers and ants are called insects

Select a committee of students to prepare and present a report about an unusual mammal or other kind of animal of their choice.

because they have two pairs of wings and three pairs of legs.

C2 **9. Chordates.** All animals in the phylum **chordata** have some kind of backbone. The more complex chordates having a true backbone are the vertebrates. The chordates are subdivided into five classes: fish, amphibians, reptiles, birds and mammals.

C4 **Fish** are the simplest type of vertebrate animal. They belong to the class **pisces.** Fish are well suited to live in water. The bodies of most fish are covered with scales. Fish have fins by which they move themselves. They breathe by gills which allow the oxygen in the water to be absorbed by the blood.
C3 Fish are **cold-blooded animals.** A cold-blooded animal is one whose body temperature changes with the temperature of its surroundings. Examples of fish are the salmon, tuna, trout, mackerel and flounder.

C4 **Amphibians,** class **amphibia,** usually spend part of their lifetime in the water and part on land. They use gills for breathing in water in the early part of their lives. Later they develop lungs that allow them to breathe air when they live on land. Although the adults live on land, they go back to the water to lay eggs. Like the fish, amphibians are cold-blooded. Examples of amphibians are the salamander, frog and toad.

C4 **Reptiles,** class **reptilia,** may live either on land or spend part of their lives in the water. Reptiles have lungs for breathing. Some examples of reptiles are the snake, lizard, alligator, turtle and crocodile. Their bodies are covered with dry scales. They, too, are cold-blooded animals. All reptiles, except the snake, have two pairs of legs.

C4 **Birds,** class **aves,** are **warm-blooded**
C3 vertebrates. A warm-blooded animal is one whose body temperature remains the same regardless of the temperature of its surroundings. Birds are covered with feathers and have scales on their legs. They breathe by lungs and have wings. They have horny beaks but no teeth. Most of them can fly. You are familiar with many birds: chickens, robins, crows, blue jays.

Mammals, class **mammalia,** are the most C4
developed group in the animal kingdom. They are warm blooded and breathe by means of lungs. Almost all mammals give birth to live young and feed their young with milk. Their bodies are covered completely or in part by
hair or fur. Human beings belong to this C5
group. So do many other animals like the elephant, whale, rabbit, bat, cat, dog and monkey.

Discover for yourself

how you could learn more about the animals around you.

PURPOSE: To determine how to classify certain members of the animal kingdom • To develop skills of observation, comparison and classification

Observe the animals in your own environment. Study them in your own home, a friend's home, or a park. Use your Work-a-

Discussion question: What is the human being's habitat? (Land, ocean, air, space.)

Text or a reference book on animals to help you classify them into phyla. The ones you are most likely to find are:

1. Cat Chordata
2. Dog Chordata
3. Bird Chordata
4. Fish Chordata
5. Earthworm Annelida
6. Snail Mollusca
7. Fly and ant Arthropoda
8. How is a bird like a fly? How are they different? Answers will vary.
9. How is a fish like a dog? How are they different? Answers will vary.
10. How is an earthworm like a snail? How are they different? Answers will vary.

Check Up C

C1. What are the two major groups of animals? Do they have a backbone or not?

Vertebrates — have a backbone.

Invertebrates — do not have a backbone.

C2. Name nine animal phyla and give an example of an animal in each phylum.

(1) Porifera — sponges.

(2) Coelenterata — examples will vary.

(3) Platyhelminthes

(4) Nematoda

(5) Annelida

(6) Echinodermata

(7) Mollusca

(8) Arthropoda

(9) Chordata

C3. What is the difference between "warm-blooded" and "cold-blooded" animals?

Warm-blooded animals have a constant body temperature regardless of the temperature of their surroundings.

C4. List the names of the five classes of chordates and give an example of an animal in each class.

(1) Pisces — examples will vary.

(2) Amphibia

(3) Reptilia

(4) Aves

(5) Mammalia

C5. Human beings belong to what kingdom, phylum and class? Kingdom — Animal;

Phylum — Chordata; Class — Mammalia

D. The Plant Kingdom

After doing this lesson you will be able:

D1 • to name the two phyla that make up the plant kingdom and give an example of a plant in each phylum

D2 • to explain how bryophytes and tracheophytes are different from each other

D3 • to explain how pteridophytes and spermatophytes reproduce

Plants are classified in the same way as animals; that is, according to likeness in

Obtain a fern plant for observation. Study the spore cases on the undersides of the leaves with a strong magnifying glass or the low power objective of a microscope.

Mosses and parasites (organisms that live on or in other organisms) grow along tree limbs in a tropical forest. How are the plants different from the tree that they grow on? How are they different from the ferns in the photo on the right?

The mosses and parasites do not need soil to grow in. They are smaller and simpler in structure than the trees and ferns.

structure. The first scientist to classify plants was the Swedish botanist, Karl von Linné (1707–1778). Since then, a more modern system of classification has been worked out. It has two phyla: **bryophytes** and **tracheophytes.** D1

1. Bryophytes. The moss plants are bryophytes. These plants have many cells. They have simple leaves and parts that look like stems and roots. They were the first plants D2
to appear on the earth. All bryophytes have chlorophyll and can make their own food by photosynthesis. The mosses and liverworts are examples of this phylum. Mosses are believed to be the first green land plants to develop in the course of evolution.

2. Tracheophytes. Plants that have true roots, stems and leaves are called **tracheo-** D2
phytes. Some tracheophytes reproduce by special cells called spores and are called D3
pteridophytes. Ferns are the most common example of this phylum.

The **spermatophytes,** or seed-bearing plants, have roots, leaves and stems. They reproduce by seeds. They are the most numerous kind of plant life. D3

One kind of spermatophytes produces seeds in cones and is a non-flowering plant. Examples are the evergreens, such as pines, spruces and firs. Another kind has flowers which produce seeds. Some examples are maple, willow and oak trees, bushes, vegetables and garden flowers.

Discuss the fact that almost all plants that students see every day are spermatophytes (seed bearers).

Maple leaf and seeds. How does the shape of the maple seed help it to be carried far from its parent tree? How is this helpful in the reproduction of maple trees?

The maple has paired, winged fruits commonly called maple keys. The seeds are inside the dry, hard fruits. Because of the wings, the fruit can be transported for long distances by the wind. If all the maple seeds fell around the parent tree, the area would become overcrowded and the new plants would not grow well.

Discover for yourself
how you could learn more about the plants around you.

PURPOSE: To determine how to classify certain members of the plant kingdom • To develop the skills of observation, comparison and classification

You need: leaves, paper, glue or tape, reference books on plants and animals

Gather leaves of plants from your own yard, a park or from your family's potted plants. Paste each one to a sheet of paper. Label the leaves with both their common and scientific names, using a reference book as a guide.

1. To what phylum do most of your leaf plants belong? Tracheophytes

2. How do leaf plants reproduce? By spores or seeds.

Check Up D

D1. List the two phyla that make up the plant kingdom and give an example of a plant in each phylum.

(1) Bryophytes — examples will vary.

(2) Tracheophytes

D2. How are bryophytes and tracheophytes different from each other? Bryophytes have simple leaves and parts that resemble stems and roots. Tracheophytes have true roots, leaves and stems.

D3. How do pteridophytes reproduce? How do spermatophytes reproduce? Pteridophytes reproduce by spores. Spermatophytes reproduce by seeds.

Unit 1, Chapter 2
REVIEW

Matching Exercise Match each term in **Column A** with its description in **Column B**. Write the letter of the description in the space beside the correct term.

	Column A	*Column B*
g	**1.** vertebrate	**a.** reproduces by seeds
j	**2.** porifera	**b.** animal with fur or hair
a	**3.** spermatophyte	**c.** fishes
d	**4.** arthropod	**d.** jointed-legged animal
h	**5.** tracheophyte	**e.** live only in salt water
b	**6.** mammal	**f.** reproduces by spores
f	**7.** pteridophyte	**g.** animal with a backbone
c	**8.** pisces	**h.** has roots, stem and leaves
e	**9.** echinoderms	**i.** flatworms
i	**10.** platyhelminthes	**j.** sponge

True/False In the space provided, write **T** if a statement is true. If it is false, change the underlined word or words to make the statement true.

Viruses	**1.** <u>Lichens</u> are the smallest living things.
T	**2.** Jellyfish, hydra and coral are examples of <u>Coelenterates</u>.
porifera	**3.** Sponges belong to the phylum <u>annelida</u>.
invertebrates	**4.** Animals that do not have a backbone are <u>vertebrates</u>.
parasites	**5.** Organisms that get their food at the cost of other living things are called <u>fungi</u>.
arthropoda	**6.** The phylum <u>chordata</u> has more different kinds of animals than any other phylum in the animal kingdom.
insects	**7.** Arthropods with two pairs of wings and three pairs of legs are called <u>chordates</u>.
Plant	**8.** Bryophytes are part of the <u>Animal</u> Kingdom.
T	**9.** <u>Tracheophytes</u> are plants that have true roots, stems and leaves.
T	**10.** Spermatophytes are <u>seed-bearing</u> plants.
two	**11.** All reptiles, except for the snake have <u>three</u> pairs of legs.
cold-blooded	**12.** An animal whose body temperature changes with the temperature of its surroundings is <u>warm-blooded</u>.
T	**13.** <u>Multicellular</u> animals are animals that are made up of more than one cell.
protists	**14.** Paramecia and amebas are <u>fungi</u>.
T	**15.** All <u>algae</u> have chlorophyll and can carry out photosynthesis.

Multiple Choice Exercise Circle the letter of the word or phrase that best completes each statement.

1. An animal whose body temperature does not change with the temperature of its surroundings is **(a)** cold-blooded **(b)** warm-blooded **(c)** an invertebrate **(d)** an amphibian.
2. An animal that does not have a backbone is **(a)** a vertebrate **(b)** a spermatophyte **(c)** an invertebrate **(d)** a bryophyte.
3. Mosses and liverworts are examples of **(a)** vertebrates **(b)** spermatophytes **(c)** invertebrates **(d)** a bryophytes.
4. Plants with true roots, stems and leaves are **(a)** vertebrates **(b)** mosses **(c)** tracheophytes **(d)** bryophytes.
5. The sea urchin, starfish and sea cucumber are examples of **(a)** echinoderms **(b)** mollusks **(c)** fish **(d)** bryophytes
6. All coelenterates have **(a)** hair **(b)** pores **(c)** stinging cells **(d)** a backbone.
7. The grouping of living things by likeness is called **(a)** kingdoms **(b)** phyla **(c)** vertebrates **(d)** classification.
8. Frogs, salamanders and toads are **(a)** amphibians **(b)** reptiles **(c)** fish **(d)** mammals.
9. Cold-blooded animals that have bodies covered by scales are **(a)** amphibians **(b)** reptiles **(c)** fish **(d)** mammals.
10. The spirogyra is **(a)** an alga **(b)** a virus **(c)** a bryophyte **(d)** a tracheophyte.

Application Exercise Give an example of an organism that belongs to each of the following phyla or classes.

*Answers will vary. Some typical responses are given.

1. Porifera sponges
2. Coelenterate jellyfish, hydra, coral
3. Platyhelminthe tapeworm, liver fluke
4. Nematoda hookworm, trichina
5. Annelida earthworm, leech
6. Echinoderm starfish, sea urchin
7. Mollusk clam, squid, octopus
8. Arthropod roach, spider, fly
9. Pisces flounder, tuna, trout
10. Amphibian frog, toad, salamander
11. Reptile snake, lizard, turtle
12. Ave robin, sparrow, finch
13. Mammal humans, monkey, whale
14. Bryophyte mosses, liverworts
15. Tracheophyte ferns

Early each spring, these apple blossoms appear on the apple tree. The blossoms are white and tinged with pink.

New science words: dormant, germinate, oxidation

Skills developed: observing, comparing

Chapter 3

FLOWERING PLANTS

There are two kinds of fruits: *fleshy* fruits which enclose seeds (apple, grape, tomato, cherry, squash). The *dry* fruits include milkweed, pea, chestnut, maple.

A. The Roots, Stem and Leaves

After doing this lesson you will be able:

- A1 • to explain the difference between monocotyledons and dicotyledons
- A2 • to give the location of root hairs and explain their function
- A3 • to name the four layers of a woody stem
- A4 • to list the two kinds of cells that make up the vascular system of a plant and tell what is carried in each
- A5 • to name the four layers of a leaf and tell what each does
- A6 • to explain what stomates and guard cells do

All green plants that bear flowers are **flowering plants.** Botanists (scientists who study plants) have grouped them according to the number of leaves found in the seed of the plant. Seeds that have only a single seed leaf
A1 are called **monocotyledons.** Seeds that have two seed leaves are called **dicotyledons.** These leaves of the seed supply the food the young plant needs until it can make its own.

Mono means one; *di* means two. A *cotyledon* is a seed leaf that contains food.

The **roots** of a plant have many branches that anchor the plant in the soil. Plant roots have **root hairs** growing out of the surface A2
cells. The root hairs take in the water and minerals that the plant needs from the soil. Water and minerals pass from the root hairs into the root and then up through the stem of the plant.

The **stem** supports the plant. It also brings up to the leaves the water and minerals taken in by the roots. When the leaves have made food, the stem carries it to all parts of the plant. Flowering plants like the petunia, marigold, carnation and tomato have soft stems. Flowering trees and bushes have hard, woody stems.

The woody stem has an outer covering A3
called **bark.** Underneath this layer is a growing region called the **cambium.** This layer makes the width of the stem increase. Under the cambium is the **wood** layer. The center of the stem is called the **pith.**

The water and minerals taken in by the roots travel to the leaves through long, thin,

Refer to the art drawing on page 24 when discussing plant roots. Note the various stages of cell division at the root tip where cell division is constantly taking place.

Root hairs increase the surface area of the root and allow much more water and minerals to be absorbed.

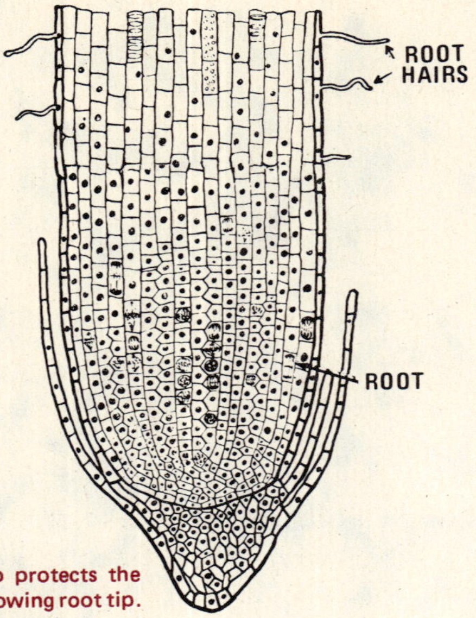

The root cap protects the constantly growing root tip.

A section of a root tip.

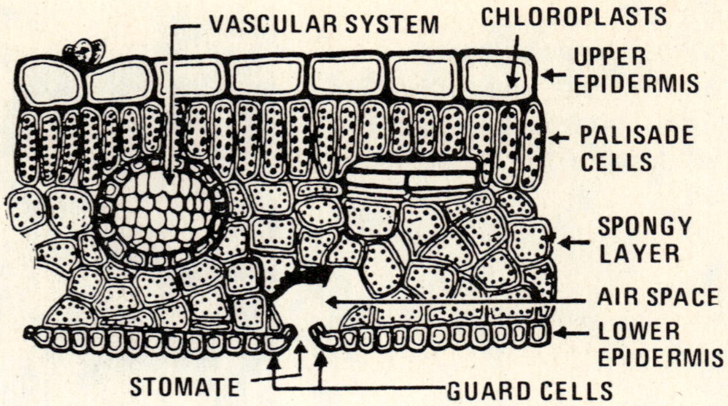

A cross section of a leaf.

A *palisade* is a fence made of logs. The palisade layer in a leaf bears a resemblance to this type of fence.

tubelike cells in the stem. These cells make up the **vascular system** which carries materials to all parts of the plant. The vascular system is made up of two kinds of cells called xylem cells and phloem cells. The **xylem cells** carry water and minerals up to the leaves from the roots.

A4 Xylem cells are formed in the spring and in late summer. The xylem cells formed in the late summer are smaller than those formed in the spring. This difference in cell size causes a ring to form in the stem. Each year a new ring appears so you can tell the age of a tree by the number of rings it has. These rings are called **annual rings.**

A4 The **phloem cells** carry food from the leaves down to the stem and roots. These bundles of cells make up the **veins** we see in the leaves of the flowering plants.

Leaves make the food that the plant uses. All green plants make their own food. The upper surface of a leaf is made up of a layer
A5 of cells called the **upper epidermis.** The upper epidermis helps stop the loss of needed water and gases from the leaf and protects other tissues in the leaf. Some leaves have a waxy material on top of the upper epidermis that also prevents the loss of water.

Just below the upper epidermis layer is the **palisade layer** of cells that contain the A5
chloroplasts. Inside the chloroplasts is a green material called **chlorophyll.** Chlorophyll gives the leaves their green color and makes it possible for the cells to use the energy from sunlight to make food.

The xylem cells, phloem cells and air spaces are found in the **spongy layer** just be- A5
low the palisade layer. This layer is used for storage. The xylem and phloem cells, as you already know, carry water and food materials to all parts of the plant. The air spaces allow gases to pass between the atmosphere and the cells of the leaf.

The bottom of the leaf is covered with a A5
layer one-cell thick called the **lower epidermis.** This layer has many tiny openings called **stomates.** The stomates allow carbon A6
dioxide and oxygen to go in and out of the leaf. They also allow water vapor to pass out of the leaf. The stomates are surrounded by **guard cells.** The guard cells regulate the size A6
of the openings of the stomates. When the guard cells fill with water, they become larger and cause the stomates to open. The stomates close when the guard cells lose their water. The amount of gas leaving and entering the leaf is thus controlled by the guard cells.

Use a microprojector to view prepared slides of plant roots, stems and leaves. Encourage students to make some slides of their own for viewing.

Check Up A

A1. What is the difference between monocotyledons and dicotyledons? Seeds that have only one seed leaf are monocotyledons. Dicotyledons have two seed leaves.

A2. Where are root hairs located and what do they do? Root hairs grow out of the surface cells of roots. They absorb water, minerals and oxygen from the soil.

A3. Name the four layers of a woody stem.

(1) Bark

(2) Cambium

(3) Wood

(4) Pith

A4. What two kinds of cells make up a plant's vascular system? What is carried by each?

(1) Xylem — water and minerals.

(2) Phloem — food.

A5. Name the four layers of a leaf and tell what each layer does.

(1) Upper epidermis — prevents water loss; protects other tissues.

(2) Palisade layer — cells contain chloroplasts where food is made.

(3) Spongy layer — storage; gas exchange; movement of water and food.

(4) Lower epidermis — passage of oxygen, carbon dioxide, water vapor in and out of leaf.

A6. What do stomates and guard cells do? Stomates allow oxygen, carbon dioxide, water vapor to go in and out of the leaf. Guard cells regulate the size of the stomates.

Words to review: radiant, energy, starch

B. How Does a Plant Make Its Food?

After doing this lesson you will be able:

B1 • to name the process by which green plants make food

B2 • to list the four things needed by a green plant to make its own food

B3 • to name the gas formed during photosynthesis

B4 • to name the process by which plants (and animals) release energy

Although a plant's roots take in water and minerals, these materials are not food for the plant. How does the plant use these materials to make food? The process by which the leaves of green plants make food is called **photosynthesis.** B1

In order to carry on photosynthesis, a green plant needs carbon dioxide, water, light and chlorophyll. The carbon dioxide is B2 taken from the air. It enters the leaves through the stomates. The water is taken in from the soil in which the plant lives through the root hairs.

These materials enter the cells through the cell wall and the plasma membrane. The light gives the **radiant energy** needed by the chlorophyll to carry on photosynthesis. The raw materials, carbon dioxide and water, are changed with the help of chlorophyll and the energy from light, into sugar which serves as food for the plant.

The sugar which contains the stored energy of the light can be used by the plant when needed. The sugar made by the leaves is carried by the phloem to every part of the plant to be used for the release of energy. Some of this unused sugar is changed into starch and fat and stored for future use. Some is changed into proteins and vitamins.

Oxygen is formed during photosynthesis. B3
Some of the oxygen is used by the plant to burn its food and release energy. This process is called **oxidation.** B4

Plants of the ocean produce 80% of the fresh oxygen that the atmosphere receives each day.

The sun is the source of energy for all plants and, indirectly, for all animals. The energy you get by burning wood, coal and oil products, is the sun's energy which was trapped long ago during photosynthesis.

Check Up B

B1. What is the process by which green plants make food? Photosynthesis

B2. What four things does a green plant need to make its own food?

(1) Carbon dioxide

(2) Water

(3) Light

(4) Chlorophyll

B3. What gas is formed during photosynthesis? Oxygen

B4. What is the process by which plants release energy? Oxidation

C. How Do Flowering Plants Reproduce?

After doing this lesson you will be able:

C1 • to name the kinds of cells complex organisms form in order to reproduce

C2 • to name the male and female reproductive organs in flowering plants

C3 • to name the male and female sex cells in flowering plants

C4 • to explain the difference between self- and cross-pollination

C5 • to define the following terms:
(1) fertilization (2) seed (3) embryo (4) cotyledons (5) fruit

C6 • to explain the difference between dormant and germinating seeds

You have learned that single-celled living things form more of their own kind by splitting or dividing in half. More complex organisms produce more of their own kind by forming special cells called **sex cells.** C1

Flowering plants produce new plants by the formation of seeds. The plants develop flowers and the flowers produce the seeds.

Have you ever noticed a flower in its bud stage? Next time, look carefully and you will see that the bud is protected by the **sepals**—petallike parts. When the bud opens you usually see colored petals. These petals often have a very pleasant odor. They protect the reproductive organs of the flower. The petals play an important part in reproduction by attracting insects to the flower.

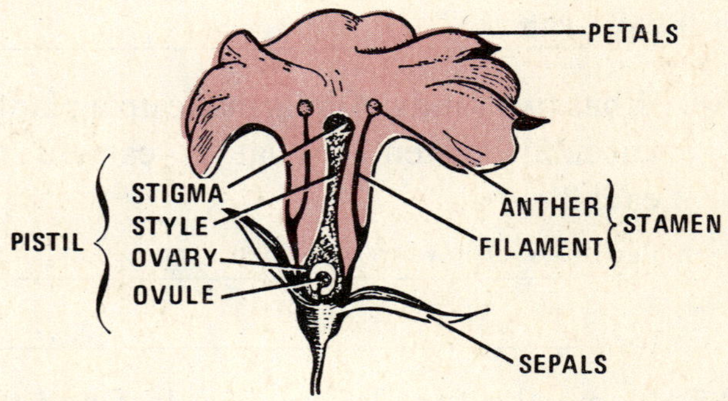

A cross section of a flower showing its main parts.

The **pistil** is the female reproductive organ. It is found in the center of the flower surrounded by the **stamens.** The top part of the pistil is the **stigma.** The larger bottom of the pistil is the **ovary.** A solid stalk called the **style** links the stigma to the ovary. If you take a close look at the inside of the ovary, you will see many small parts called **ovules.** The ovules contain female sex cells, sometimes called egg cells. C2 C3

The **stamens** are the male reproductive organs. Each stamen consists of a long, thin stalk called the **filament.** At the upper end of C2

Certain plants, such as the apple, could not produce without the cross-pollination provided by such insects as the honey bee.

each stamen there is a boxlike part called an **anther.** The anthers produce pollen grains which make male sex cells, sometimes called C3 sperm cells.

For reproduction to take place, pollen grains must be brought from the anther of one flower to the stigma of the same flower, or to the stigma of another flower of the same kind. This transfer of pollen is called **pollination.** Pollination is usually carried out by wind, water and animals. Bees and hummingbirds are the most common carriers. If the pollen is carried to the stigma of the same flower, the process is called **self-pollination.** If it is C4 carried to a flower of another plant, the process is called **cross-pollination.**

After a pollen grain has landed on the stigma of a pistil, it grows a long, thin tube called the **pollen tube.** This tube goes down through the style until it reaches the ovule. When the nucleus of the sperm cell combines with the nucleus of the ovule, C5(1) **fertilization** takes place. Fertilization is the joining of a sperm cell with an egg cell to produce a fertilized egg cell. A new organism can then grow from the fertilized egg cell.

After fertilization, the egg cell divides and forms many cells. Not all the cells are the same. This division into different kinds of cells forms the young plant, within a seed, C5(3) called the **embryo.** It is covered by a protective coating. The embryo with its stored food C5(2) is called a **seed.**

Have you ever examined a lima bean closely? The outside covering is a protective coating called the **seed coat.** If you split the bean in half you will find stored food inside C5(4) the white seed leaves, or **cotyledons.** You will also see the embryo with a root tip and a stem tip. These parts will grow into the root and stem of the new plant.

As the seed is being formed, the ovary en- C5(5) larges and becomes the **fruit** of the plant. Tomatoes, apples, beans and peaches all are produced this way.

Words to review: tissue, cycle

How are new plants formed? When you cut open an apple, you find many seeds inside. These seeds have stopped growing. They are **dormant.** Plants cannot grow from C6 dormant seeds. The right conditions must exist for a seed to **germinate,** or grow into a C6 new plant. Seeds need warmth and water in order to germinate.

How do water and warmth cause the seeds to germinate? The water causes the seed coat to split apart. The root tip then can grow down into the soil. The stem tip starts to grow above the ground. The seed leaves, or cotyledons, supply the young plant with food until it can make its own. As the food is used up in the cotyledons, they begin to shrink. Leaves appear on the young plant. The plant is then able to make its own food. When the plant is fully grown, the **life cycle** will start all over again. Many new healthy plants may be produced.

Some beans, *packaged for food use*, are treated to prevent or retard sprouting. For best results, use beans from a packet that is intended for *garden use*.

Discover for yourself

how you could compare the seeds of two different kinds of flowering plants.

PURPOSE: To study some common monocotyledons and dicotyledons • To develop the skills of observation and comparison

You need: lima bean seeds, corn seeds, dish, warm water

a. Soak a few lima bean seeds and a few corn seeds in warm water. When their seed coats are soft, peel them off.

b. Look at each seed carefully. Try to split the seeds in two with your fingers.

1. Does the lima bean split easily in two? ____
   Yes
2. Does the corn seed split easily in two? ____
   No

Some plants reproduce in other ways: rhizomes (iris), runners (strawberries), tubers (potatoes), bulbs (onions, tulips), corms (gladiola), cuttings (coleus, geranium).

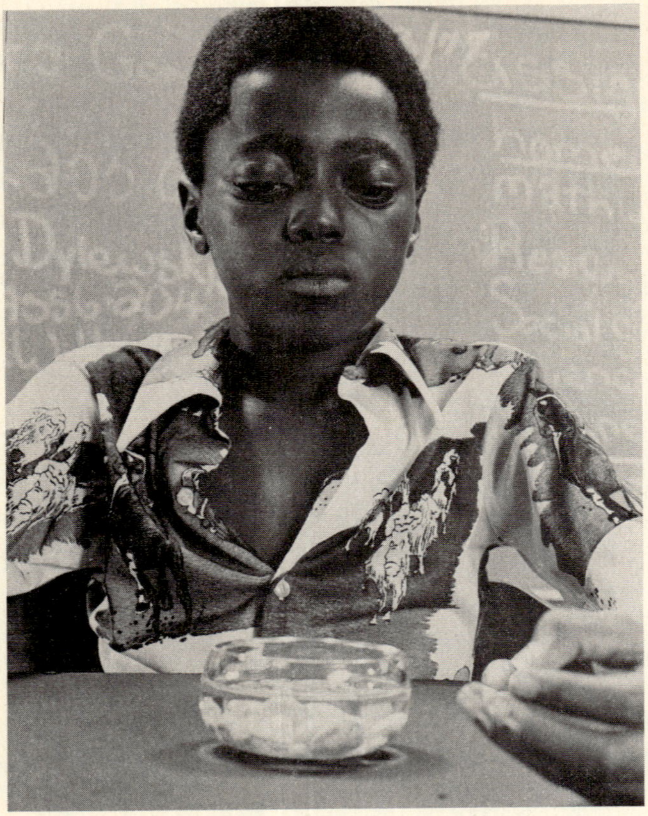

c. Examine the embryo of each seed. Look for a root tip and a stem tip.

1. Can you see the embryo plants? Yes
2. What color are they? Most are white.

Check Up C

C1. What kind of cells do complex organisms form in order to reproduce? Sex cells.

C2. What are the male and female reproductive organs in a flowering plant called? The male reproductive organs are called stamens. The female reproductive organ is called the pistil.

C3. What are the male and female sex cells in a flowering plant called? The male sex cell is the pollen grain or sperm. The female sex cell is the ovule or egg cell.

C4. What is the difference between self- and cross-pollination? Self-pollen is carried from the anther to the stigma of the same flower. Cross-pollen is carried from the anther of one flower to the stigma of another.

C5. Define the following terms:

(1) Fertilization The uniting of a sperm cell with an egg cell.

(2) Seed A plant embryo with its stored food.

(3) Embryo A young plant within a seed.

(4) Cotyledons Seed leaves containing stored food.

(5) Fruit An enlarged and developed ovary.

C6. What is the difference between a dormant and a germinating seed? A dormant seed is not developing. A germinating seed is developing.

Unit 1, Chapter 3
REVIEW

Matching Exercise Match each term in **Column A** with its description in **Column B**. Write the letter of the description in the space beside the correct term.

	Column A	*Column B*
f	**1.** ovules	**a.** seed with one leaf
b	**2.** pistil	**b.** female reproductive organ in flowering plants
j	**3.** pollen grains	**c.** control the size of the stomates
g	**4.** stamen	**d.** seed with two leaves
i	**5.** stomates	**e.** carries food to stems and roots
c	**6.** guard cells	**f.** female sex cells in flowering plants
e	**7.** phloem	**g.** male reproductive organ in flowering plants
h	**8.** xylem	**h.** carries water and minerals to the leaves
a	**9.** monocotyledon	**i.** openings in the lower epidermis
d	**10.** dicotyledon	**j.** male sex cells in flowering plants

True/False In the space provided, write **T** if a statement is true. If it is false, change the underlined word or words to make the statement true.

water **1.** Green plants need carbon dioxide, oxygen, radiant energy and chlorophyll to carry out photosynthesis.

larger **2.** The xylem cells that are formed in the summer are smaller than those formed in the spring.

Phloem cells **3.** Stomates carry food from the leaves of a plant to its stems and roots.

female **4.** As a seed in a flowering plant forms, part of the male reproductive organ becomes its fruit.

wood **5.** The four layers of the woody stem are the bark, cambium, palisade layer and pith.

T **6.** Seeds of flowering plants may have one or two cotyledons.

spongy layer **7.** The xylem cells, phloem cells and air spaces are found in the upper epidermis of a leaf.

T **8.** Petals play an important part in reproduction by attracting insects to a flower.

male **9.** The pistil of a plant is surrounded by the female reproductive organs.

T **10.** Phloem and xylem cells are found in both the stems and the leaves.

Oxidation **11.** Photosynthesis is the process in which oxygen is used by a plant to burn its food and release energy.

Fertilization **12.** Germination occurs when the nucleus of the sperm cell combines with the nucleus of the ovule.

T 13. Self-pollination occurs when pollen is carried from the anther of a flower to the stigma of the same flower.

T 14. A seed is an embryo with stored food.

dormant 15. Seeds that have stopped growing are germinating.

Multiple Choice Exercise Circle the letter of the word or phrase that best completes each statement.

1. The transfer of pollen from one plant to another plant is **(a)** germination **(b)** cross-pollination **(c)** self-pollination **(d)** photosynthesis.
2. The food-making process of photosynthesis uses **(a)** carbon **(b)** carbon dioxide **(c)** oxygen **(d)** pollen.
3. The joining of a nucleus of a sperm cell with the nucleus of an egg cell is **(a)** pollination **(b)** germination **(c)** fertilization **(d)** photosynthesis.
4. Water and minerals are carried from the roots of a plant to the stems and leaves by the **(a)** xylem **(b)** phloem **(c)** cambium **(d)** pith.
5. The fruit of a plant is **(a)** a plant embryo with stored food **(b)** a young plant within a seed **(c)** a dormant seed **(d)** an enlarged and developed ovary.
6. The process by which green plants make food is **(a)** germination **(b)** pollination **(c)** fertilization **(d)** photosynthesis.
7. The stamens of a flowering plant are the **(a)** female reproductive organs **(b)** female sex cells **(c)** male reproductive organs **(d)** male sex cells.
8. In woody plants, the layer of cells that makes the width of the stem increase is the **(a)** cambium **(b)** bark **(c)** wood layer **(d)** pith.
9. Water and minerals from the soil are taken into a plant by the **(a)** stem **(b)** root hairs **(c)** cambium **(d)** pith.
10. Leaves get their green color from **(a)** chloroplasts **(b)** photosynthesis **(c)** chlorophyll **(d)** xylem.

Application Exercise Label the parts of the flower in the diagram below. Write your answers next to each line.

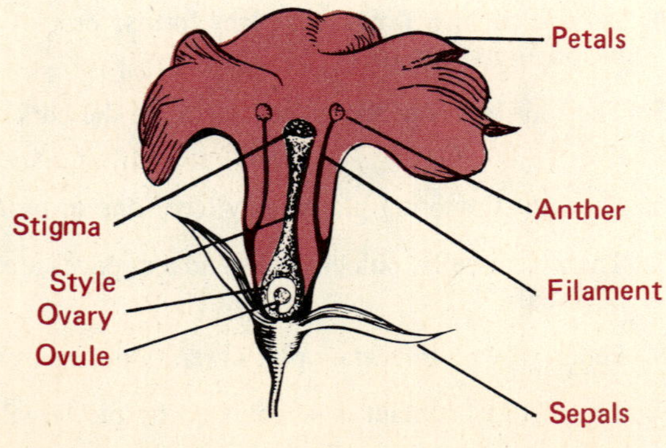

Scientists believe that the earth looked like this 200 million years ago.

New science words: era, evolution, fossil

Skills developed: observing, comparing, stating a conclusion

Chapter 4

HOW LIVING THINGS CHANGE

After students read about the Mesozoic Era on page 35, have them compare what they read to what they see in this photograph.

In this chapter you will find out what happened after the earth was formed. Try to picture the earth without any kind of plant or animal life on it—nothing around except rocks and water. How and when did life appear on the earth? What did the first living things look like? Paleontology is the study of life of past geological periods as known from the remains of fossils.

A. Fossils and the Past

After doing this lesson you will be able:

A1 • to tell what is meant by the term "evolution"

A2 • to explain what a fossil is

A3 • to tell the type of rock in which most fossils are found

A4 • to explain how the position of a fossil in sedimentary rock helps to tell its age

A5 • to tell what a geologist is

Some of the living things that were on earth millions of years ago are much like things that live today. Others no longer exist. Very slow changes have been going on in living things for hundreds of millions of years and are still going on. These changes are called **evolution,** the process of very slow change as simple organisms become more **complex.** A1

Much of what is known about the past comes from **fossils.** Fossils are the remains or imprints of living things that existed millions of years ago. By studying fossils, scientists can tell us something about what kinds of life existed on the earth long ago. A2

Scientists think that the first kind of life existed in the oceans. They believe that carbon, oxygen, hydrogen and nitrogen combined to form a proteinlike substance.

The simplest living things on earth today are the one-celled protists. So scientists assume that the first life on earth also must have been a one-celled organism. These one-celled organisms lived in the oceans over two billion years ago. No fossil remains of these organisms have been found.

There are no fossil remains of these early forms of life because they had no hard parts, such as shells. They were also very small. When the cells died, they fell to the bottom

Obtain some fossils or a fossil set for display, examination and discussion.

Graphite is grayish-black carbon: soft, greasy and having a metallic luster. Graphite is used widely: one use is in pencils. The lead in pencils is a mixture of graphite and clay.

of the ocean and decayed. Eventually they were covered with sand and rock fragments.

All living things, in part, are made of carbon. Scientists have found large amounts of **graphite,** a form of carbon, in sedimentary rock over two billion years old. Therefore, they think that simple forms of plant life may have existed at that time.

A5 **Geologists** are scientists who study the history of the earth and its life forms, especially as found in rocks. They think that it takes between 4,000 and 10,000 years to form one foot of sedimentary rock. Scientists can figure out the age of a fossil found in the
A3 layers of sedimentary rock. The lowest layers of sedimentary rock would be the
A4 oldest. These would contain fossils of the simplest forms of life. The upper layers would contain fossils of more advanced life forms. The first plants and animals are thought to have been protists or something like them.

After thousands of years, the single cells began to come together. It probably was easier for them to get food and to protect themselves. This was the beginning of many-celled organisms.

The oldest fossils found in sedimentary rock are the remains of the **trilobites,** which are animals that lived in the water about 550 million years ago. These animals had a hard shell. After they died, their shells sank to the bottom and left an imprint in the sedimentary rock. From these imprints scientists have a good idea what these animals looked like.

The trilobite, named for its 3-part body, was the ancestor of the lobster and crab.

Discover for yourself
how you could make your own fossil print.

PURPOSE: To show how fossils are formed • To develop the skills of observation, comparison and stating a conclusion

You need: plaster of paris, water, spoon, coffee can, shoe box, petroleum jelly, leaf, shell or bone

a. Mix plaster of paris with water in a coffee can. Pour a layer into a shoe box.

b. Cover a leaf, shell or bone with petroleum jelly and place it on the wet plaster.

c. When the plaster hardens, spread a thin layer of petroleum jelly on it. Mix more plaster of paris in the coffee can and pour another layer of plaster into the box.

d. When the plaster hardens, tear away the box and gently take the two layers apart. Observe the fossil imprint.

1. How is your fossil imprint like a real fossil? It leaves the same shape as a real fossil.

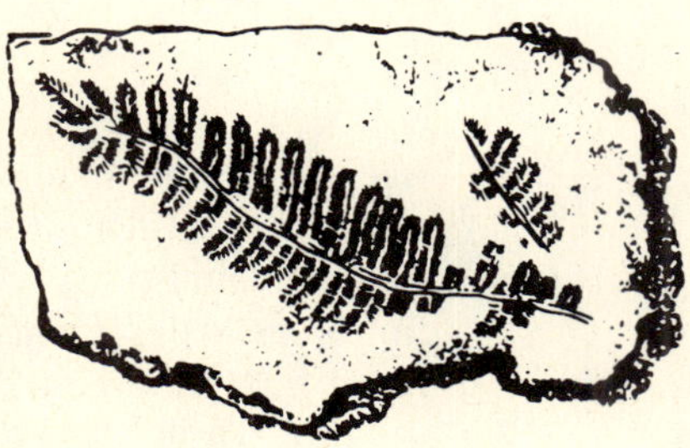

2. How does the way you made your own fossil differ from the way a real fossil is made? A real fossil takes thousands of years to form.

3. Compare your fossil imprint with those made by others in class. Which objects formed the clearest imprints? Why? Answers will vary.

Check Up A

A1. What is meant by evolution? The process of gradual change as simple organisms become more complex.

A2. What is a fossil? Remains or imprints of living things that existed millions of years ago.

A3. In what type of rock are most fossils found? Sedimentary rock.

A4. How does a fossil's location in this rock help tell its age? The lowest layers of sedimentary rock would be the oldest and would contain the oldest fossils.

A5. What is a geologist? A scientist who studies the history of the earth and its life as shown by rock formations.

Words to review: invertebrate; vertebrate

B. The Earth's Timetable

After doing this lesson you will be able:

B1 • to tell about how old the earth is thought to be

B2 • to name the earth's four geological eras and tell about how long each lasted

B3 • to give one change that took place during each of the earth's four geological eras

B4 • to give the scientific name of a giant lizard, called the "terrible lizard," and tell the era in which it lived

B5 • to give the name of the largest flying reptile and tell the era in which it existed

B6 • to give the name of an early reptilelike bird with feathers and claws, and tell the era in which it lived

B7 • to give the name of the earliest fossil of a human being found in Germany in 1856

B8 • to give the name of the fossil of a human being found in Java

B9 • to give the name of a fossil of a human being formed in Peking, China

B10 • to give the name of the people who inhabited 30,000 years ago an area in Europe, which is now France

Geologists have divided the age of the earth into long spaces of time called **eras.** They have done this in order to better know what the earth was like and what kind of life there was at different times in its history.

1. The **Precambrian Era** was the first era of
B1 the earth's geological history. It began when the earth was formed, about 4.5 billion years ago. During this era, oceans, lakes, rivers and mountains were formed. Protists began to appear in waters. Blue-green algae are thought to have started at this time. Later on in this era came simple sea animals, such as sponges and other invertebrates. The Precambrian Era lasted about 4 billion years.

2. The **Paleozoic Era** saw many changes in plants and animals. Many invertebrates lived in the seas. The first vertebrates appeared during this era. They looked like fish. However, they had a hard shell covering the body but no jaw. Their mouths were always open in order to catch food. Later during this period, true (real) fish appeared. This is known because scientists have discovered fossils in sedimentary rocks of this era.

Eras have been broken down into smaller spans of time called periods and epochs.

Plants began to appear on the land. It is thought that the ocean's tides left these plants on the land. Some of these plants died since they were unable to live out of the water. Some, however, were able to live on land. Many-celled plants came from these simple plants. It was easier for these many-celled plants to live on land.

GEOLOGICAL TIMETABLE B2

Era	Forms of Life	B3
PRECAMBRIAN 4 billion years	One-celled life in the sea; no fossils; simple sea invertebrates and algae; very few remains of life exist; blue-green algae may have existed	
PALEOZOIC 345 million years	Invertebrates with shells; trilobites most numerous; starfish and fishlike animals; primitive sharks, lungfish, amphibians; forests and sporebearing plants	
MESOZOIC 160 million years	Age of reptiles; link between birds and reptiles; egg-laying mammals and birds with teeth; dinosaurs become extinct (disappeared); flowering plants and modern insects	
CENOZOIC 65 million years	Age of mammals; primitive monkeys, horses, elephants; number of mammals increases; modern plants; early mammals become extinct; early ape forms; human beings	

Toward the close of this era, air-breathing animals began to live on land as well as in water. Some of them developed lungs that allowed them to breathe air. These animals are **amphibians.** Eventually, some animals developed legs and were able to crawl out of the water onto the land. They lived on land as well as in water.

At the same time, many plants were beginning to grow on the land. Much of the area that is now the United States was a huge swamp. It was hot. Plants grew very large. When the plants died they fell. The swamps eventually dried up. Over a long period of time layers of materials, pressure from the layers and the heat of the earth changed the plants into **coal.**

Fossils of spiders, dragon flies, fish and snails have been found. In this era fossils of plants look like mosses and ferns. The Paleozoic Era lasted about 345 million years.

Brontosaurus means thunder lizard; *Pteranodon* means winged lizard.

3. The **Mesozoic Era** started about 225 million years ago. This era is known as the "age of reptiles." The reptiles probably developed from the amphibians. The reptiles were very small at first. As millions of years passed, some of them developed into very large reptiles called **dinosaurs,** or "terrible lizards." B4

Many of the dinosaurs were vegetarians, or plant eaters. The brontosaur, which was a vegetarian, was the largest reptile to live on land.

Other dinosaurs were meat eaters. They ate the smaller and weaker animals. Tyrannosaur, the largest of the meat eaters, was about 15 meters long, and 6 meters high. It had great hind legs and a powerful tail.

During this era, animals took to the air. At first there were the flying reptiles. Present day birds are related to the flying reptiles which had wings but no feathers. The largest flying reptile was the **pteranodon.** B5

Measure the length of this model of a dinosaur, Stegosaurus. Then measure the length of its head. Do you think this dinosaur had a large brain?

Stegosaurus was a plant eater and had bony plates covering its back and spiked tail, probably for protection against meat-eating dinosaurs, like Tyrannosaurus rex. Compare the size of their mouths referring to the terms carnivore and herbivore.

Other possible reasons for the dinosaur's extinction could have been a combination of factors among which were geographic and climatic changes. The natural habitat of many dinosaurs, the swamp, was being drained away due to crustal movements. Colder climates evolved with the formation of mountains; dinosaurs could not thrive in this weather.

The first true bird appeared toward the end of this era. Many changes took place in the flying reptile before the appearance of the first bird. The animal developed feathers and became warm-blooded. The **archae-**
B6 **opteryx** was a reptilelike bird. It was one of the first birds that had feathers and claws.

From fossils, scientists know that the first mammals evolved during the end of this era. They were small in size.

Dinosaurs became extinct (disappeared) before the end of the Mesozoic Era. Because of their size they needed a lot of food. They probably ate everything in sight, even one another. They ate the eggs they laid. Without eggs, there could be no new life. Why didn't they eat the other forms of life? Birds could fly out of their reach. Small mammals could run faster than the large, clumsy dinosaurs. Mammals were better able to protect themselves because they had a more highly-developed brain.

Plants during the Mesozoic Era were also advancing. A damp, warm climate helped cone-bearing trees and simpler plants to grow everywhere. The Mesozoic Era lasted about 160 million years.

4. The **Cenozoic Era** began about 65 million years ago. It is the era we live in. It has many different kinds of plant and animal life. The climate in the early part of this era was hot and damp. Plant life was like the plant life of hot and damp climates of today, but the mammals were small.

During this era the climate changed from hot and damp to cool and dry. The plants and animals had to live through this change to survive. Flowering plants began to look as they do today and grasses started to grow.

Animals had to change because their food changed. Some of the mammals grew larger. Monkeys, horses and elephants evolved during this era.

B10

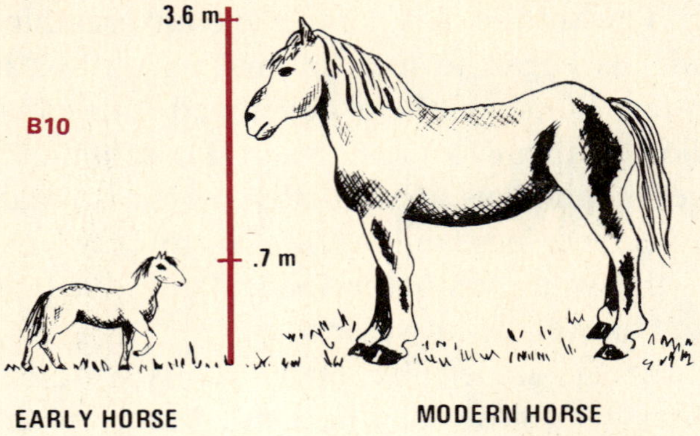

Measure the height of these horses. Can you give roughly the difference in size of the modern horse as compared with the early horse?

During the Cenozoic Era the first human beings appeared. The earliest fossil of a human being was found in Neanderthal, a valley in western Germany, in 1856. Although this fossil was somewhat different from living people, it was a race of modern people. It is B7
called **Neanderthal Man.** In the 1890s, another fossil was found in Java, from which it B8
got its name, **Java Man.** About 1929 some more human fossils, including skulls, were found near Peking, China. One skull is called B9
Peking Woman. Other fossils of human beings have been found since then.

The most recent of these early human beings is the **Cro-Magnon** people. They lived some 30 thousand years ago in the area of Europe which is now France.

These early human beings made simple tools and it is believed that they could talk. However, they are different from present-day people as you can see from the pictures. For example, it is known that Neanderthal people were shorter than present-day people, averaging about 1.5 meters in height. Their faces were full with deep-set eyes and they had a stocky build. Their brains were slightly smaller than those of modern people.

You have seen how living things have slowly become more advanced by slowly changing through the eras of the earth. It is

Recent fossil evidence indicates that bipedalism (walking erect) evolved before the development of a brain. Skulls and face bones recently unearthed are less than one-third the size of the modern brain. Bipedalism is the key to separation between humans and apes.

Neanderthal man lived between 100,000 and 400,000 years ago; they used fires, lived in caves, and were the first to have burials and possibly religious rites.

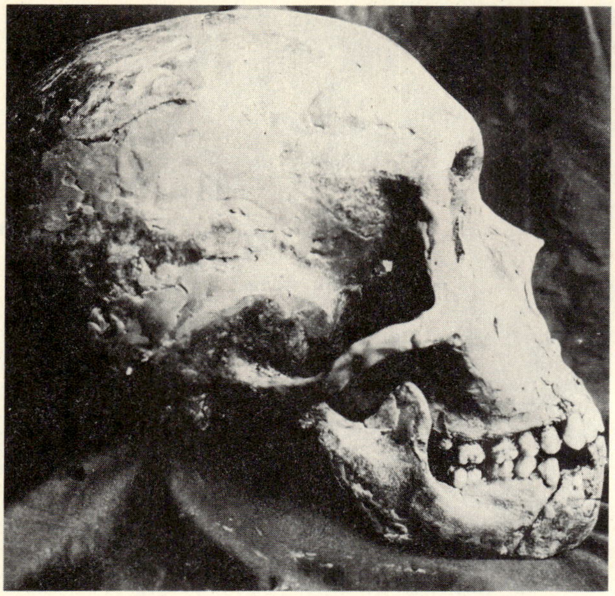

A rebuilt Neanderthal skull. Compare it with the pictures of other skulls in this section.

A rebuilt skull of Java Man.

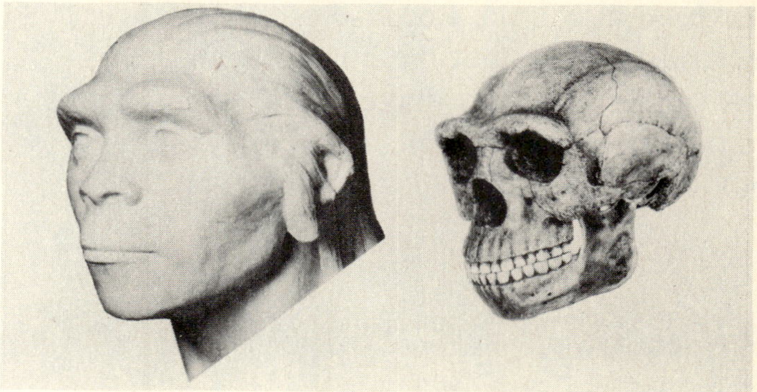

A rebuilt skull of Peking Woman.

Java man and Peking woman both belong to an extinct species of humans called Homo erectus. These humans existed from 300,000 to a million years ago, had erect posture and used fires and primitive tools.

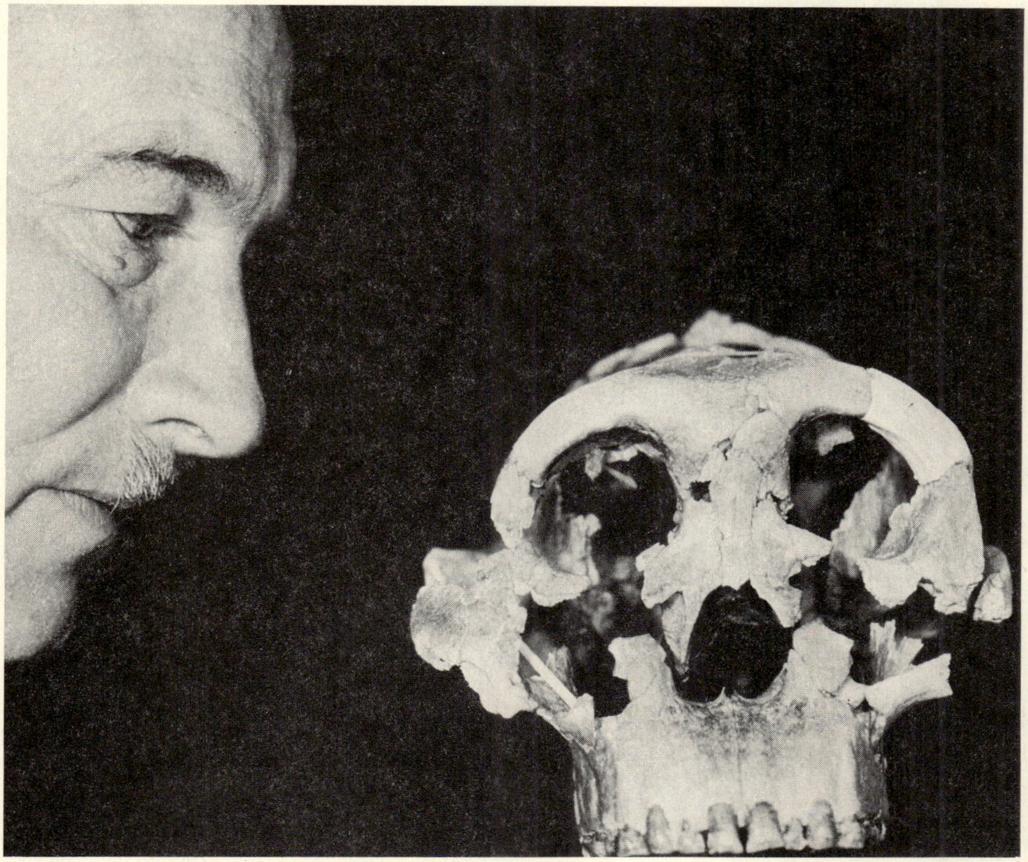

Dr. Louis Seymour Bazett Leakey (1903–1972), a famous British archeologist, with the skull of Zinjanthropus. *This is the skull of an 18 year old man who lived over a million years ago. Its name combines two words,* zinj *meaning East Africa, where the skull was found, and* anthropus, *from a Greek word meaning human being.*

Students may want to find out more about the interesting work of Louis Leakey and Mary Leakey who specialized in the early ancestry of humans.

Zinjanthropus actually belongs to the genus *Australopithecus*. Most Australopithicines lived between 1 million and 2.5 million years ago. Skeletal features usually indicate that this genus resembled modern humans but had a less developed brain and less dexterity.

only reasonable to think that living things will keep on changing.

In 1972, a skull thought to be 2.3 million years old, was found in Africa. In 1976, Mary Leakey, a British archeologist (a scientist who studies how ancient peoples lived) found some footprints that are thought to be 3.6 million years old. They were made by a creature that walked like a human being—on two legs. In 1979, some jawbones that are more than 10 million years old were discovered in Burma. Because of this discovery scientists think that the human being's earliest ancestors may have come from Asia rather than from Africa. But they are still not certain.

Check Up B

B1. How old is the earth thought to be? 4.5 billion years old.

B2. What are the four geological eras and about how long did each last?

(1) Precambrian — 4 billion years

(2) Paleozoic — 345 million years

(3) Mesozoic — 160 million years

(4) Cenozoic — 65 million years

B3. Give one event that took place during each of the four geological eras.

(1) Answers will vary.

(2)

(3)

(4)

B4. What is the scientific name of a giant lizard called the "terrible lizard"? When did it exist? Dinosaurs; Mesozoic Era

B5. What is the name of the largest flying reptile? In which era did it exist? Pteranodon; Mesozoic Era

B6. What is the name of a reptilelike bird with feathers and claws? When did it exist? Archaeopteryx; Mesozoic Era

B7. What is the name given to the earliest fossil of a human being found in Germany in 1856? Neanderthal Man

B8. What is the name given to the fossil of a human being found in Java? Java Man

B9. What is the name given to the fossil of a human being found in Peking, China? Peking Woman

B10. What were the people called who lived 30,000 years ago in an area of Europe which is now France? Cro—Magnon people

Unit 1, Chapter 4
REVIEW

Matching Exercise Match each term in **Column A** with its description in **Column B**. Write the letter of the description in the space beside the correct term.

	Column A	*Column B*
f	**1.** fossils	**a.** type of rock in which fossils are found
j	**2.** evolution	**b.** the era the Earth is in now
a	**3.** sedimentary rock	**c.** the earliest era of the Earth
c	**4.** Precambrian era	**d.** scientist who studies rocks
i	**5.** Archaeopteryx	**e.** earliest known human being
e	**6.** Neanderthal man	**f.** remains or imprints of early life
b	**7.** Cenozoic era	**g.** largest flying reptile
g	**8.** pteranodon	**h.** an early form of modern people
h	**9.** Cro-Magnon	**i.** reptilelike bird with feathers and claws
d	**10.** geologist	**j.** process of gradual change

True/False In the space provided, write **T** if a statement is true. If it is false, change the underlined word or words to make the statement true.

highest or uppermost **1.** Fossils of the most complex forms of life are usually found in the lowest layers of sedimentary rock.

Mesozoic **2.** Dinosaurs lived during the Paleozoic Era.

T **3.** One of the reasons the dinosaurs became extinct was because they ate their own eggs.

ferns **4.** Fossils of plants from the Paleozoic Era look like the mosses and flowering plants of today.

Precambrian **5.** Sponges, blue-green algae and protists first appeared during the Cenozoic Era.

T **6.** The early horse was about the same size as many small dogs of today.

more complex **7.** Evolution is the process of changes as simple organisms become less complex.

one-celled protists **8.** The simplest living things on earth today are the fishes.

trilobites **9.** The oldest fossils found in sedimentary rocks are of the remains of dinosaurs.

T **10.** The climate in the early part of the Cenozoic Era was hot and damp.

T **11.** Dinosaurs were poorly-adapted for survival.

Neanderthal **12.** The earliest fossil of a human being was found in Java.

fossils **13.** The remains or imprints of living things that existed long ago are called evolutions.

Cro-Magnon **14.** The most recent of the early human beings are Peking people.

Cenozoic **15.** The earth is now in the Precambian Era.

Multiple Choice Exercise Circle the letter of the word or phrase that best completes each statement.

1. Animals that can breathe on land and in water are **(a)** amphibians **(b)** reptiles **(c)** mammals **(d)** trilobites.
2. Remains or imprints of life from long ago are called **(a)** sediments **(b)** minerals **(c)** fossils **(d)** graphite.
3. The Age of Mammals is the **(a)** Precambrian era **(b)** Paleozoic era **(c)** Mesozoic era **(d)** Cenozoic era.
4. The oldest fossils found in sedimentary rocks are of **(a)** dinosaurs **(b)** trilobites **(c)** flowering plants **(d)** mammals.
5. The first era in the earth's geological history was the **(a)** Precambrian era **(b)** Paleozoic era **(c)** Mesozoic era **(d)** Cenozoic era.
6. Dinosaurs lived during the **(a)** Precambrian era **(b)** Paleozoic era **(c)** Mesozoic era **(d)** Cenozoic era.
7. The largest flying reptile was the **(a)** trilobite **(b)** pteranodon **(c)** stegosaurus **(d)** tyrannosaur.
8. All living things are made in part of **(a)** carbon **(b)** graphite **(c)** fossils **(d)** coal.
9. The process of very slow change as living things become more complex is **(a)** graphite **(b)** fossilization **(c)** sedimentation **(d)** evolution.
10. The earliest human beings are known as **(a)** Java Man **(b)** Peking Woman **(c)** Neanderthal Man **(d)** Cro-Magnon people.

Application Exercise Indicate the era described by each of the occurrences below. Choose from these eras: **Precambrian, Paleozoic, Mesozoic,** and **Cenozoic.** Write your answer in the space provided.

Cenozoic 1. Earth's present era
Precambrian 2. one-celled life lived in the seas
Cenozoic 3. human beings appear on Earth
Mesozoic 4. the Age of Reptiles
Paleozoic 5. trilobites are very common
Precambrian 6. no fossils exist for this era
Paleozoic 7. forests and sporebearing plants appear
Mesozoic 8. flowering plants and modern insects appear
Mesozoic 9. dinosaurs become extinct
Precambrian 10. oceans, lakes, mountains and rivers form
Mesozoic 11. first true bird appears
Cenozoic 12. Cro-Magnon people live on Earth
Precambrian 13. many invertebrates live in the seas
Paleozoic 14. amphibians begin to appear.
Cenozoic 15. climate changed from hot and damp to cool and dry

Get Ready To Read UNIT 2

LOOKING FOR RELATIONSHIPS

In Unit 1, you learned that it is important to make comparisons among things that are the same and things that are different. Sometimes it is helpful also to take information that is new to you and compare it to something you already know. Relating new information to what you already know helps you to understand and remember what you read.

The astronaut in this drawing is trying to explain what the soil on the planet Zeno feels like by comparing it to something familiar. If you know what wet cement looks and feels like, you have some idea of what the soil on the planet Zeno is like.

What does the astronaut compare the soil to? wet cement

Do you think the soil is heavy or light in weight? heavy

Do you think the soil is thin and watery or thick and lumpy? thick and lumpy

Is the soil gray or brown in color? gray

Have students make other comparisons. For example, they could describe a new pair of slacks, or a coat, or a dress by comparing it to similar clothing or objects of the same color or fabric.

In Unit 2, you will be reading about how the human body works. You will be learning new information that you can relate to what you already know. For example, you will find out that the circulatory system can be compared to a railroad system. Use what you already know about a railroad system to do the following exercise. Draw a line to show which part of the circulatory system best matches the part of a railroad system in terms of the job each part does.

PARTS OF THE CIRCULATORY SYSTEM	PARTS OF A RAILROAD SYSTEM
Blood vessels carry blood through your body	train
Blood travels through the blood vessels	engine
The heart is the power that pumps blood through the blood vessels	railroad tracks

You will also read that the body's nervous sytem is the *communication* system of your body. To which of the following systems might you compare the nervous system? a heating system? the solar system? a telephone system?

As you read Unit 2, think about how the new information you are learning relates to what you already know.

USING GRAPHIC AIDS

In this Science Work-A-Text you will be asked to read charts, diagrams, and graphs. These are called **graphic aids.** Graphic aids help to explain the topic, or subject that is written about in the text. They also give the reader more information about the topic.

Now, read the following paragraph about teeth. Then label the diagram of a tooth based on what you read.

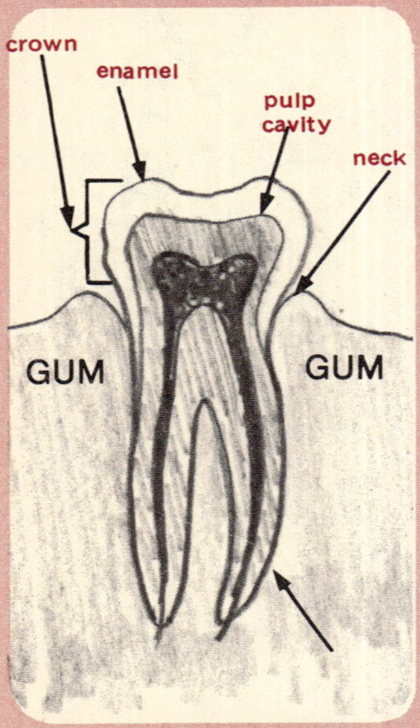

Have you ever looked carefully at one of your teeth? Each tooth is made up of three parts. The part that sticks out of the gums is called the *crown.* The crown is covered by a very hard substance called *enamel.* Directly next to the crown is the tooth's *neck,* the narrow part of the tooth that is surrounded by the gums. The roots of a tooth extend down into the jawbone. Inside the tooth is the *pulp cavity.* This part of a tooth contains nerves and blood vessels.

Chapter 8 in Unit 2 is called *Keeping Healthy.* A chart, such as the one on snacks, gives the reader more information about the topics in Chapter 8.

Ask the students to select one diagram (with labels) from any page in their science books. Have them write a brief paragraph describing the object and telling where each labelled part is located. These should be underlined in their paragraph. They should exchange diagrams, being sure not to get the same diagram they selected. Have them label the "new" diagram using the descriptive paragraph as an aid.

COMPARE AND CONTRAST THIS INFORMATION

Nutritional Value of Some Common Snacks

	fresh orange juice	soda pop	malted milk	raw carrot	pizza	chocolate bar
amount	1 cup	1 cup	1 cup	about 8 strips	1 slice	1.5 oz bar
calories	110	96	245	20	185	160
protein	2 g	0	11 g	1 g	7 g	2 g
calcium	27 g	0	317 g	18 g	107 g	33 g
iron	.5 ml	0	.7 ml	.4 ml	.7 ml	.4 ml
Vitamin C	12.4 ml	0	2 ml	4 ml	4 ml	00
Riboflavin	.07 ml	0	.49 ml	.3 ml	.12 ml	.05 ml

Based on percentage of RDA—Recommended Daily Allowance

1. Which snacks are the most nutritious? orange juice, malted milk
2. Which two snacks contain the same amount of Vitamin C?

   carrot and pizza

3. Which snack offers nothing but calories? soda pop
4. According to the chart, the highest amount of calcium can be found in malted milk.

One type of graphic aid is called a bar graph. Anita used a bar graph to show which activities ten students in her sixth grade class like to do best. Use Anita's bar graph to answer questions 1 through 5.

Write *T* if the statement is true and *F* if the statement is false.

F 1. More students enjoy sports of some type more than any other activity.

F 2. Watching TV and reading comic books are enjoyed by the same number of students.

T 3. Few students enjoy reading poetry in their spare time.

T 4. For about 8 students camping is their favorite activity.

F 5. The same number of students who enjoy seeing movies enjoy listening to music.

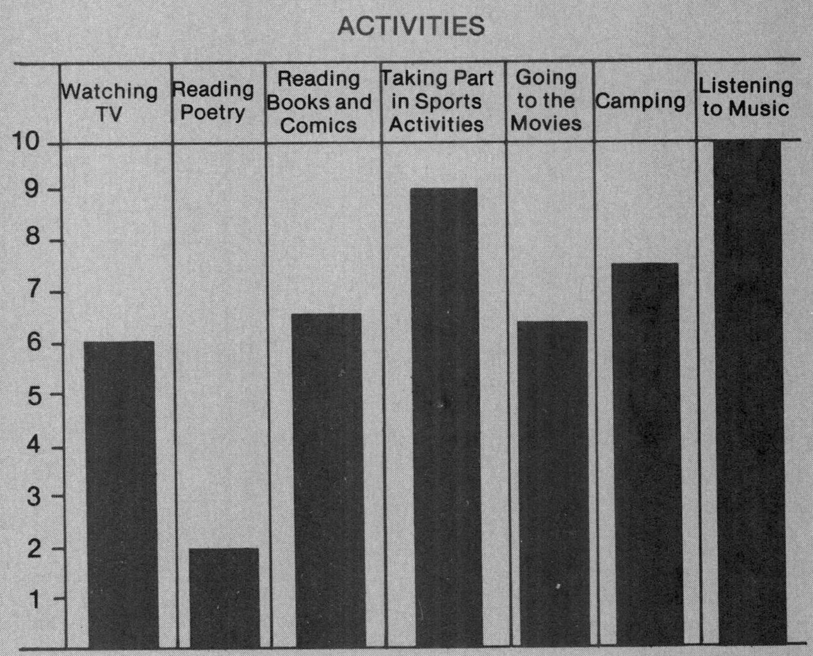

Now, use what you have just learned about graphic aids to help you read the charts, graphs and diagrams you will see in this text.

Have each student select a chart or graph from any of their textbooks. Have them make up some questions using information from the chart they chose. The answers should be found on the chart or graph. (Some students may want to make up their own chart or graph.)

UNIT 2

Introduction to Human Biology

Looking through a microscope, this student is studying a sample of blood tissue.

New science words: tissue, organ, system

Skills developed: observing, comparing

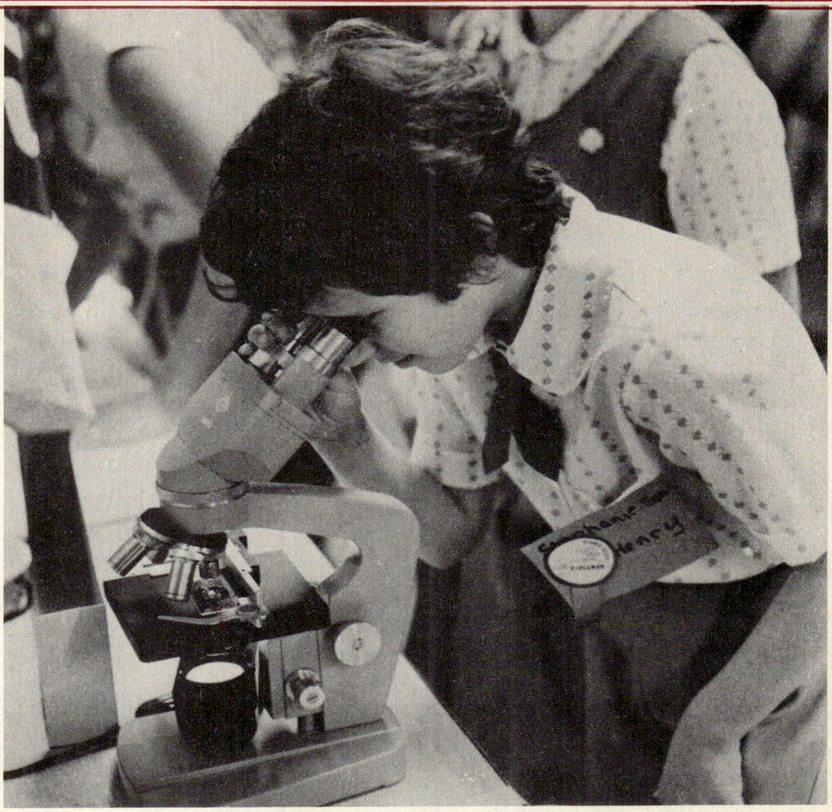

Chapter 1

WHAT IS YOUR BODY MADE OF?

You have learned that all living things are made up of tiny parts of living material called **cells.** Your body has trillions of cells. Earlier you learned that one-celled organisms are able to carry on all the life activities. However, in many-celled plants and animals, cells have become special and carry on special functions. They work together for the good of the entire body. A group of cells that carry on a special function is called a **tissue.** Tissues that work together to perform a special function are called **organs.** Organs that work together are known as a **system.**

A. Tissues

After doing this lesson you will be able:

A1 • to list five main types of tissues that make up your body

A2 • to tell where epithelial tissue is found and give its two main functions

A3 • to name three kinds of muscle tissue and tell one place where each kind is found

A4 • to explain the difference between voluntary and involuntary muscles

A5 • to list the three kinds of connective tissue and give one function of each kind

A6 • to explain the difference between ligaments and tendons

A7 • to describe the two functions of the blood

A8 • to list the four parts of the blood

Human beings, like all higher animals, are made up of five main types of tissues: (1) A1
epithelial, (2) **muscle,** (3) **nerve,** (4) **connective** and (5) **blood** tissues.

1. **Epithelial tissue** makes up the skin and A2
lines almost all of the inner parts of the body. Epithelial tissue also makes up the glands

Have the students observe various types of tissues in prepared slides under a microscope or microprojector.

which secrete (give off) chemical substances needed for body activities, such as digestion. Its main functions are protection and secretion.

2. **Muscle tissue** makes it possible for you to move and work. Your body has three differ-
A3 ent types of muscle tissue: (a) **voluntary,** (b) **involuntary** and (c) **cardiac** (heart) muscle.

A4 **a.** "Voluntary" means you can control these muscles. For example, you can control the muscles in your arms and legs.

b. "Involuntary" means you have no control. The muscles in your stomach are involuntary. They act on their own.

A4 **c.** A third kind of muscle tissue is found only in the heart. It is called **cardiac muscle.** Your heart is also an involuntary muscle.

3. **Nerve tissue** carries messages from one part of the body to other parts of the body. Messages travel along the nerves and cause muscles in your body to act.

4. **Connective tissue** helps to support, shape and protect the body. Some kinds of connec-
A5 tive tissue are: (a) **fatty tissue,** (b) **fibrous tissue** and (c) **cartilage** and **bone.**

a. Fatty tissue helps protect the body against the cold. It is stored by the body in the deep layers of the skin and can be used as food if it is ever needed.

b. Fibrous tissue helps hold the body together. Some examples of fibrous tissue are **tendons** and **ligaments.** The place where two bones meet is called a **joint.** You have many joints in your body. If you did not have them, you would not be able to walk, swim or pick up things. In fact, you would not be able to
A6 move. **Ligaments** hold the bones together at the joint. **Tendons** connect muscle to bone.

c. Cartilage and **bone** help support the body. They also protect inner organs and give the body its shape. Because of the minerals it contains, bone is hard. Cartilage is tough and elastic.

Words to review: secretion, flexible

There are about 206 bones in your body. These bones make up your skeleton. The bones act as machines that make it possible for you to move and do work.

5. **Blood tissue** is often referred to as blood. Although it is a liquid, blood is considered to be a tissue because it is made up of similar
cells doing a special job. The blood carries A7
food and oxygen to all parts of the body. It also takes away waste materials made by the cells of the body. Blood tissue is made up of
plasma, red blood cells, white blood cells and A8
platelets.

Check Up A

A1. What are the five main types of tissues that make up your body?

(1) Epithelial

(2) Muscle

(3) Nerve

(4) Connective

(5) Blood

A2. Where is the epithelial tissue found? What are its two main functions? Makes up the skin. Its two main functions are protection and secretion.

A3. Name the three types of muscle tissue and tell one place where each is found.

(1) Voluntary – arms and legs

(2) Involuntary – stomach

(3) Cardiac – heart

 A tendon that students can easily find on their own bodies is the "Achilles" tendon on the back of the leg, right above the heel.

Blood tissue differs from skin tissue and heart muscle tissue in that blood is a fluid tissue. The plasma, the liquid part of the blood, carries dissolved and suspended materials as it circulates throughout the body via the blood vessels.

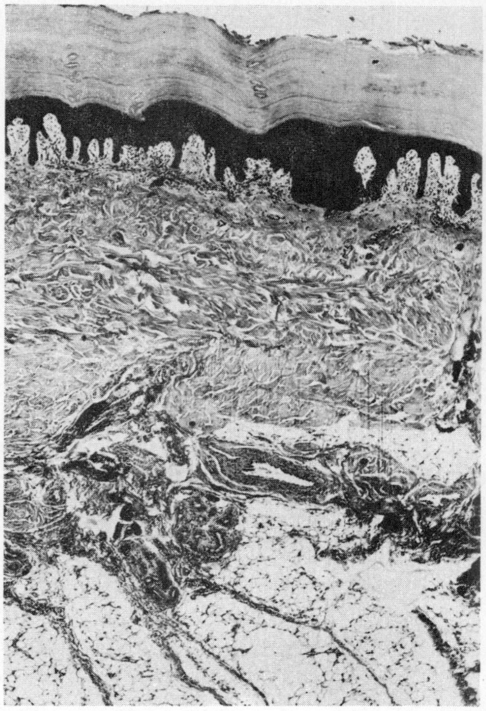

(a)

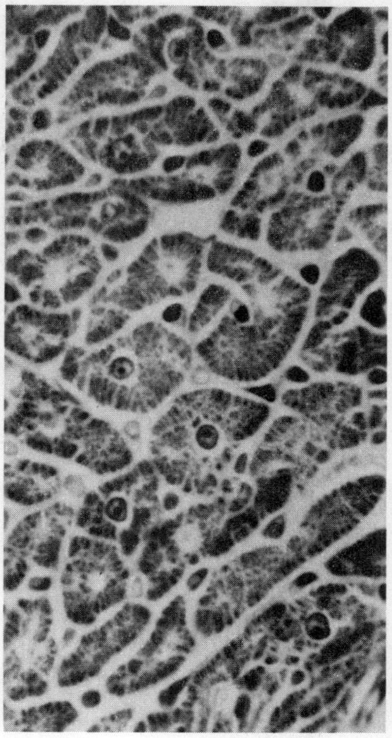

(b)

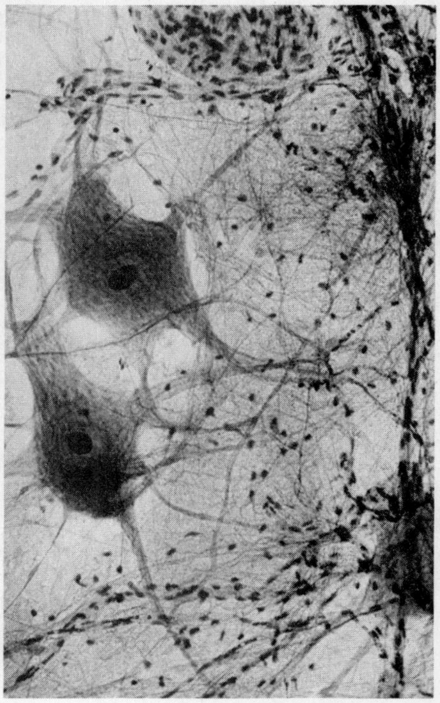

(c)

(a) Cross section of skin tissue. The outer layer is dead skin tissue. (b) Cross section of heart muscle tissue. Note the central location of nuclei in the muscle fibers. (c) Nerve cells. (d) A section of human bone. (e) Blood tissue of a human being. The white blood cells are stained and look darker than the red blood cells.

(d)

(e)

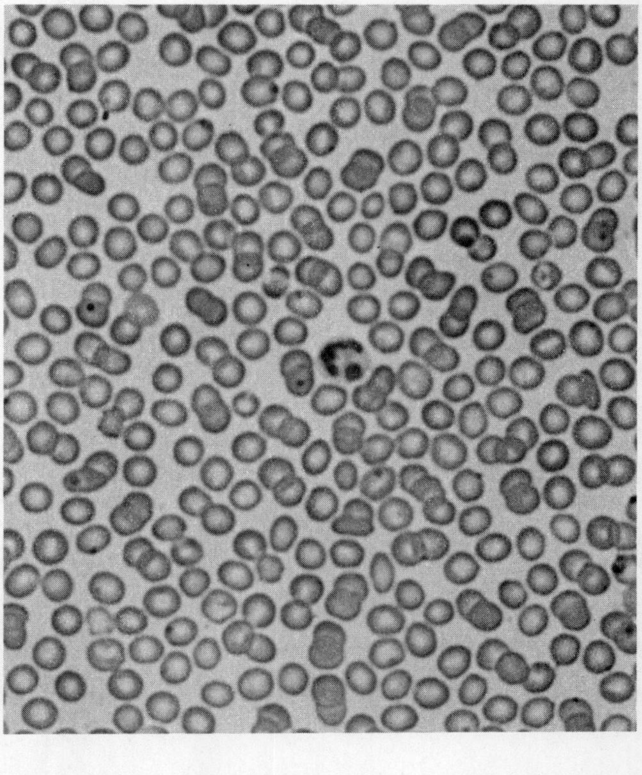

Photos (a & d) courtesy of Carolina Biological Supply Co.

A4. Describe the difference between voluntary and involuntary muscles. Voluntary muscles can be controlled. Involuntary muscles cannot be controlled.

A5. Name the three kinds of connective tissue and give one function of each kind.

(1) Fatty — insulates the body against cold.

(2) Fibrous — helps hold the body together.

(3) Cartilage and bone — support the body.

A6. What is the difference between a ligament and a tendon? Ligaments hold bones together at joints. Tendons connect muscles to bones.

A7. What are the two functions of the blood?

(1) Carries food and oxygen to all parts of the body;

(2) Removes waste materials.

A8. List the four parts of the blood.

(1) Plasma

(2) Red blood cells

(3) White blood cells

(4) Platelets

THINK ABOUT THIS

The World Food Problem

The world population has been increasing by about 2% each year. At this rate, the population will be doubled by the year 2000. The earth's surface that is above the sea will not be able to supply enough food, energy, living space and natural resources to support all these people.

The oceans may be able to help with these problems. The sea produces five to ten times as much living matter as does the land. Scientists are finding ways to make use of these possible food resources. Some advances have been made.

In limited numbers, oysters, shrimp, mussels and eels are being raised by "sea farming" methods. Mineral-rich water pumped from deep parts of the ocean into coastal bays has increased plant growth. This method supplies much food for marine (sea) animals.

Algae and plankton (microscopic plants and animals) as a source of food for human beings has had very little success. People just don't want to eat plankton!

Unit 2, Chapter 1
REVIEW

Matching Exercise Match each term in **Column A** with its description in **Column B**. Write the letter of the description in the space beside the correct term.

	Column A		*Column B*
i	**1.** bone	**a.**	connects muscle to bone
h	**2.** cells	**b.**	cardiac muscle
f	**3.** tissue	**c.**	holds bones together at joints
e	**4.** system	**d.**	carries messages throughout the body
g	**5.** organ	**e.**	organs that work together
b	**6.** heart muscle	**f.**	group of similar cells
j	**7.** joint	**g.**	tissues working together to perform a special function
d	**8.** nerve tissue	**h.**	building blocks of all living things
a	**9.** tendon	**i.**	gives the body shape and support
c	**10.** ligament	**j.**	place where two bones meet

True/False In the space provided, write **T** if a statement is true. If it is false, change the underlined word or words to make the statement true.

a tissue ______ **1.** A group of cells that carry on a special function is called <u>an organ</u>.

Epithelial ______ **2.** <u>Muscle</u> tissue makes up the skin and lines almost all of the inner parts of the body.

epithelial ______ **3.** The glands of the body are made up of <u>muscle</u> tissue.

T ______ **4.** <u>Muscle</u> tissue makes it possible for the body to move.

connective ______ **5.** The body get its support, shape and protection from <u>epithelial</u> tissue.

T ______ **6.** Blood tissue is made up of plasma, red blood cells, <u>platelets</u> and white blood cells.

heart ______ **7.** Cardiac muscle is found only in the <u>stomach</u>.

an involuntary ______ **8.** Cardiac muscle is <u>a voluntary</u> muscle.

cartilage ______ **9.** The body gets its shape from <u>ligaments</u> and bone.

T ______ **10.** Nerve tissue carries <u>messages</u> from one part of the body to another.

Fatty ______ **11.** <u>Fibrous</u> tissue helps protect the body against the cold.

Ligaments ______ **12.** <u>Tendons</u> hold the bones together at the joints.

Tendon ______ **13.** <u>Cartilage</u> connects muscle to bone.

T ______ **14.** Tendons and ligaments are examples of <u>fibrous</u> connective tissues.

voluntary ______ **15.** The muscles in the arms and legs are <u>involuntary</u> muscles.

Multiple Choice Exercise Circle the letter of the word or phrase that best completes each statement.

1. The human skeleton is made up of **(a)** 102 bones **(b)** 206 bones **(c)** 352 bones **(d)** 406 bones.
2. All of the following are parts of the blood except **(a)** plasma **(b)** platelets **(c)** adipose **(d)** white blood cells.
3. Protection and secretion are the two main functions of **(a)** connective tissue **(b)** epithelial tissue **(c)** blood tissue **(d)** cartilage and bone.
4. Messages are carried from one part of the body to other parts of the body by **(a)** connective tissue **(b)** epithelial tissue **(c)** nerve tissue **(d)** blood tissue.
5. Muscle that can be controlled is **(a)** involuntary muscle **(b)** cardiac muscle **(c)** epithelial muscle **(d)** voluntary muscle.
6. Bones are held together at the joints by **(a)** ligaments **(b)** tendons **(c)** cartilage **(d)** nerve tissue.
7. Muscle is connected to bone by **(a)** ligaments **(b)** tendons **(c)** cartilage **(d)** nerve tissue.
8. A group of cells that carry on a special function is called **(a)** an organ **(b)** a system **(c)** a tissue **(d)** a cartilage.
9. Cardiac muscle is **(a)** voluntary **(b)** involuntary **(c)** epithelial tissue **(d)** nerve tissue.
10. Cartilage and bone are types of **(a)** fatty tissue **(b)** nerve tissue **(c)** muscle tissue **(d)** connective tissue.

Short Answer Answer each of the following in complete sentences.

1. Why is blood considered to be a tissue? Blood is a tissue because it is made up of a group of similar cells that work together to perform a specific function.

2. What are the three types of muscle tissue that are found in the body? Name one place where each type of muscle tissue can be found. The three types of muscle tissue found in the body are voluntary muscle, involuntary muscle and cardiac muscle. Voluntary muscle is found in the arms and legs. Involuntary muscle is found in the stomach. Cardiac muscle is found only in the heart. (Examples may vary).

New science words: alimentary, digestion, enzyme

Skills developed: observing, comparing

Do you know how their digestive systems will change the food so their bodies can use it?

Chapter 2

HOW THE BODY DIGESTS FOOD

Four to six hours after a meal is eaten, some of that food is in the blood and on its way to the cells.

The cells of your body cannot use foods as they are when you eat them. To be usable, foods must be broken down into a form that can be used by the cells. The process by which foods are broken down into a usable liquid form is **digestion.** The system in which the process of digestion is carried on is the **digestive system.**

A. Digestion

After doing this lesson you will be able:

A1 • to list the two types of digestion and explain the difference between them

A2 • to name the substances that change the chemical makeup of food and tell where they are produced

A3 • to explain how enzymes work

There are two kinds of digestion: (1) mechanical digestion and (2) chemical digestion.

1. Mechanical digestion happens when food is A1
chewed, mixed and broken down into smaller pieces. This process helps food pass more easily through the digestive system.

2. Chemical digestion. Some proteins called A2
enzymes change the chemical makeup of the food you eat. Many different kinds of enzymes are given off by organs called **glands.** A3
These enzymes act upon food particles to break them down into simple liquid form. Enzymes do not become part of the chemical change; they just make the chemical reaction A3
go faster. Thus, they are not used up in digestion, but are used over and over again.

Digestion begins in the *mouth* and ends in the small intestine. Saliva begins to break down starches into sugar.

On a chart of the digestive system help students identify the organs.

Check Up A

A1. Name the two types of digestion. What is the difference between them? (1) Mechanical — food is broken up into small pieces. (2) Chemical — enzymes change the chemical structure of food.

A2. What substances change the chemical makeup of food? Where are they produced? Enzymes; they are produced in glands.

A3. How do enzymes work? Enzymes act on food particles to break them down into simple liquid form.

B. The Alimentary Canal

After doing this lesson you will be able:

- B1 • to tell what the alimentary canal is
- B2 • to list the five organs that make up the alimentary canal
- B3 • to name two organs that are not part of the alimentary canal but help in digestion
- B4 • to tell how many permanent teeth adults have
- B5 • to name the glands in the mouth that produce a digestive juice and name this juice
- B6 • to name the enzyme found in saliva and explain what it does
- B7 • to name the glands in the lining of the stomach and name the digestive juice they give off
- B8 • to tell what rennin, pepsin and hydrochloric acid do in digestion
- B9 • to name the thick liquid that leaves the stomach

The rectum opens to the exterior through the anus or anal opening. Ask students if they can remember the structure in the paramecium through which solid wastes are released (anal spot).

- B10 • to give the average length of the small and large intestines
- B11 • to explain the function of the large intestine

Food passes through a long tube called the **alimentary canal,** a path for food between the mouth and the rectum. B1 The organs that make up the alimentary canal include the **mouth, esophagus, stomach, small intestine** and **large intestine.** B2 The digestive system includes the alimentary canal through which food passes and organs which give off enzymes that carry on digestion in the canal. These organs are the **liver** and **pancreas.** B3

1. The mouth. Food first enters the alimentary canal through the mouth. Digestion, therefore, begins in the mouth.

Your teeth grind and cut the food into small pieces. The enzymes work better when the pieces of food are small. An adult should have 32 permanent teeth. There are four dif- B4

The digestive system.

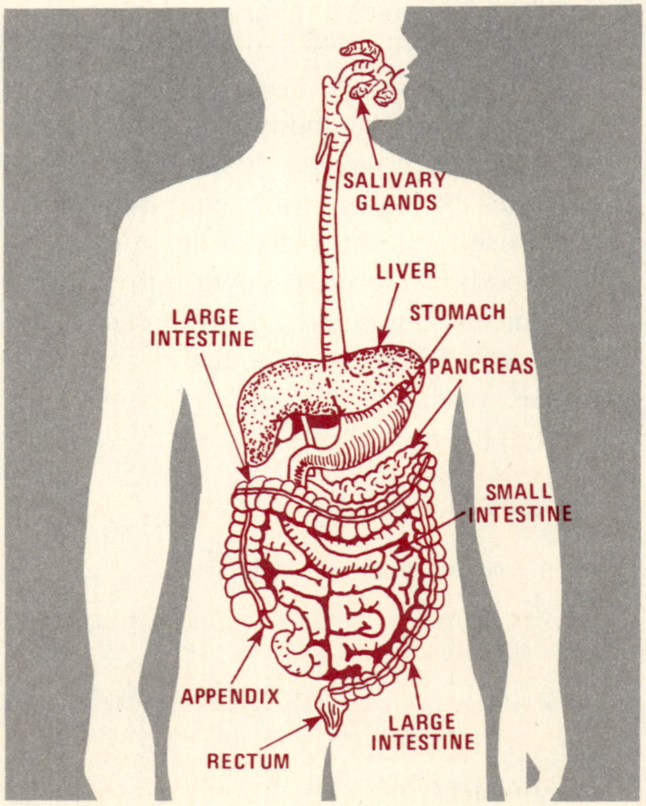

The alimentary canal is a food tube approximately 9 meters long. The small intestine which absorbs the end products of digestion makes up about 2/3 of this length.

Many sixth grade students will still have only 24 teeth at this time. By the end of the school year, most of them will probably have 28. The final four (the wisdom teeth) come in between the ages of 17 and 21 years.

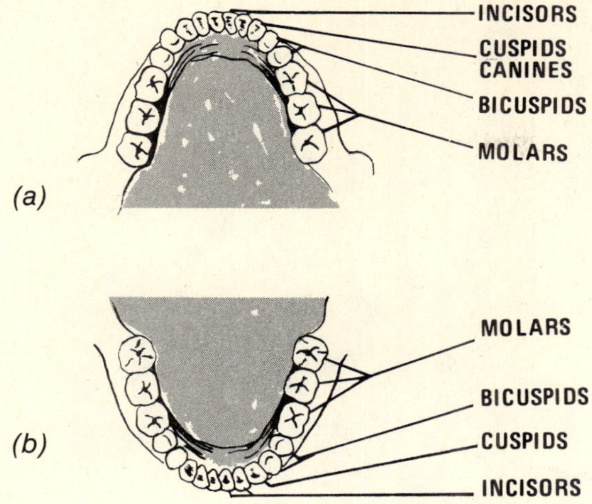

Take care of your permanent teeth. They are the last you can grow! (a) upper teeth of an adult person; (b) lower teeth of an adult person. The last four teeth to appear in your mouth are the "wisdom teeth".

ferent types of permanent teeth: **incisors, cuspids, bicuspids** and **molars**.

The **incisors** are the teeth found in the front of your upper and your lower jaw. The eight incisors have straight edges and are used for cutting food. On either side of the incisors you have a sharp pointed tooth. These four teeth, called the **cuspids,** or **canines,** tear food apart. Behind the cuspids on both sides of each jaw there are two **bicuspids.** Each of these eight bicuspids has two flat surfaces. They are used for grinding food. Behind the bicuspids are the **molars** which are also used for grinding. You should have eight molars. This gives you 28 teeth altogether. The last four molars are called the "wisdom teeth." Look at your teeth in a mirror and see if you can locate the four different kinds of teeth.

The part of a tooth that you can see is the **crown.** The narrow part of the tooth inside the gum is the **neck.** The **roots** extend from the neck of the tooth to the jawbone and hold the tooth in place.

The outside of the crown consists of a hard material containing calcium. This material is **enamel.** Underneath the enamel is **dentin.** Dentin makes up the mass of the tooth. It is somewhat like bone tissue but harder. Within the tooth there is a hollow space called the **pulp cavity.** Inside the cavity there are many blood vessels and nerves. This is why drilling into the cavity may cause pain.

Words to review: contraction, abdomen, diaphragm

The **tongue** helps push the food around so it can be chewed more easily. It also helps push the food toward the back of the mouth so it can be easily swallowed. The tongue has tiny **taste buds** that pick up the flavor of food.

In addition to the mechanical digestion carried out by your teeth, chemical digestion also takes place in the mouth. Three pairs of **salivary glands** give off a digestive juice B5
called **saliva.** Saliva contains the enzyme B6
amylase. As the food is chewed, it mixes with the amylase which changes some of the starch to sugar.

2. The esophagus. As the food is swallowed, it passes from the mouth into the throat. Then it passes into the **esophagus,** or **gullet.** The esophagus is a long, muscular tube that connects the throat with the stomach. Wavelike movements squeeze the food down the esophagus and into the stomach.

3. The stomach is a pear-shaped organ located in the upper left side of the abdomen under the diaphragm. The walls of the stomach have a great deal of muscle tissue. The inner lining of the stomach has many tiny **gastric glands** that give off the digestive juice B7
called **gastric juice.** Gastric juice contains two enzymes, **rennin** and **pepsin,** as well as a small amount of **hydrochloric acid.** Rennin B8
digests the protein in milk. Pepsin and hydrochloric acid act together to start the digestion of other proteins.

After remaining in the stomach a few hours, the food is only partly digested. It is still not broken down enough to enter the bloodstream or be used by the cells. The food that leaves the stomach is a thick liquid called **chyme.** B9

Peristalsis is the term given to these wavelike contractions of the involuntary muscle that lines the alimentary canal.

The large intestine absorbs excess water from the food wastes, making them a soft solid rather than a liquid.

4. The small intestine. The partly digested food then passes into the **small intestine.** It is a narrow tube about 7 meters long. The process of digestion is completed in the small intestine. B10

5. The large intestine. Food substances that the body does not digest are passed on through the **large intestine.** It is wider than the small intestine and about $1\frac{1}{2}$ meters long. B10
The undigested wastes pass to the **rectum** at the end of the large intestine and then out of the body through an opening called the **anus.** B11

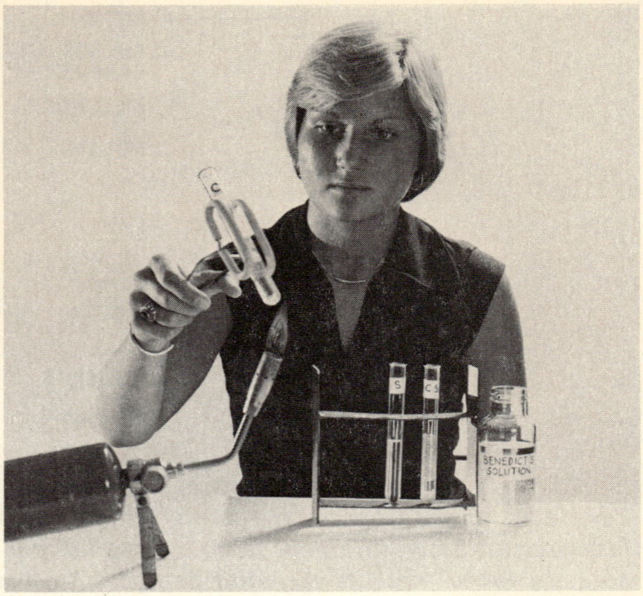

Discover for yourself

how you could show that digestion begins in the mouth.

PURPOSE: To observe the effects of ptyalin on starch • To develop the skills of observation and comparison

You need: unsalted soda crackers, 3 test tubes, test tube rack, Benedict's solution, Bunsen burner or alcohol lamp, matches

a. Put a small piece of cracker in a test tube. Mark the test tube C. Fill the tube about one-third full with Benedict's solution. Place test tube C on a test tube rack.

b. Pour some saliva into a test tube and mark it S. Add Benedict's solution. Put test tube S on the test tube rack.

c. Chew a piece of soda cracker and put it in a test tube. Mark it CS. Add Benedict's solution. In turn, heat all three test tubes slightly over a Bunsen burner or an alcohol lamp.

1. What happens to test tubes C and S when heated? They remain the same.

2. What happens to test tube CS when heated? It becomes brick-red.

d. Benedict's solution turns brick red in presence of sugar. The change of color proves that sugar is present in test tube CS.

1. Where did the sugar come from? The chewed cracker.

2. What enzyme changes starch to sugar? Amylase

Check Up B

B1. What is the alimentary canal? A passageway for food between the mouth and the rectum.

B2. List, in order, the five organs that make up the alimentary canal.

(1) Mouth

(2) Esophagus

(3) Stomach

(4) Small intestine

(5) Large intestine

B3. What two organs help in digestion, but are not part of the alimentary canal?

(1) Liver

(2) Pancreas

B4. How many permanent teeth do adults have? 32

B5. What are the glands in the mouth that give off a digestive juice? What is this digestive juice called? The salivary glands give off the digestive juice saliva.

B6. What is the enzyme found in saliva? What does it do? Amylase — changes starch to sugar.

B7. What are the glands in the lining of the stomach? What digestive juice do they give off? The gastric glands give off gastric juice.

B8. What do rennin, pepsin and hydrochloric acid do in digestion?

(1) Rennin causes digestion of protein in milk.

(2) Pepsin } work together to digest protein

(3) Hydrochloric acid }

B9. What is the thick liquid that leaves the stomach called? Chyme

B10. How long is the small intestine? How long is the large intestine? The small intestine is about 7 meters long. The large intestine is about 1 1/2 meters long.

B11. What does the large intestine do? Accumulates undigested food wastes for elimination from the body.

C. The Villi

After doing this lesson you will be able:

C1 • to describe the villi

C2 • to explain how the villi function

Before the cells of the body can use the digested food, it must enter the bloodstream. Food passes through the walls of the small intestine and into the blood vessels.

The walls of the small intestine are covered with thousands of fingerlike projec- C1
tions called **villi** (singular: villus). The villi
increase the surface of the small intestine. C2
Also they allow for faster movement of di-

Villi cover the inside wall of the small intestine. There are so many they look like fuzz. The villi you see in the photo have been greatly magnified.

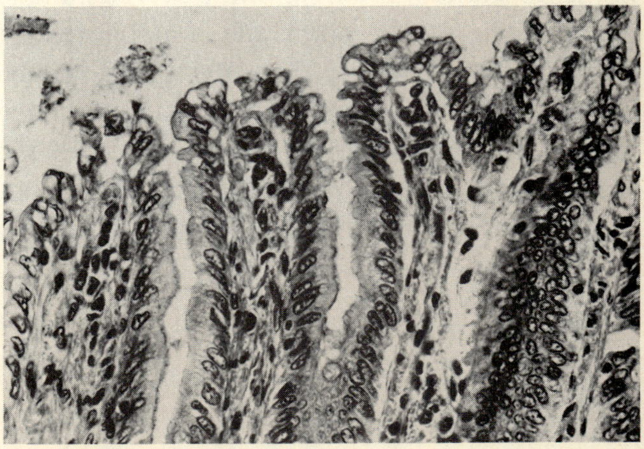

Photo courtesy of Carolina Biological Supply Co.

Refer back to the art on the digestive system which appears on page 52 when studying the pancreas and liver. Note on the art that although it is not labeled, the gall bladder is illustrated; it is the sac underneath the liver and next to the stomach.

gested food into the bloodstream. In addition, the villi have many blood vessels. Digested food, nutrients and water pass through the villi into the bloodstream and are carried to all cells of the body.

Check Up C

C1. What are villi? Villi are fingerlike projections from the wall of the small intestine.

C2. What do they do? They increase the surface area of the small intestine and allow for more rapid passage of food substances into the bloodstream.

D. Helper Glands

After doing this lesson you will be able:

- D1 • to name the largest gland in the body and tell what it does
- D2 • to explain where bile is stored and what it does
- D3 • to name the digestive juice produced by the pancreas and tell what it does
- D4 • to tell where the digestion of starches, fats and proteins is completed

Two glands, the liver and pancreas, make substances which enter the small intestine
D1 through small tubes. The **liver,** the largest gland in the body, makes a substance called **bile.** The bile is stored until needed in the
D2 **gall bladder,** a small sac attached to the liver. Bile is not an enzyme, but helps in the digestion of fat. Bile breaks fat down into small
drops which can be digested better by en- D2
zymes.

The **pancreas** is located below and behind the stomach. The digestive juice it makes is
called **pancreatic juice.** Pancreatic juice has D3
a number of enzymes which continue the digestion of starches, fats and proteins.

The glands of the small intestine make
intestinal juice. The enzymes in intestinal D4
juice complete the digestion of starches, fats and proteins. The food is now in a simple, liquid form that can pass into the bloodstream and be used by the cells of the body.

Check Up D

D1. What is the largest gland in the body? What does it do? The liver; it produces bile.

D2. Where is bile stored until it is needed? What does bile do? Gall bladder; bile helps digest fats.

D3. What digestive juice is made by the pancreas? What does it do? Pancreatic juice; it contains a number of enzymes that continue the digestion of starches, fats, proteins.

D4. Where is the digestion of starches, fats and proteins completed? Small intestine

Unit 2, Chapter 2
REVIEW

Matching Exercise Match each term in **Column A** with its description in **Column B**. Write the letter of the description in the space beside the correct term.

	Column A	*Column B*
g	**1.** enzymes	**a.** the breaking down of food
i	**2.** rennin	**b.** glands in the mouth that give off saliva
j	**3.** pepsin	**c.** largest gland in the body
d	**4.** amylase	**d.** enzyme in saliva
h	**5.** glands	**e.** small gland that sends juices to the small intestine
f	**6.** villi	**f.** fingerlike projections in the small intestines
b	**7.** salivary glands	**g.** substances that change the chemical makeup of food
c	**8.** liver	**h.** organs that give off enzymes
e	**9.** pancreas	**i.** enzyme that digests milk protein
a	**10.** digestion	**j.** enzyme that works with hydrochloric acid to digest proteins

True/False In the space provided, write **T** if a statement is true. If it is false, change the underlined word or words to make the statement true.

T	**1.** The process by which foods are broken down into usable liquid form is digestion.
Mechanical	**2.** Chemical digestion happens when food is chewed, mixed and broken down into smaller pieces.
cuspids or canines	**3.** Food is torn apart by four teeth called the incisors.
roots	**4.** Teeth are held in place by the crown.
T	**5.** The flavor of food is picked up by taste buds located on the tongue.
villi	**6.** Thousands of fingerlike projections called chymes cover the walls of the small intestine.
mouth	**7.** The alimentary canal is the passageway for food between the stomach and the rectum.
Enzymes	**8.** Pancreas are substances that change the chemical makeup of food.
T	**9.** The gall bladder is a small sac attached to the liver.
made	**10.** Bile is stored in the liver.
salivary	**11.** Saliva is produced by the gastric glands.
pepsin	**12.** Rennin and chyme are two enzymes given off by the gastric glands.
anus	**13.** Undigested wastes pass out of the body through an opening called the rectum.
liver	**14.** The pancreas is the largest gland in the body.
T	**15.** Glands are organs that give off enzymes.

Multiple Choice Exercise Circle the letter of the word or phrase that best completes each statement.

1. Mechanical digestion takes place in the **(a)** mouth **(b)** stomach **(c)** large intestine **(d)** small intestine.
2. The glands in the stomach that give off gastric juice are the **(a)** villi **(b)** pancreas **(c)** gastric glands **(d)** alimentary canal.
3. The enzyme that digests milk protein is **(a)** pepsin **(b)** rennin **(c)** amylase **(d)** chyme.
4. No digestion takes place in the **(a)** large intestine **(b)** small intestine **(c)** mouth **(d)** stomach.
5. The outside of the crown of a tooth is called the **(a)** dentin **(b)** pulp **(c)** enamel **(d)** rennin.
6. Small finger-like projections that cover the walls of the small intestine are **(a)** pulp **(b)** gastric glands **(c)** dentin **(d)** villi.
7. Undigested wastes pass out of the body through the **(a)** small intestine **(b)** large intestine **(c)** anus **(d)** arms.
8. All of the following are involved in the digestion of proteins except **(a)** ptyalin **(b)** hydrochloric acid **(c)** pepsin **(d)** rennin.
9. The thick liquid that leaves the stomach is called **(a)** ptyalin **(b)** chyme **(c)** hydrochloric acid **(d)** pepsin.
10. The two organs that help in digestion, but are not part of the alimentary canal are the **(a)** esophagus and liver **(b)** esophagus and tongue **(c)** liver and pancreas **(d)** pancreas and esophagus.

Application Exercise Label the parts of the digestive system indicated by the arrows in the diagram below. Write the name of the part on the line with the same number.

1. salivary glands
2. liver
3. large intestine
4. stomach
5. pancreas
6. small intestine
7. appendix
8. rectum

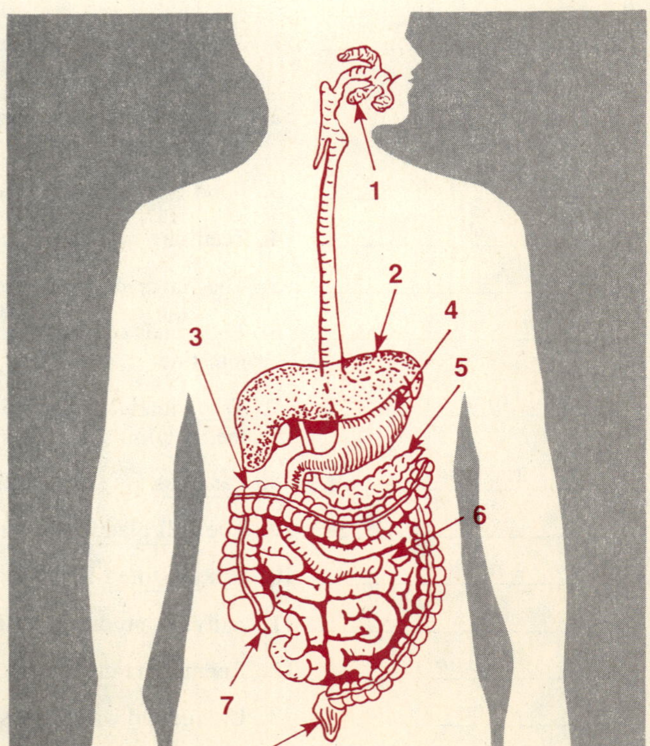

The blood this donor is giving will help stock a blood bank for the American Red Cross.

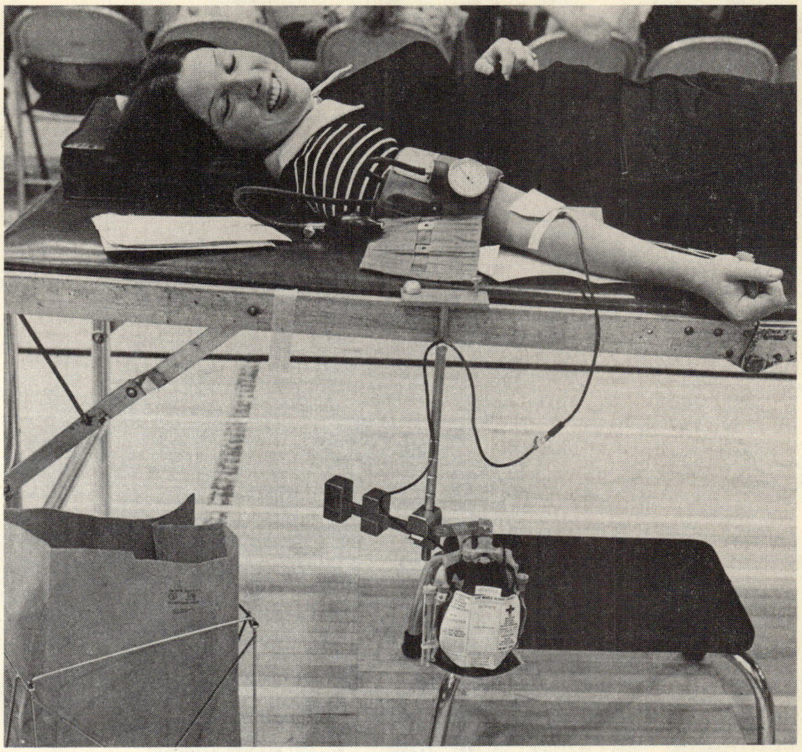

New science words: plasma, pulse

Skills developed: comparing, inferring, measuring, observing

Chapter 3

THE CIRCULATORY SYSTEM

In order to understand what happens to food after it has been digested, you must know the parts of the circulatory system and how it works.

The circulatory system can be compared to a railroad system. Trains travel back and forth on tracks in order to reach certain locations. Your blood travels through the blood vessels much like the trains travel back and forth on the tracks. Your heart can be compared to the engine since it has the power to force the blood through the blood vessels.

A. The Heart

After doing this lesson you will be able:

- A1 • to describe two functions of the circulatory system
- A2 • to list the three parts of the circulatory system
- A3 • to name the upper chambers of the heart and describe their function
- A4 • to name the lower chambers of the heart and describe their function
- A5 • to explain what happens to the blood when it reaches the lungs
- A6 • to explain when the blood is rich in oxygen

The circulatory system carries digested food and oxygen to all cells of the body. It also removes all waste materials that form as the cell carries out its life activities. The circulatory system is made up of the **heart,** the **blood** and the **blood vessels.** A1 A2

If you put your hand on the upper center area of your chest between the lungs, you will feel your heart beat. Your **heart** is a hollow, muscular organ. It is rather small, only about the size of your fist. It continues to grow as you grow. It is constantly pumping so that the blood can supply your cells with food and oxygen. There are two sides to the heart. A wall called the **septum** separates the right side from the left. Each side is divided

Refer back to the photograph of heart muscle tissue on page 47 when discussing the heart.

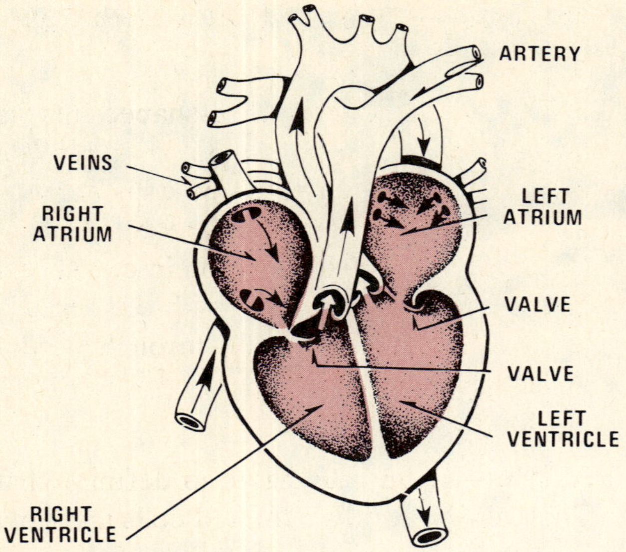

Your heart.

into upper and lower chambers. The **upper**
A3 chambers are called the right and left **atria**
(singular, atrium). The **lower** chambers are
A4 called the right and left **ventricles.**

A3 The atria receive the blood brought back to the heart. They have thin walls with very little muscle in them. The ventricles pump
A4 the blood to all parts of the body. They have thick muscular walls. Located between the atria and ventricles are **valves** which permit the blood to flow in one direction *only*. Thus, when the walls of the ventricles contract to pump the blood from the heart, the valves close and stop the blood from flowing back into the atria.

A5 When the blood reaches the lungs, it *picks up* oxygen and *gets rid of* the waste materials:
A6 carbon dioxide and water. The blood, now rich in oxygen, flows from the lungs back to the left atrium. The left atrium contracts and forces the blood through a valve into the left ventricle. The left ventricle contracts and pumps the blood containing oxygen to all the body's cells.

The right atrium receives the blood brought back to the heart from all parts of the body. This blood, low in oxygen, carries the wastes made by the cells. The walls of

The wall of the left ventricle is the most highly muscular; it pump[s] blood through the aorta to all parts of the body.

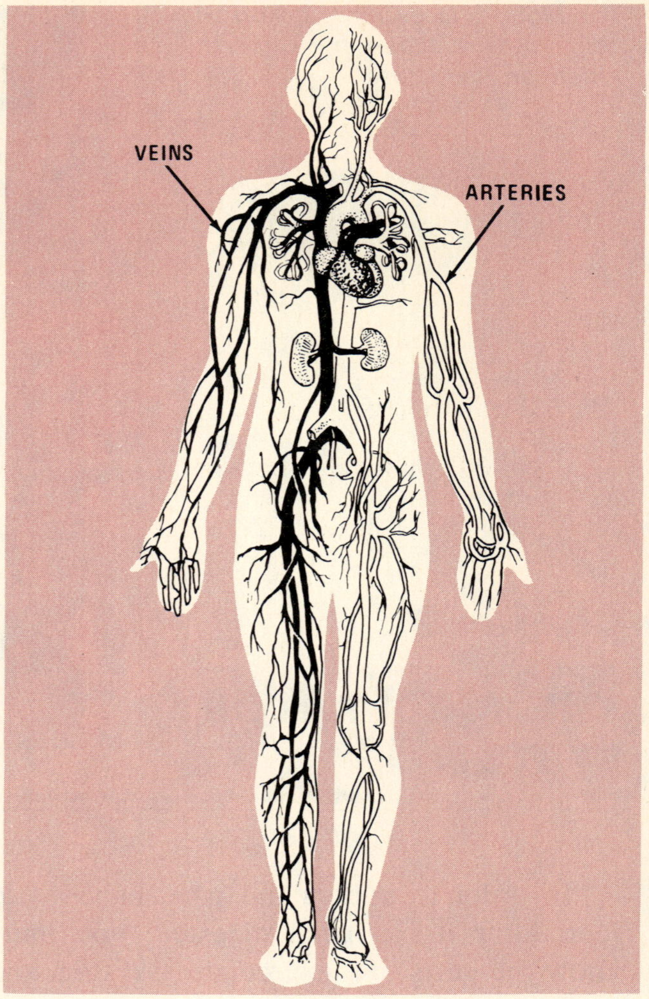

The circulatory system.

the right atrium contract and force the blood through a valve into the right ventricle. The right ventricle contracts and forces the blood out of the heart and to the lungs. The chambers of the heart are like one-way streets because the blood flows in one direction only.

Check Up A

A1. Describe two functions of the circulatory system. (1) Carries digested food and oxygen to all parts of the body; (2) Removes waste materials from the cells.

Point out that there are blood vessels on both sides of the body. The diagram shows only one side for simplicity. Mention that although the kidney plays an extremely important role in filtering the blood, it is *not* considered an organ of the circulatory system.

A2. List the three parts of the circulatory system. (1) Heart, (2) blood, (3) blood vessels.

A3. What are the upper chambers of the heart called? What do they do? Atria — receive the blood brought back to the heart.

A4. What are the lower chambers of the heart called? What do they do? Ventricles — pump blood to all parts of the body.

A5. What happens to the blood when it reaches the lungs? Blood picks up oxygen and gets rid of carbon dioxide and water.

A6. When is the blood rich in oxygen? When it leaves the lungs.

B. The Blood

After doing this lesson you will be able:

- **B1 • to name the four parts of the blood and give a function of each part**
- **B2 • to explain what hemoglobin is and what it does**

The blood carries all the materials needed by the cells of the body. Blood is made up of **plasma,** and three kinds of cells: **red blood cells, white blood cells** and **platelets.** The cells float in the plasma.

B1 **1. Plasma** is the liquid part of the blood. It is about 90% water. Dissolved in the plasma are digested foods and the wastes from the cells.

White blood cells have the ability to move like ameba. They can surround and digest invading bacterial cells.

2. Red blood cells are disc-shaped and have B1
no nucleus. There are millions of red blood cells in the body. Red blood cells **carry oxygen** to the cells of the body. Red blood cells contain a red, iron-rich protein called **hemo-** B2
globin. Hemoglobin joins with oxygen in the lungs. Then, as blood flows through the body, the hemoglobin gives the oxygen to the cells of the body.

3. White blood cells have no definite shape. B1
They are larger than the red cells and have a nucleus. The white blood cells fight infection and destroy bacteria that get into the body. There are fewer white blood cells than red blood cells in the body.

4. Platelets have no definite shape, no nu- B1
cleus and are small in size. They help stop the flow of blood when a part of the body is cut. Platelets are tiny pieces of living cells.

There are many more red blood cells than white blood cells.

Note the one white blood cell (stained) among the numerous red blood cells.

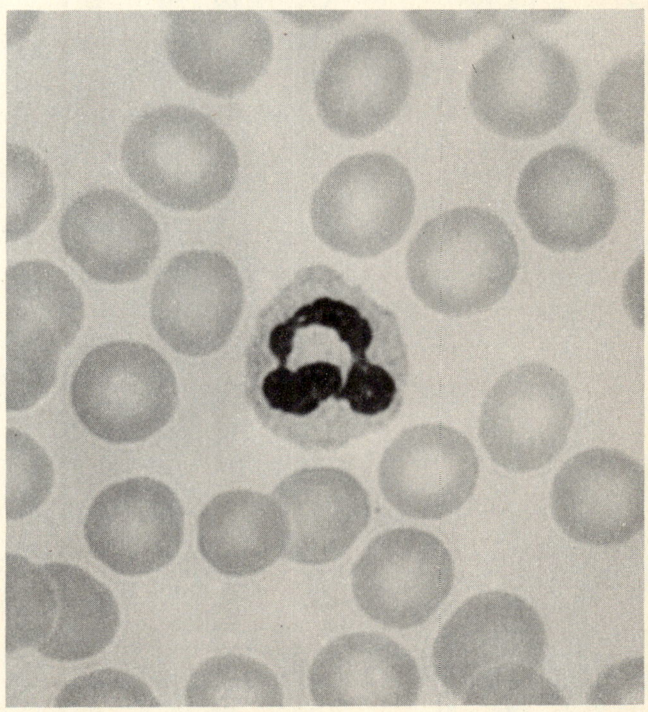

Human blood contains approximately 5 million red blood cells per cubic mm; 7,000–8,000 white blood cells per cubic mm.

Discover for yourself
how you could compare frog blood with human blood.

PURPOSE: To look at the blood cells of different animals • To develop the skills of observation and comparison

You need: prepared slides of frog blood and human blood, two microscopes

Your teacher will set up the slides under the microscopes. Observe the shapes of both kinds of blood cells.

1. How are the frog and human blood cells alike? Answers will vary.

2. How are they different? Answers will vary.

Check Up B

B1. Name the four parts of the blood and give the function of each part.

(1) Plasma — transports all materials needed by cells.

(2) Red blood cells — carry oxygen to the cells of the body.

(3) White blood cells — fight infection and destroy bacteria.

(4) Platelets — help stop the flow of blood from a cut.

B2. What is hemoglobin? What does it do?
A red iron-rich protein in red blood cells that picks up oxygen in the lungs. It releases oxygen as it flows through the body.

C. The Blood Vessels

After doing this lesson you will be able:

C1 • to name the three kinds of blood vessels and describe a function of each

C2 • to explain what a pulse is

C3 • to name the kind of blood vessel through which food, oxygen and waste materials pass and why this is so

Blood flows through the body in tubes called **blood vessels.** There are three main kinds of blood vessels: (1) arteries, (2) veins and (3) capillaries.

Arteries carry blood directly *from* the heart to all parts of the body. This blood is rich in food and oxygen. The wall of the arteries are thick and muscular. If you cut an artery, blood will spurt out. Cutting an artery can be dangerous. C1

As blood is pushed from the heart into the main artery, the artery gets larger and longer. This continues to happen as each "push" of blood flows through all arteries in the body. As each wavelike "push" stretches an artery, it produces a **pulse.** C2

If you listen to your heart beating, it might be very hard for you to count the number of beats. You can easily count your heartbeat by taking the **pulse** of an artery that lies close to the surface of the skin, as in the wrist. Each pulse beat is produced by a single heart beat.

Discover for yourself
how you could show that your heart beats faster when you run than when you sit.

PURPOSE: To show the effects of exercise on the rate of heartbeat • To develop the skills of measurement, comparison and explanation

You need: paper and pencil, watch or clock with a second hand

You may wish to vary this activity by comparing the pulse rate of all the students in the class at rest.

a. Your pulse can be felt on the inner part of your wrist. Use your fingers not your thumb, which also has a pulse. Count your pulse for 30 seconds, using a watch or a clock with a second hand to keep time.

Word to review: network

1. How many times does your pulse beat in 30 seconds? Answers will vary. __________

2. Figure out your pulse rate for 1 minute (60 seconds) using your answer to question 1. Answers will vary. __________

b. Now run around for 1 minute. Take your pulse for 30 seconds.

1. What is your pulse rate for 30 seconds? Answers will vary.

   For one minute? Answers will vary. __________

2. What is the difference between your sitting pulse rate and your running pulse rate? Answers will vary.

3. Why did your pulse rate increase after you ran? Running made your heart rate increase, so pulse rate increased also. __________

Veins have thinner walls than arteries. C1
The veins carry the blood *back* to the heart. This blood does not travel in "pushes." It flows evenly. When a vein has been cut, the blood flows out slowly and in a steady stream. If you should accidentally cut a blood vessel, you can tell whether you have cut a vein or an artery by observing how the blood flows from the cut.

Three types of blood vessels. Identify each.

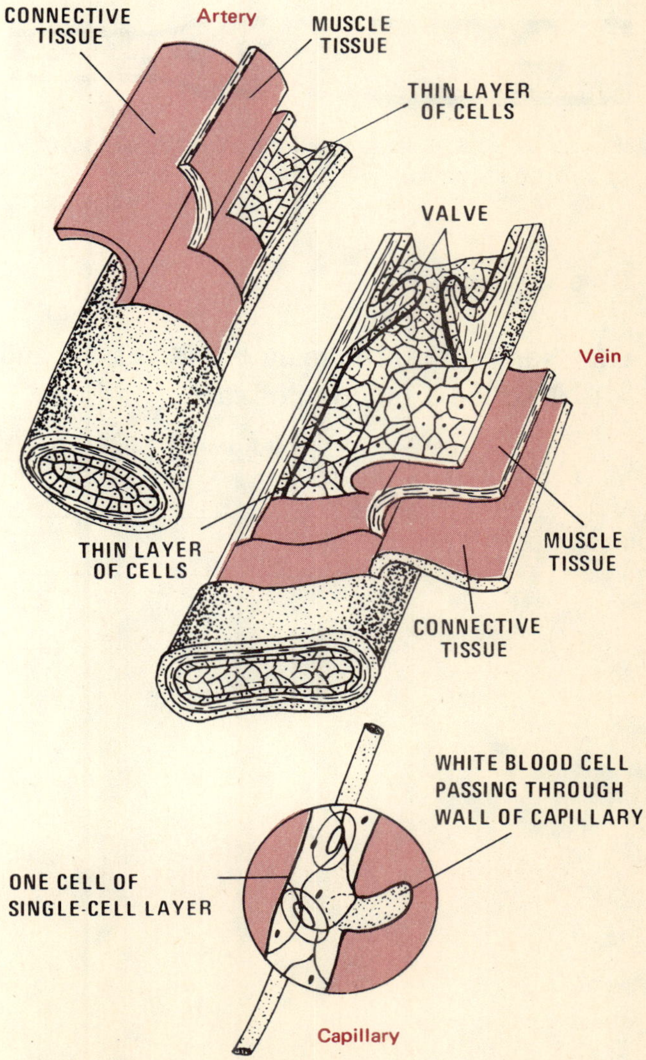

On the average, the pulse rate for an adult at rest is 70 to 90 pulsations a minute; for children at rest, generally, the pulse rate is 90 to 120 pulsations a minute.

C1 **Capillaries.** Although the blood is in the arteries or veins, it still cannot reach the cells. The walls of both the arteries and veins are too thick to let oxygen and digested food pass through them. Arteries and veins are connected by tiny blood vessels that are
C3 called **capillaries.** These very tiny, thin-walled blood vessels are only one cell thick. Thus digested food and the oxygen can pass through the walls of the capillaries, and enter the cells of the body. Waste materials can leave the cells and enter the capillaries.

The blood vessels are one-way streets. The blood travels in only one direction. It travels first away from and then back toward the heart.

Veins have many valves which stop the blood from flowing backward.

Check Up C

C1. Name the three main blood vessels and describe a function of each.

(1) Arteries – carry blood from the heart to all parts of the body.

(2) Veins – carry blood back to the heart.

(3) Capillaries – connect arteries and veins.

C2. What is a pulse? A wavelike surge of blood through an artery.

C3. What kind of blood vessel makes it possible for food, oxygen and waste materials to pass in and out of cells? Explain why this can be done. Capillaries – they are only one cell thick.

Unit 2, Chapter 3
REVIEW

Matching Exercise Match each term in **Column A** with its description in **Column B**. Write the letter of the description in the space beside the correct term.

	Column A	*Column B*
i	**1.** veins	**a.** carries digested food and wastes
d	**2.** arteries	**b.** upper chambers of the heart
e	**3.** capillary	**c.** lower chambers of the heart
a	**4.** plasma	**d.** carry blood away from the heart
j	**5.** platelets	**e.** smallest kind of blood vessel
f	**6.** white blood cells	**f.** fight infection
c	**7.** ventricles	**g.** carry oxygen to cells of the body
b	**8.** atria	**h.** red protein found in blood cells
h	**9.** hemoglobin	**i.** carry blood to the heart
g	**10.** red blood cells	**j.** help stop bleeding from a cut

True/False In the space provided, write **T** if a statement is true. If it is false, change the underlined word or words to make the statement true.

Answer	Statement
cells	**1.** The circulatory system carries digested food and oxygen to the heart.
septum	**2.** A wall called the atrium separates the right side of the heart from the left side of the heart.
White blood cells	**3.** Platelets fight infection and destroy bacteria that get into the body.
T	**4.** About 90% of plasma is made up of water.
ventricles	**5.** The atria pump blood to all parts of the body.
T	**6.** The atria are the upper chambers of the heart.
Valves	**7.** Platelets between the atria and ventricles permit the blood to flow in only one direction.
Red	**8.** White blood cells are disc-shaped and have no nucleus.
T	**9.** Hemoglobin is the red, iron-rich protein found in red blood cells.
Platelets	**10.** Red blood cells have no definite shape, no nucleus and are small in size.
Plasma	**11.** Hemoglobin is the liquid part of the blood.
lungs	**12.** When blood reaches the heart, it picks up oxygen and gets rid of waste materials.

T 13. Arteries carry blood directly from the heart to all parts of the body.

Veins 14. Arteries carry blood from the body back to the heart.

Capillaries 15. Arteries are tiny, thin-walled vessels that are only about one cell thick.

Multiple Choice Exercise Circle the letter of the word or phrase that best completes each statement.

1. The wall that separates the right side of the heart from the left side of the heart is the **(a)** ventricle **(b)** atrium **(c)** septum **(d)** lung.
2. The red, iron-rich protein found in red blood cells is **(a)** hemoglobin **(b)** platelet **(c)** oxygen **(d)** plasma.
3. The liquid part of the blood is **(a)** hemoglobin **(b)** platelet **(c)** oxygen **(d)** plasma.
4. Red blood cells **(a)** help to stop the flow of blood from a cut **(b)** fight infection and destroy bacteria **(c)** carry oxygen to the cells of the body **(d)** transport all materials needed by the cells.
5. The arteries carry blood **(a)** from the heart to all parts of the body **(b)** from the body back to the heart **(c)** from the veins to the cells of the body **(d)** to the heart from the lungs.
6. A pulse can only be felt in the **(a)** veins **(b)** arteries **(c)** capillaries **(d)** heart.
7. Bleeding from a cut is stopped by **(a)** plasma **(b)** red blood cells **(c)** white blood cells **(d)** platelets.
8. Blood containing wastes produced by the cells enters the heart through the **(a)** right ventricle **(b)** left ventricle **(c)** left atrium **(d)** right atrium.
9. Blood is pumped to all parts of the body by the **(a)** atria **(b)** ventricles **(c)** veins **(d)** capillaries.
10. The valves between the atria and ventricles **(a)** allow the blood to flow in only one direction **(b)** supply the heart with oxygen **(c)** carry blood out of the heart **(d)** help to fight infection and destroy bacteria.

Application Exercise On the diagram of the heart below, label the following parts: **right ventricle, right atrium, left ventricle, left atrium,** a **valve,** a **vein,** and an **artery.**

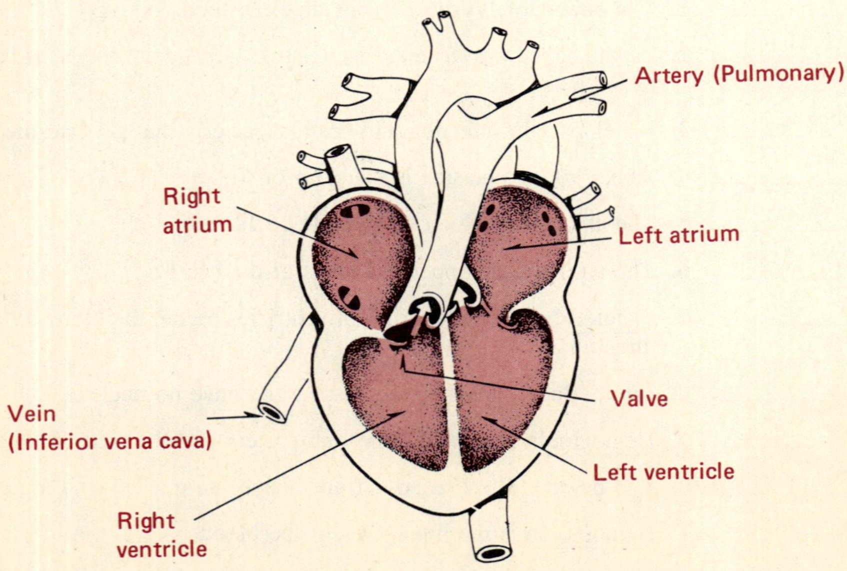

Humans take in oxygen by diffusion through air sacs in the lungs.

Even someone who swims "like a fish" must come up for air. How do you think you take in oxygen from the air?

New science words: diaphragm, respiration

Skills developed: observing, inferring, stating a conclusion

Chapter 4

TAKING IN AND USING OXYGEN

Have you ever heard the term **artificial respiration** used? Do you know what it means? Do you know what a **respirator** is? A respirator is a tool to help you breathe. Artificial respiration is the rhythmic forcing of air in and out of the lungs. It is often given to persons who have been overcome by smoke in a fire; or, who have been rescued from nearly drowning. When you have finished this chapter, discuss this question with your class: Do the words respiration and respirator describe what they are meant to describe?

A. What Is Respiration?

After doing this lesson you will be able:

A1 • to explain what respiration is

A2 • to explain what breathing is and name the two parts of the breathing process

A3 • to tell how the lungs are protected

A4 • to name the sheet of muscle that separates the chest from the abdomen

A5 • to explain what happens to the pressure in the chest cavity when you inhale and when you exhale

Respiration is the life activity by which A1
the body takes in oxygen and uses it to oxidize food and let out energy. Carbon dioxide and water are given off as waste products of this process.

Before oxygen can enter the blood and then the cells, air that contains oxygen must be taken into the lungs. **Breathing** is taking A2
air into the lungs and pushing out carbon dioxide from the lungs.

Breathing is something you do without thinking. Breathing *in* is called **inhaling** while A2
breathing *out* is called **exhaling.** Stop reading for a few seconds and inhale deeply. Now exhale. You just filled your lungs with oxygen and exhaled some carbon dioxide. Put your hands on your ribs and inhale again. What did you feel?

Before you can explain what you felt, you must find out more about the lungs. Because they are such important organs, they are well protected. The ribs form what is called the **rib cage** to protect the lungs. The two lungs A3
are located within the rib cage. A sheet of muscle called the **diaphragm** separates the A4
chest from the abdomen, the lower part of the body. The diaphragm is important in the breathing process.

Discussion topic: People can go without food for about 3 weeks and without water for about 3 days. What is it that we can do without for only a matter of minutes? (oxygen)

The lungs of a sixth grader hold from 2 to 3 liters of air. An adult's lungs hold from 3 to 4 liters of air.

Have students inhale and exhale. Discuss the changes they feel as described in these paragraphs.

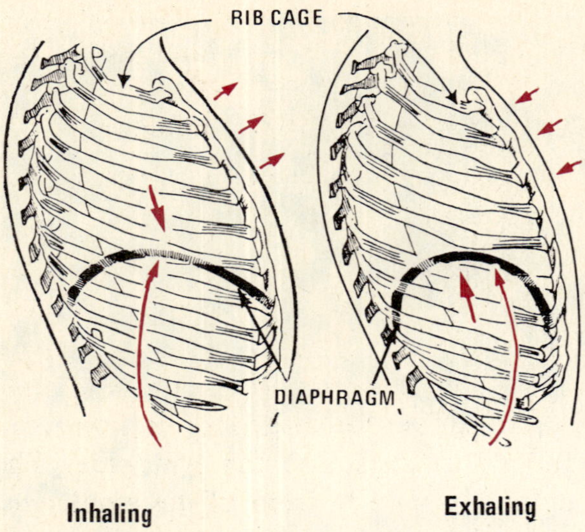

Note the position of the diaphragm as air is inhaled and exhaled.

Hale means breathe.

1. Inhaling. When you breathe in, muscles move the ribs upward and outward. The diaphragm moves downward. The chest cavity becomes larger. Because the cavity is larger, the **pressure** in it becomes **lower** than the air
A5 pressure outside the body. Air rushes into the lungs to equalize the pressure.

2. Exhaling. After the lungs are filled with air, the rib muscles relax and the ribs move back in place. The diaphragm moves upward. As a result of these actions, the chest cavity becomes *smaller* and the **pressure** in the chest
A5 is **greater** than the air pressure outside the body. Air, therefore, rushes out of the lungs.

Discover for yourself
what would happen if you exhaled into limewater.

PURPOSE: To show that exhaled air contains carbon dioxide • To develop the skills of observation and inference

You need: limewater (a water solution of calcium hydroxide), a beaker, a drinking straw

Half fill a beaker with limewater. Put a straw in the beaker. Exhale through the straw. Repeat this two or three times.

Note: Limewater is used to test for the presence of carbon dioxide. Carbon dioxide will turn limewater milky white.

1. What happens to the color of the limewater? It becomes milky white.
2. What does this tell you about the air you exhaled? It contains carbon dioxide.

Check Up A

A1. What is respiration? The life activity by which the body takes in oxygen and uses it to oxidize food and release energy.

A2. What is breathing? Name the two parts of the breathing process. Breathing is the taking in of air into the lungs and the pushing out of carbon dioxide from the lungs. Inhaling and exhaling.

A3. How are the lungs protected? By the ribs.

A4. What separates the chest from the abdomen? The diaphragm

A5. Explain what happens to the pressure in the chest cavity when you inhale and when you exhale. Inhale — the pressure becomes lower than the air pressure outside the body. Exhale — the pressure becomes greater than the pressure outside the body.

B. The Respiratory System

After doing this lesson you will be able:

- B1 • to list, in order, the five parts of the respiratory system
- B2 • to explain how air is filtered, warmed and moistened in the nasal passages
- B3 • to explain what the epiglottis is
- B4 • to explain what the bronchi and bronchial tubes are
- B5 • to tell what the alveoli are and the kind of blood vessels that surround them

The system involved with getting the air into and out of the lungs is called the **respiratory system.** The organs of the respiratory
B1 system are (1) the **nose,** (2) the **pharynx,** (3) the **trachea,** (4) the **bronchi** and (5) the **lungs.**

1. The nose. When you inhale, the air enters the openings in your nose called **nostrils.** Then it passes into long passageways called the **nasal passages.** The nostrils are lined with **cilia** (hairs) which filter the air; that is, they
B2 remove dirt and dust. The nasal passages are lined with a moist **mucous membrane** that also catches dust. Because of the cilia and the mucous membrane, air that reaches your lungs is fairly free of dirt and dust. The
mucous membrane of the nasal passages also B2
warms and moistens the incoming air.

2. The pharynx. The clean, warm air passes through the **pharynx.** The pharynx is a cavity found behind the mouth.

3. The trachea. After leaving the pharynx, and before entering the **trachea,** also called
the **windpipe,** the air must pass the epiglottis. B3
The **epiglottis** is a cartilage flap that opens to permit air to pass into the windpipe. The windpipe lies just in front of the **esophagus,** the muscular tube that leads from the pharynx to the stomach. The epiglottis closes the opening into the windpipe when you swallow food. Otherwise, food would pass into the windpipe and down into your lungs.

Near the top of the trachea is your **voice box** which is called the **larynx.** The voice box has vocal cords which vibrate (move back and forth) when air passes them. These vibrations make sounds. The inner wall of the trachea is lined with tiny hair-like **cilia** which are always in motion. The cilia catch dust particles. Then they push them

The respiratory system.

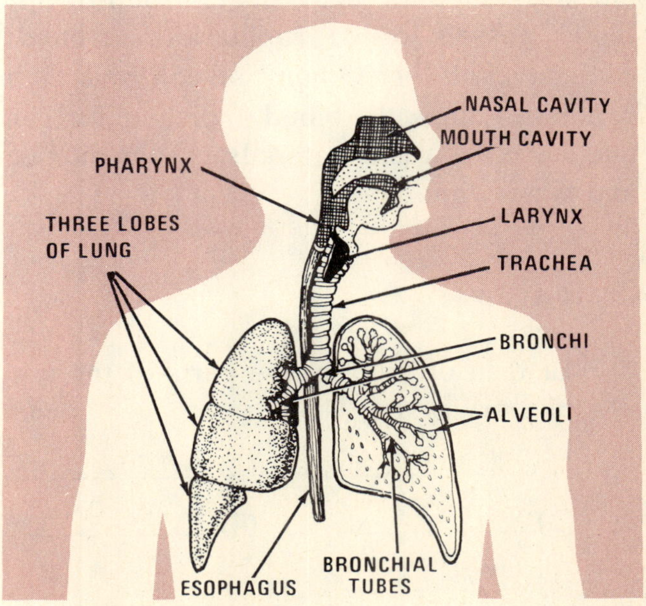

Prepare a diagram of the respiratory system. Have students label the organs.

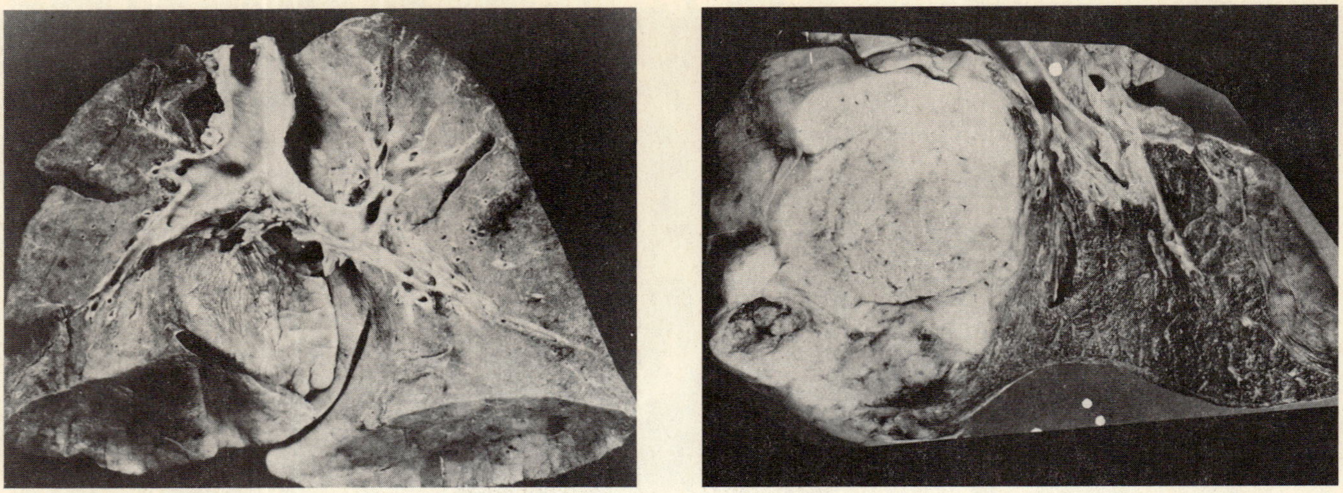

The lungs of a nonsmoker are shown in the photo on the left. On the right are the lungs of a cancer victim. Note the "black lung" caused by coal tars from cigarette smoke on the right side and the large white tumor on the left side.

Remind students that capillary walls are only one cell thick. The walls of the alveoli are also only one cell thick.

out of the trachea and up toward your throat and mouth.

4. The bronchi. At its lower end the trachea branches out into two large tubes called **bronchi.** The bronchi enter the lungs where
B4 they divide into smaller tubes called **bronchial tubes** (sometimes called **bronchioles**). Air sacs are at the ends of these smaller tubes.

5. The lungs. The lungs are made up of tiny
B5 air sacs called **alveoli.** A large group of *tiny* blood vessels called **capillaries** surrounds each alveolus. The many alveoli and their capillaries allow the blood to pick up oxygen and leave the waste materials, carbon dioxide and water vapor.

The lungs contain approximately 1 billion alveoli.

Check Up B

B1. List, in order, the five parts of the respiratory system.

(1) Nose (2) Pharynx

(3) Trachea (4) Bronchi

(5) Lungs

B2. How is the air filtered, warmed and moistened in the nasal passages? Air is filtered by the cilia, and warmed and moistened by the mucous membrane that lines the nasal passages.

B3. What is the epiglottis? What are its two functions? A cartilage structure that opens to permit air to pass into the windpipe. It also partially closes to keep food from entering the windpipe.

B4. What are the bronchi and bronchial tubes? The bronchi are two large tubes that branch out from the lower end of the trachea. They divide into smaller tubes called bronchial tubes.

B5. What are the alveoli? What kind of blood vessels surround them? Tiny air sacs in the lungs. They are surrounded by networks of capillaries.

Unit 2, Chapter 4
REVIEW

Matching Exercise Match each term in **Column A** with its description in **Column B**. Write the letter of the description in the space beside the correct term.

	Column A	*Column B*
d	**1.** respiration	**a.** windpipe
j	**2.** pharynx	**b.** stops food from entering the trachea
f	**3.** larynx	**c.** breathing out
a	**4.** trachea	**d.** life activity that releases energy
h	**5.** diaphragm	**e.** air sacs
b	**6.** epiglottis	**f.** voice box
i	**7.** bronchial tubes	**g.** breathing in
e	**8.** alveoli	**h.** muscle between the chest and abdomen
g	**9.** inhaling	**i.** branches of the bronchi
c	**10.** exhaling	**j.** cavity behind the mouth

True/False In the space provided, write **T** if a statement is true. If it is false, change the underlined word or words to make the statement true.

Breathing	**1.** Respiration is taking air into the lungs and pushing carbon dioxide out of the lungs.
T	**2.** The rib cage protects the lungs.
downward	**3.** When you inhale, the diaphragm moves upward.
larynx	**4.** The pharynx is located near the top of the trachea.
nostrils	**5.** The trachea and pharynx are lined with cilia.
T	**6.** The esophagus is a muscular tube that leads from the pharynx to the stomach.
bronchi	**7.** The alveoli are the two branches at the lower end of the trachea.
pharynx	**8.** The larynx is a cavity behind the mouth.
T	**9.** The larynx is the voice box.
lower	**10.** When air enters the lungs, the pressure in the chest cavity is higher than the air pressure outside the body.
water vapor	**11.** Carbon dioxide and oxygen are given off as waste products of respiration.
T	**12.** Artificial respiration is the rhythmic forcing of air into and out of the lungs.
bronchial tubes or bronchioles	**13.** Air sacs are located at the ends of the bronchi.
capillaries	**14.** The alveoli are surrounded by a large group of arteries.
alveoli	**15.** Blood picks up oxygen and gives off waste materials in the capillaries surrounding the bronchioles.

Multiple Choice Exercise Circle the letter of the word or phrase that best completes each statement.

1. The tiny air sacs in the lungs are the **(a)** bronchial tubes **(b)** alveoli **(c)** bronchi **(d)** cilia.
2. The nostrils and trachea are lined with tiny hair-like structures called **(a)** mucous membranes **(b)** alveoli **(c)** bronchi **(d)** cilia.
3. The trachea is also called the **(a)** larynx **(b)** pharynx **(c)** windpipe **(d)** epiglottis.
4. The cartilage flap that opens to permit air to pass through the trachea is the **(a)** larynx **(b)** pharynx **(c)** windpipe **(d)** epiglottis.
5. The voice box is also called the **(a)** larynx **(b)** pharynx **(c)** windpipe **(d)** epiglottis.
6. At its lower end, the trachea branches into two large tubes called the **(a)** pharynx **(b)** bronchi **(c)** alveoli **(d)** bronchial tubes.
7. The tiny blood vessels that surround the alveoli are **(a)** capillaries **(b)** bronchial tubes **(c)** veins **(d)** arteries.
8. The chest is separated from the abdomen by a sheet of muscle called the **(a)** rib cage **(b)** alveoli **(c)** bronchi **(d)** diaphragm.
9. The life activity by which the body takes in oxygen and uses it to oxidize food and release energy is **(a)** circulation **(b)** digestion **(c)** respiration **(d)** inhalation.
10. The waste products of respiration are **(a)** carbon and water **(b)** oxygen and water **(c)** carbon dioxide and oxygen **(d)** carbon dioxide and water vapor.

Application Exercise Eight parts of the respiratory system are listed below. Write the number of each listed part next to the arrow in the diagram that points to that part.

1. nasal cavity
2. alveoli
3. bronchi
4. pharynx
5. trachea
6. bronchial tubes
7. larynx
8. mouth cavity

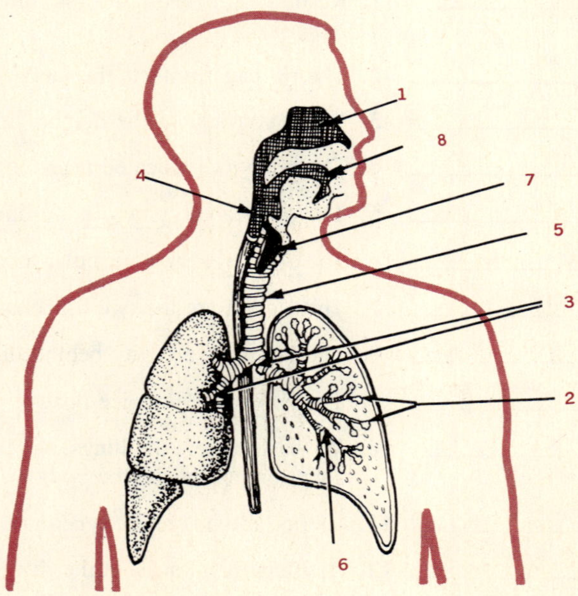

Vinnie is receiving on-the-job management training in industrial relations at a Bethlehem steel plant. Why is it so important for her to wear a hard hat and safety glasses?

The hard hat and safety glasses will help prevent injury to Vinnie's head and eyes.

New science words: impulse, reflex

Skills developed: observing, describing, inferring

Chapter 5

THE NERVOUS SYSTEM

What keeps you from getting killed when you cross a busy street? What makes you back away from a hot fire? Why do you jump when a car backfires? Why can you hit a ball with a bat or tie your shoe laces? What makes it possible for you to learn? The answer is the nervous system, one of the most important and best-developed systems in the body.

A. What Are the Main Functions of the Nervous System?

After doing this lesson you will be able:

- A1 • to list the three main functions of the nervous system
- A2 • to list three sense organs of the nervous system

The **nervous system** receives and sends out messages much as a telephone system
A2 does. Sense organs, such as the eyes, ears and skin, are specialized structures which receive messages from the external environment. These messages are then sent on to other parts of the body.

Your nervous system makes it possible for you to think and reason. It allows you to A1
adjust to all kinds of changes in your environment.

Check Up A

A1. What are the three main functions of the nervous system? (1) To think, (2) to reason, (3) to adjust to changes in the environment.

A2. List three sense organs of the nervous system. (1) Eyes, (2) ears, (3) skin.

There are about 10 billion neurons in the nervous system of the human. About half of these neurons are located in the brain.

B. A Nerve Cell

After doing this lesson you will be able:

B1 • to give another name for a nerve cell

B2 • to name three parts of a neuron and describe their functions

B3 • to tell what an impulse and a response are

B4 • to explain the difference between sensory and motor neurons

B5 • to tell what a stimulus is

B6 • to name the two parts of the central nervous system

B1 The **neuron,** or nerve cell, is the basic unit of structure and function of the nervous sys-
B2 tem. Every neuron is made up of a **cell body** which controls the functions of the neuron.
B2 Branching from the cell body are **dendrites.** They pick up "messages" and carry them to the nerve cell body of which they are a part. These messages are called **impulses.** In many
B2 cells, another branch called the **axon** extends from the nerve cell body. The axon carries impulses to other neurons, to muscles and to glands. You cannot see a neuron without a microscope. But you can see a nerve. It looks like a thin white thread. There are many bundles of nerves in the body. The bundles are usually made up of many axons wrapped in connective tissue. They look much like telephone cables, which have many wires wrapped into a bundle.

Nerves react to **stimuli** (singular, stimu-
B5 lus). A **stimulus** is a change in the environment (surroundings) that can be picked up by a sense organ. A stimulus can be heat, cold, light, taste, smell or sound. A stimulus starts
B3 an **impulse** along a nerve fiber. An impulse is a "message" that causes something to happen. The action that results when an impulse is received is called a **response.**

Neurons carry impulses to all the different parts of the body. **Sensory neurons**
B4 carry messages from the sense organs to various parts of the nervous system.

Nerve impulses travel at approximately 100 meters/second.

Dendrite means *branching, tree-like*. Axon means *main part*.

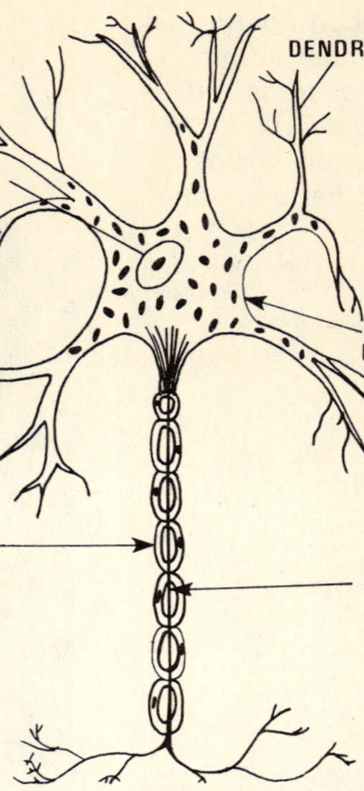

A motor neuron.

Besides sensory neurons, the body also has **motor neurons.** These nerve cells carry B4
impulses from the brain and spinal cord to all the muscles and other glands. Impulses reaching the muscles make them contract (get shorter). Sensory neurons react to stimuli; motor neurons stimulate the muscles and glands to respond to impulses.

Responding to a stimulus. Suppose you are walking along and you suddenly step on a sharp object. The sharp object is the external **stimulus.** The **sensory** neurons carry the impulse to the spinal cord. There the impulse is sent to the set of **motor** neurons that extends to your foot. The impulse is then carried to the muscles in your foot. The muscles react by pulling your foot away from the stimulus. The time between receiving the stimulus and removing your foot from it, is only a fraction of a second.

Many nerve cells make up a **nerve.** All nerves come together either in the spinal cord or the brain. The spinal cord and the brain B6
make up the **central nervous system.**

Word to review: coordinate

Check Up B

B1. What is another name for a nerve cell?

Neuron

B2. Name three parts of a neuron and describe the function of each.

(1) Cell body — controls the functions of the neuron.

(2) Dendrites — pick up impulses and conduct them to the cell body.

(3) Axon — carries impulses to other neurons.

B3. What is an impulse? a response?

An impulse is a "message" that travels through the nervous system. A response is a reaction to an impulse.

B4. What is the difference between sensory and motor neurons?

Sensory — carry impulses from sense organs to various structures of the nervous system. Motor — conduct impulses from brain and spinal cord to muscles and other organs that respond to stimuli.

B5. What is a stimulus?

A change in the environment that can be picked up by the sense organs.

B6. What are the two parts of the central nervous system? (1) Brain, (2) spinal cord.

C. The Central Nervous System

After doing this lesson you will be able:

- C1 • to tell where the spinal cord is and how it is protected
- C2 • to tell what the spinal cord is made up of
- C3 • to tell how the brain is protected
- C4 • to name the three parts of the brain and a function of each part
- C5 • to explain what reflexes are and what they do for the body

The **spinal cord** is made up of many nerve C2
cells. It is located in the spinal column C1
(backbone) which protects it. The nerves which make up the spinal cord extend from the brain to the very end of the spinal column. From the spinal cord the nerves reach out to all parts of the body. Thus, the nerves of the spinal cord coordinate (put in order) the actions of the body by carrying messages from all parts of the body to the brain, and from the brain to all parts of the body.

The **brain** is located inside the skull cavity
and protected by the hard bones of the skull. C3
Like the spinal cord, the brain is made up of many neurons and nerves. The brain is divided into three parts: the **cerebrum,** the **cerebellum** and the **medulla.**

1. The **cerebrum** is the largest part of the C4
brain. The outer part of the cerebrum contains many, many nerve cells which are found in creases or folds called **convolutions.** The cerebrum controls certain activities of
the body. There are special centers for sight, C4
taste, smell, speech and muscular activity. Reasoning, memory and thought are also centered in the cerebrum.

2. The **cerebellum** is a smaller portion of the C4
brain found below and in back of the cerebrum. It coordinates the muscles which per-
form such actions as walking and running. C4

The spinal cord can also control certain reflex actions without conducting impulses to the brain. The knee jerk reflex is an example of this.

Use a model of the human brain or obtain preserved specimens of sheep or calf's brains. Discuss the parts of the brain.

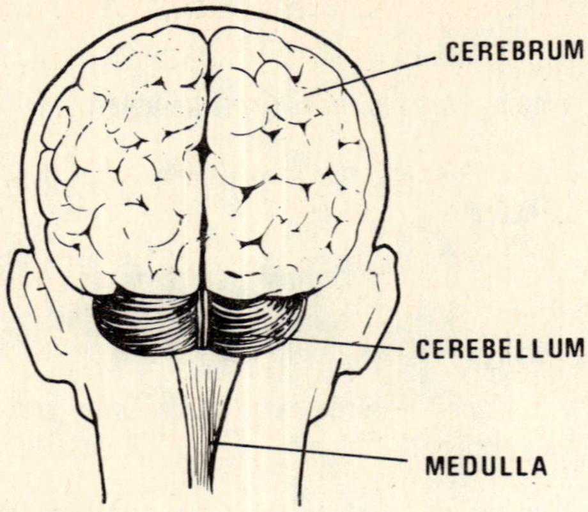

Three sections of the brain: the cerebrum, cerebellum and medulla.

C4 3. The **medulla,** a very small part of the brain connects the brain to the spinal cord. It controls all the involuntary (not under your
C4 control) actions of the body, such as heartbeat and breathing. The medulla also controls other involuntary actions, called **re-**
C5 **flexes.** Coughing, sneezing and blinking are reflex actions. You are born with reflexes. They are automatic responses. They help to protect the body.

Check Up C

C1. Where is the spinal cord and how is it protected? The spinal cord is located within the spinal column which protects it.

C2. What is the spinal cord made up of? The spinal cord is made up of many nerves.

C3. How is the brain protected? The brain is protected by the hard bones of the skull.

C4. List the three parts of the brain and give a function of each part.

(1) Cerebrum — centers of sight, taste, smell, speech, reasoning, memory, thought.

(2) Cerebellum — coordinates muscles.

(3) Medulla — controls involuntary actions such as heartbeat and breathing.

C5. What are reflexes? What do they do for the body? Automatic involuntary actions that help to protect the body.

D. What Are Sense Organs?

After doing this lesson you will be able:

- D1 • to name the five sense organs
- D2 • to give the functions of the following parts of the eye: cornea, pupil, iris, lens, retina, optic nerve
- D3 • to name the three parts of the ear
- D4 • to give the function of the outer ear
- D5 • to explain what each of the following middle ear parts are: hammer, anvil, stirrup, Eustachian tube
- D6 • to explain what each of the following inner ear parts do: cochlea, auditory nerve, semicircular canals
- D7 • to list the four stimuli the nerve endings in the skin are sensitive to
- D8 • to explain what taste buds do

The body has certain parts that can help you become aware of various happenings around you. These parts, called **sense organs,** D1
are your eyes, ears, nose, skin and tongue. After picking up the stimuli, messages are brought from the sense organs to the brain where they are interpreted (explained).

1. The eye. Your eyes make it possible for you to see. Let us find out how. The eyes are

Words to review: focus, transparent, vibration

Refer to diagram on page 78 when discussing the topic of the ear.

small, round organs set in the bony sockets of the skull which protect them. The outer covering of the eye is called the **sclerotic coat.** It is a tough, protective material. The front part of the sclerotic coat is called the **cornea.** It is transparent (clear) and protects D2 the lens of the eye. The back part that does not allow light to enter is the **sclera.** The colored part of the eye, the **iris,** is just behind the cornea. What color is your iris?

The **pupil** is found in the center of the iris. D2 The pupil is an opening that permits light to D2 enter the eye. The iris controls the size of the pupil. When bright light shines into the eye, the iris causes the pupil to become smaller. When the light is dim, the pupil enlarges, allowing more light to enter the eye.

D2 Behind the pupil is the **lens.** The lens focuses the light coming from the object you are looking at. It focuses (brings to a point) light on the inner surface of the eye, called D2 the **retina.** In the retina there are many nerve cells. When an image forms on the retina, an D2 impulse is formed in the **optic nerve** and travels to the brain. The brain then interprets the message and tells you what you are looking at. Muscles attached to the eyeball allow you to move your eyes around.

The part of the eyeball behind the lens is filled with a jellylike material called the

The parts of the eye.

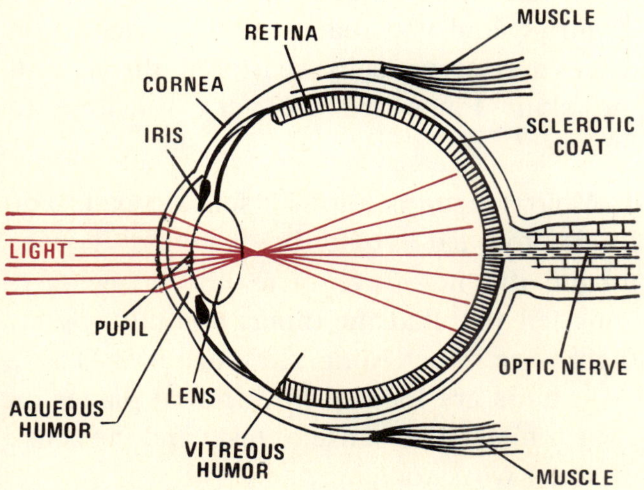

vitreous humor. The liquid material between the lens and the cornea is called the **aqueous humor.** Both these materials help give the eyeball its shape.

Your eyes really do not "see." They are **sensitive** (react) to light. The light causes impulses to travel along the optic nerve to the brain. The brain translates (explains) the impulses and you then "see" what you are looking at.

2. The ear. You are able to hear because sound waves received by your ear are changed into impulses. The impulses are carried to the brain. The brain changes the impulses into sounds. The ear is divided into three parts: the **outer ear,** the **middle ear,** and D3 the **inner ear.**

The **outer ear** is shell-shaped so that it can D4 pick up sound waves. It is made up of the ear lobe, the ear canal and the eardrum. The outer ear catches the sound waves and carries them through the **ear canal** into the middle ear. The ear canal has a sticky wax and tiny hairs that catch dust particles. The outer ear is separated from the middle ear by a membrane called the **eardrum.** Sound waves strike the eardrum and cause it to vibrate (shake).

In the **middle ear** vibrations from the ear- D5 drum are carried to three tiny bones called the **hammer, anvil** and **stirrup.** The stirrup carries the vibrations into the inner ear.

The middle ear is filled with air. The air comes from a tube called the **Eustachian tube.** D5 One end of the Eustachian tube leads to the throat and the other end leads into the middle ear. You can sometimes feel this air when you swallow. Often when you are in an area of greater pressure (in a rapidly descending elevator, for example), you feel a push in your ears. If you swallow, the ear pressure changes, and the push on the eardrums disappears.

The main part of the **inner ear** is a snail-like coil called the **cochlea.** It is filled with

The range of human hearing is from 15 to 20,000 cycles (vibrations) per second. Most children can hear higher-pitched sounds than adults can.

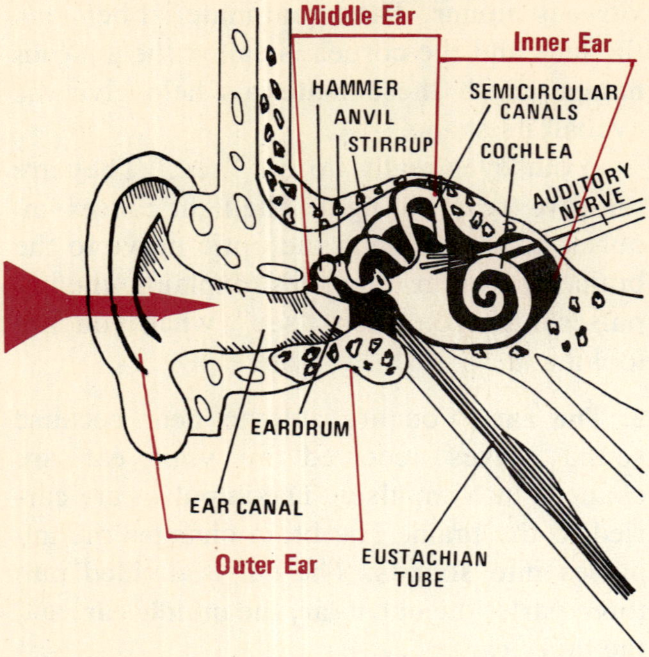

The parts of the ear.

When the heat went off in their school, these pupils put on hats, coats and gloves. Why don't they feel as cold now?

fluid. There are many nerve endings in the cochlea. The fluid is set in motion by the vibrations that come from the stirrup of the middle ear. The nerves in the cochlea carry the impulses to the **auditory nerve** which D6
then carries them to the brain to be interpreted.

The vibrations are also carried to the semicircular canals. The **semicircular canals** are three small bony tubes which are filled with fluid. At the end of the canals there are nerve cells that are connected to the brain. The fluid in the semicircular canals helps you D6
to keep your balance.

How do you hear? When someone speaks to you, the air vibrates, or shakes. Sound waves reach your ear and are picked up by the outer ear. The sound waves then pass through the ear canal, hit the eardrum and cause it to vibrate. The hammer begins to vibrate, then the anvil and finally the stirrup. The vibrations are passed on to the fluid within the cochlea. Vibrations in the cochlea cause an impulse in the nerve endings located there. The impulse is carried to the brain by the auditory nerve. The brain interprets the impulses and you hear.

3. Your **skin** also has nerve endings. They are sensitive to the stimuli of heat and cold, D7
pain and pressure. The nerves in the skin send impulses to the brain. The brain then translates these impulses into sensations (feelings) and you react. For each sensation there are nerve cells scattered throughout the skin. Fingertips are very sensitive to feeling.

4. Your **tongue** is another sense organ. The foods you eat would be tasteless if your tongue did not have taste buds. The **taste buds** tell you that the things you put in your mouth are sweet, sour, salty or bitter. These D8
taste buds are located in different places on your tongue. The sense of taste and the sense of smell work together.

 Clothing acts as insulation; it helps to slow down the loss of body heat to the surrounding air.

5. The **nose** is sensitive to odors. The nerves in the nose react to different smells, and impulses are set up. These impulses are carried to the brain. Your eyes also play a part in recognizing what you eat. Your memory helps you to know what it should taste like. Foods are often recognized by smell rather than by taste.

Discover for yourself
how you could observe how your eyes control the amount of light entering them.

PURPOSE: To show how the iris controls the size of the pupil • To develop the skills of observation, comparison and influence

You need: a partner, clock or watch

a. Choose a partner and look into his/her eyes. Observe the size of the pupil. The pupil is the dark spot in the center of the eye.

b. Ask your partner to close his/her eyes and to cover them with his/her hands for about one minute. Then ask your partner to uncover the eyes.

c. Look at the pupil right away.

1. Has the size of the pupil changed? ______
   It becomes larger.
2. Is there any connection between the amount of light entering the eyes and the size of the pupils? ______
   The less light entering the eyes, the larger the pupils become.

Check Up D

D1. Name the five sense organs. ______
(1) Eyes, (2) ears, (3) nose, (4) skin, (5) tongue.

D2. Give a function of each of the following parts of the eye.

(a) Cornea: Protects the lens.

(b) Pupil: Allows light to enter the eye.

(c) Iris: Changes the size of the pupil.

(d) Lens: Focuses the light on the retina.

(e) Retina: Sends messages to the optic nerve.

(f) Optic nerve: Transfers impulses to the brain.

D3. What are the three parts of the ear? (1) Outer ear, (2) middle ear, (3) inner ear.

D4. What does the outer ear do? Picks up sound waves.

D5. Describe each of the following middle ear parts.

(a) hammer, anvil, stirrup: Three tiny bones in the middle ear

(b) Eustachian tube: A tube leading from the throat to the middle ear.

D6. Give a function of each of the following inner ear parts.

(a) cochlea: Vibrations of the fluid in the cochlea stimulate the auditory nerve.

(b) auditory nerve: Carries impulse to the brain.

(c) semicircular canals: Fluid in the canals helps to maintain balance.

D7. What four stimuli can the nerve endings in skin pick up? (1) Heat, (2) cold, (3) pressure, (4) pain.

D8. What do the taste buds do? Tell whether things are sweet, salty, sour or bitter.

THINK ABOUT THIS

Is Hunger "All In Your Head"?

The hypothalamus is a tiny clump of nerve cells located at the base of the brain. Scientists are learning more and more about the part it plays in controlling the appetite.

Researchers now know that if the cells in the center part of the hypothalamus are low in sugar, people tend to overeat and get fat. Certain chemicals in the body, called hormones, also affect the hypothalamus. One of these is called adrenalin. It is made by the adrenal glands which fit like little caps right on top of the kidneys. When a person is upset, angry or frightened, the adrenal glands give off more adrenalin than usual. This causes the sugar intake of the cells of the hypothalamus to be lowered. Then hunger is felt and the person tends to overeat.

The hypothalamus also controls body temperature, blood pressure, metabolism of fats and the sugar level in the blood.

Unit 2, Chapter 5
REVIEW

Matching Exercise Match each term in **Column A** with its description in **Column B**. Write the letter of the description in the space beside the correct term.

	Column A	*Column B*
d	**1.** cell body	**a.** center of reasoning, memory and thought
h	**2.** axon	**b.** automatic responses that help to protect the body
i	**3.** neuron	**c.** carries impulses to the cell body
c	**4.** dendrite	**d.** controls the functions of the neuron
g	**5.** central nervous system	**e.** messages that travel through the nervous system
a	**6.** cerebrum	**f.** controls breathing and heartbeat
f	**7.** medulla	**g.** brain and spinal cord
j	**8.** stimuli	**h.** carries impulses from one neuron to another
b	**9.** reflexes	**i.** nerve cell
e	**10.** impulses	**j.** changes in the environment

True/False In the space provided, write **T** if a statement is true. If it is false, change the underlined word or words to make the statement true.

Sensory	**1.** Motor neurons carry impulses from sense organs to various structures of the nervous system.
cerebrum	**2.** The cerebellum is the largest part of the brain.
medulla	**3.** Breathing and heartbeat are controlled by the cerebrum.
pupil	**4.** The lens is an opening in the iris that permits light to enter the eye.
stirrup	**5.** Vibrations from the eardrum are carried to three tiny bones called the hammer, anvil and cochlea.
T	**6.** The cochlea is the main part of the inner ear.
T	**7.** Impulses are carried from the retina to the brain by the optic nerve.
smell	**8.** The senses of taste and sight work together.
iris	**9.** The size of the pupil is changed by the cornea.
stimulus	**10.** A response is a change in the environment that can be picked up by a sense organ.
T	**11.** The dendrites pick up messages and carry them back to the nerve cell body.
cerebrum	**12.** Reasoning, thought and memory occur in the cerebellum.
T	**13.** Axons carry impulses to other neurons.
spinal cord	**14.** The central nervous system is made up of the brain and the eyes.
An impulse	**15.** A stimulus is a message that is carried along a nerve fiber.

Multiple Choice Exercise Circle the letter of the word or phrase that best completes each statement.

1. Impulses are carried from sense organs by the **(a)** motor neurons **(b)** dendrites **(c)** sensory neurons **(d)** cerebellum.
2. The part of the brain that is responsible for coordinating the muscles is the **(a)** cerebrum **(b)** medulla **(c)** spinal cord **(d)** cerebellum.
3. The part of the brain that is responsible for controlling breathing and the heartbeat is the **(a)** medulla **(b)** cerebrum **(c)** spinal cord **(d)** cerebellum.
4. The opening that allows light to enter the eye is the **(a)** pupil **(b)** lens **(c)** retina **(d)** iris.
5. Nerves that stimulate the auditory nerve to carry messages to the brain are in the **(a)** eye **(b)** ear canal **(c)** Eustachian tube **(d)** cochlea.
6. The tiny bones in the middle ear are the **(a)** hammer, anvil and Eustachian tube **(b)** hammer, anvil and stirrup **(c)** anvil, stirrup and Eustachian tube **(d)** hammer, stirrup and Eustachian tube.
7. Impulses are carried from the eye to the brain by the **(a)** auditory nerve **(b)** Eustachian tube **(c)** optic nerve **(d)** cornea.
8. The parts of the ear that help to keep balance are the **(a)** semicircular canals **(b)** Eustachian tubes **(c)** hammer and anvil **(d)** eardrums.
9. A change in the environment that can be picked up by a sense organ is a **(a)** dendrite **(b)** stimulus **(c)** impulse **(d)** neuron.
10. The central nervous system consists of **(a)** only the brain **(b)** only the spinal cord **(c)** only the nerve cells **(d)** the brain and the spinal cord.

Application Exercise Label the parts of the neuron indicated by the arrows in the diagram below. Write your answers in the provided spaces.

1. nucleus
2. dendrites
3. cell body
4. sheath
5. axon

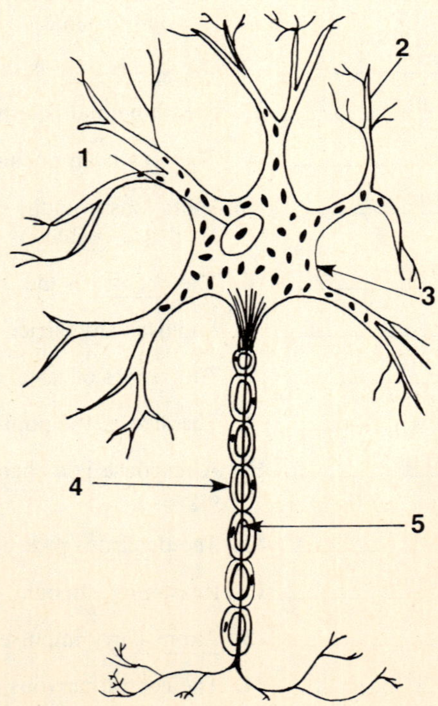

This man is perspiring. Perspiration is one of the ways your body gets rid of wastes.

New science words: excretion, perspiration

Skills developed: observing, comparing

Chapter 6

HOW YOUR BODY GETS RID OF WASTES

Excretion is a function of the skin. Why is this person perspiring? Through perspiration the human body excretes water and salts to cool the overheated body.

You have learned that the lungs get rid of the wastes, carbon dioxide and water vapor, from the body. The large intestine removes the solid wastes of the body. What happens to the liquid wastes made by the cells of the body?

A. The Excretory System

After doing this lesson you will be able:

- A1 • to name the life activity by which all waste material is removed from the body
- A2 • to list the four organs of waste excretion in the human body
- A3 • to tell where the kidneys are located
- A4 • to explain why wastes are forced out of the blood in the capillaries of the kidney
- A5 • to give the function of ureters, urinary bladder and the urethra

The life activity by which the waste ma- A1
terials are removed from the body is called **excretion.** The organs of excretion, in addi- A2
tion to the lungs and the large intestine, are the kidneys and the skin.

The **kidneys** are two bean-shaped organs found on each side of your body just be-
low your waist. They are very close to the A3
back wall of your body. Each kidney is made up of millions of tiny tubes that have cup-shaped ends. In each of the cups is a tiny knot of capillaries. When the blood flows through these capillaries, it is under great A4
pressure and the liquid wastes are forced out. The wastes are caught in the cups and carried out of the kidneys by the tubes. All of these tubes come together in a hollow part of the kidney called the **pelvis.**

A small tube called a **ureter** carries liquid A5
wastes from the pelvis of each kidney into the

Filtration in the kidney is so efficient that a person can survive with only half a kidney functioning.

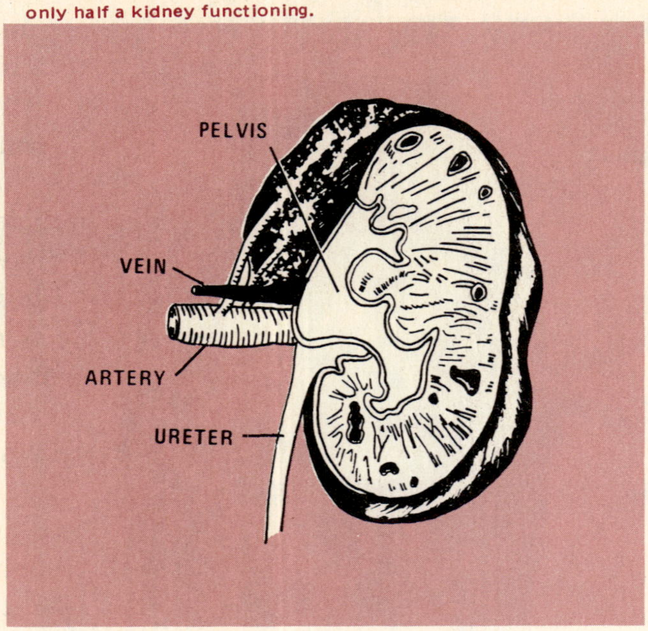

A cross section of the kidney.

The excretory system.

The functional unit of the kidney is the nephron, a microscopic clump of coiled capillaries and tubules. There are approximately 1 million nephrons in each kidney.

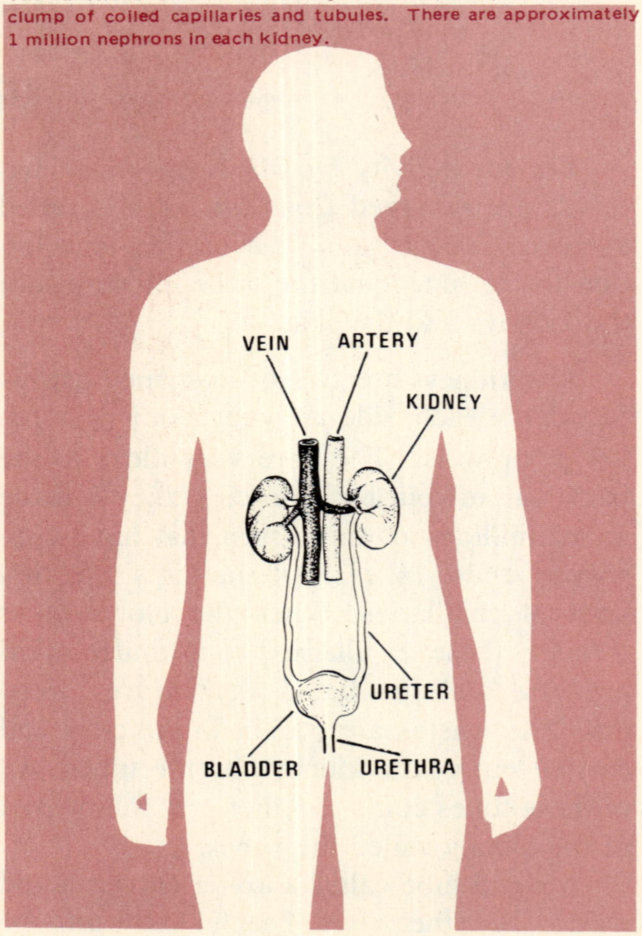

urinary bladder. The **urinary bladder** stores the liquid wastes until they are ready to leave the body. They leave through a tube called the **urethra** that leads from the urinary bladder to the outside.

The adult bladder has a capacity of approximately 0.5 liter.

Check Up A

A1. Which life activity removes wastes from your body? Excretion

A2. Name the four organs of waste excretion in the human body. (1) Lungs, (2) large intestine, (3) kidneys, (4) skin.

A3. Where are the kidneys located? Just below the waist on each side of the body, close to the back wall.

A4. Why are wastes forced out of the blood in the capillaries of the kidneys? Blood flows through the capillaries under great pressure which forces the liquid wastes out of it.

A5. Give a function of each of the following:

(a) ureters: Takes liquid wastes from the kidneys to the bladder.

(b) urinary bladder: ______

Stores liquid wastes until they are ready to leave the body.

(c) urethra: ______

Tube that leads from the urinary bladder to the outside.

B. The Skin

After doing this lesson you will be able:

B1 • to name the two layers of the skin

B2 • to describe and give the function of the epidermis

B3 • to name four structures found in the dermis

B4 • to give the function of the sweat glands

B5 • to tell what perspiration is

The skin is made up of two layers. The
B1 outer layer is called the **epidermis.** This layer
is made of dead cells. As the dead cells are worn away, new cells from lower layers take
their place. The epidermis helps protect the B2
inner layer of the skin.

The inner layer of the skin is the **dermis.** B1
It is thicker than the epidermis. Here you
find the sweat glands, oil glands, hair follicles B3
and nerve endings. The sweat glands have
tiny tubes, called ducts, that bring liquid B4
wastes to the surface (top part) of the skin.
The liquid wastes are called **perspiration.** B5
Perspiration passes through openings (pores) in the skin and evaporates into the air. Why is it important to wash often?

You have learned about many of the systems in the body, but the body is so complicated that it is impossible in this book to cover all the systems. However, you now have a better understanding of how your body works and why it is important to take good care of it. You will learn more interesting things about your body if you study biology.

Bathing removes salt and wastes from the skin and prevents the sweat ducts from becoming clogged with dirt.

SUMMARY OF THE ORGANS OF EXCRETION

Organ	Special structure	Wastes removed	Leave body
Kidneys	Tubes with cup-shaped ends	Water; salt; most protein wastes	Through ureter to bladder; leaves bladder through the urethra
Skin	Sweat glands	Water; salt; small amounts of protein wastes	Through pores (openings) of sweat glands
Lungs	Alveoli	Carbon dioxide; water vapor	Through nose and mouth
Large intestine		Solid undigested food after part of water has been taken back into blood	Through rectum

In the photograph the separation between epidermis and dermis is as clear as it is in the drawing. Have students use the drawing labels to locate the same parts in the photo.

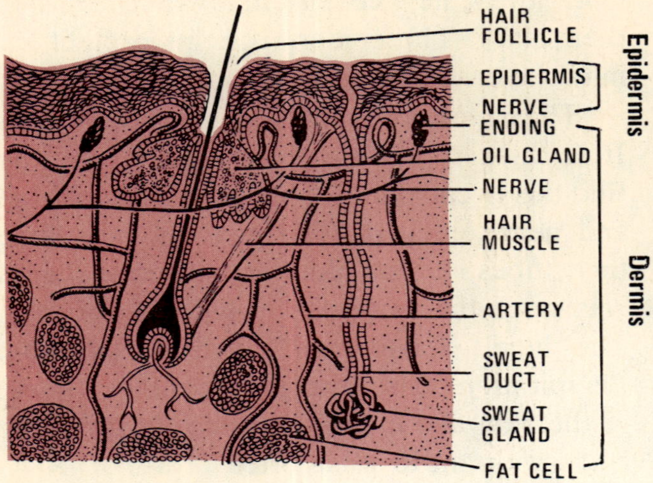

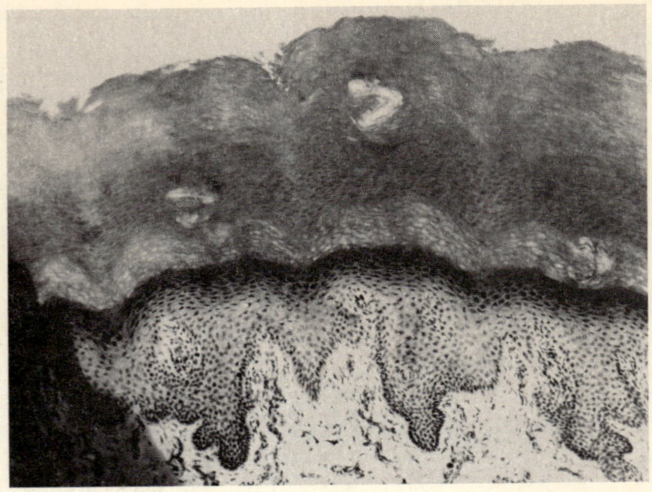

On the left is a drawing of a cross section of skin. Compare the drawing with an actual photo of skin on the right.

Check Up B

B1. What are the two layers of the skin? (1) Epidermis, (2) dermis.

B2. Describe the epidermis. What does it do? The epidermis is made of dead cells. It protects the inner layers of the skin.

B3. Name four structures found in the dermis. (1) Sweat glands, (2) oil glands, (3) hair follicles, (4) nerve endings.

B4. What do the sweat glands do? Bring liquid waste materials to the surface of the skin.

B5. What is perspiration? Liquid waste materials given off by the sweat glands.

Unit 2, Chapter 6
REVIEW

Matching Exercise Match each term in **Column A** with its description in **Column B**. Write the letter of the description in the space beside the correct term.

	Column A		*Column B*
j	**1.** epidermis	**a.**	inner layer of skin cells
a	**2.** dermis	**b.**	removes solid wastes
d	**3.** urinary bladder	**c.**	carries liquid wastes to the urinary bladder
h	**4.** pores	**d.**	storage place for liquid wastes
f	**5.** pelvis	**e.**	remove carbon dioxide and water vapor
b	**6.** large intestine	**f.**	hollow part of the kidney
i	**7.** sweat glands	**g.**	carries liquid wastes out of the body
c	**8.** ureter	**h.**	tiny openings in the skin
g	**9.** urethra	**i.**	excrete perspiration
e	**10.** lungs	**j.**	dead layer of skin cells

True/False In the space provided, write **T** if a statement is true. If it is false, change the underlined word or words to make the statement true.

epidermis	**1.** The <u>dermis</u> of the skin is made up of a layer of dead cells.
urinary bladder	**2.** The body stores liquid wastes in the <u>urethra</u>.
T	**3.** The liquid waste excreted by the skin is <u>perspiration</u>.
ureter	**4.** Liquid wastes are carried from the pelvis of the kidney to the urinary bladder by the <u>urethra</u>.
urethra	**5.** The <u>ureter</u> carries liquid wastes from the urinary bladder to the outside of the body.
large	**6.** Solid wastes are removed from the body by the <u>small</u> intestine.
T	**7.** The <u>lungs</u> are responsible for excreting carbon dioxide and water vapor.
epidermis	**8.** The upper layer of the skin is the <u>dermis</u>.
dermis	**9.** The sweat glands, oil glands, hair follicles and nerve endings are found in the <u>pelvis</u>.
T	**10.** The dermis is <u>thicker</u> than the epidermis.
excretion	**11.** The life activity by which waste materials are removed from the body is <u>perspiration</u>.
T	**12.** The <u>kidneys</u> are responsible for removing water, salt and most protein wastes from the body.

Multiple Choice Exercise Circle the letter of the word or phrase that best completes each statement.

1. The life activity by which wastes are removed from the body is **(a)** secretion **(b)** respiration **(c)** circulation **(d)** excretion.
2. Liquid wastes are carried from the urinary bladder to the outside of the body by the **(a)** pelvis **(b)** ureter **(c)** urethra **(d)** large intestine.
3. The outer layer of skin is the **(a)** dermis **(b)** epidermis **(c)** urethra **(d)** urinary bladder.
4. Liquid wastes are carried from the kidneys to the urinary bladder by the **(a)** dermis **(b)** epidermis **(c)** urethra **(d)** ureter.
5. All of the following are organs of excretion, except the **(a)** small intestine **(b)** large intestine **(c)** lungs **(d)** kidneys.
6. The inner layer of the skin is the **(a)** dermis **(b)** epidermis **(c)** urinary bladder **(d)** urethra.
7. The liquid wastes of the body are stored in the **(a)** large intestine **(b)** small intestine **(c)** ureters **(d)** urinary bladder.
8. The kidneys are located **(a)** in the neck **(b)** beneath the skin **(c)** in the back below the waist **(d)** near the heart.
9. The lungs are responsible for the removal of **(a)** oxygen **(b)** water **(c)** salt and water **(d)** carbon dioxide and water vapor.
10. The liquid waste that is excreted by the sweat glands is **(a)** perspiration **(b)** carbon dioxide **(c)** undigested food **(d)** epidermis.

Application Exercise Label the parts of the excretory system in the diagram on the left. Then label the parts of the kidney in the diagram on the right. Write your answers next to the lines that point to each part.

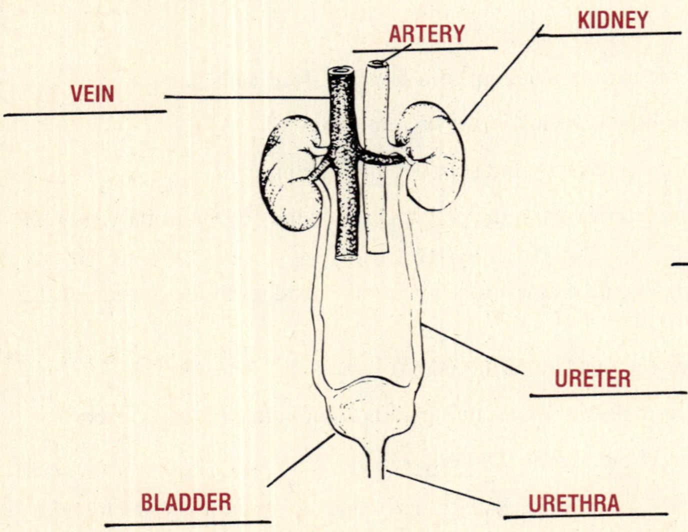

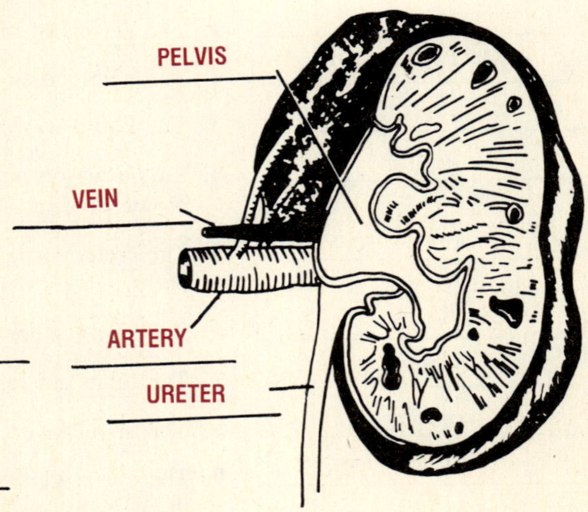

The joints in certain parts of their bodies allow these bikers to pedal. Where do you think these joints are located? in their knees

New science word: skeleton

Skills developed: observing, describing, comparing, inferring

Chapter 7

THE SKELETAL SYSTEM

Discuss the importance of exercise. Ask: "What happens to muscles and joints that are seldom used? Why is outdoor exercise particularly healthful?"

Try to picture a set of clothing standing by itself in a clothing store. It can't stand by itself. It needs the model or it would fall to the floor. Without the skeleton, your body would also collapse and fall to the floor.

A. The Functions of the Skeleton

After doing this lesson you will be able:

A • to list four functions of the skeleton

The human skeleton has four important functions.

A **1.** The skeleton gives the body **shape** and **support.** Without this support you could not stand, walk or sit.

A **2.** The skeleton *protects delicate organs* such as the brain, heart and lungs. The bone on top of the head must be very strong to protect the brain.

A **3.** The skeleton is jointed in many places and so *makes movement possible*. The skeleton is the place where muscles are attached. The muscles pull on the bones to make you move.

4. In the center of many bones of the skeleton there is a cavity (space) that contains **marrow.** This soft inner part of the bone is able to *manufacture red blood cells*. A

A local hospital may loan old x-rays for you to use.

Discover for yourself

how you could find out what the inner part of long bones look like.

PURPOSES: To observe marrow • To develop the skills of observation, description and inference

You need: the large leg bone of an animal

Ask a butcher for a leg bone cut in half. Look at it.

1. Describe the material you see in the center. Soft; pink and white in color.
2. What is it called? Marrow
3. What are the red threads? Blood vessels

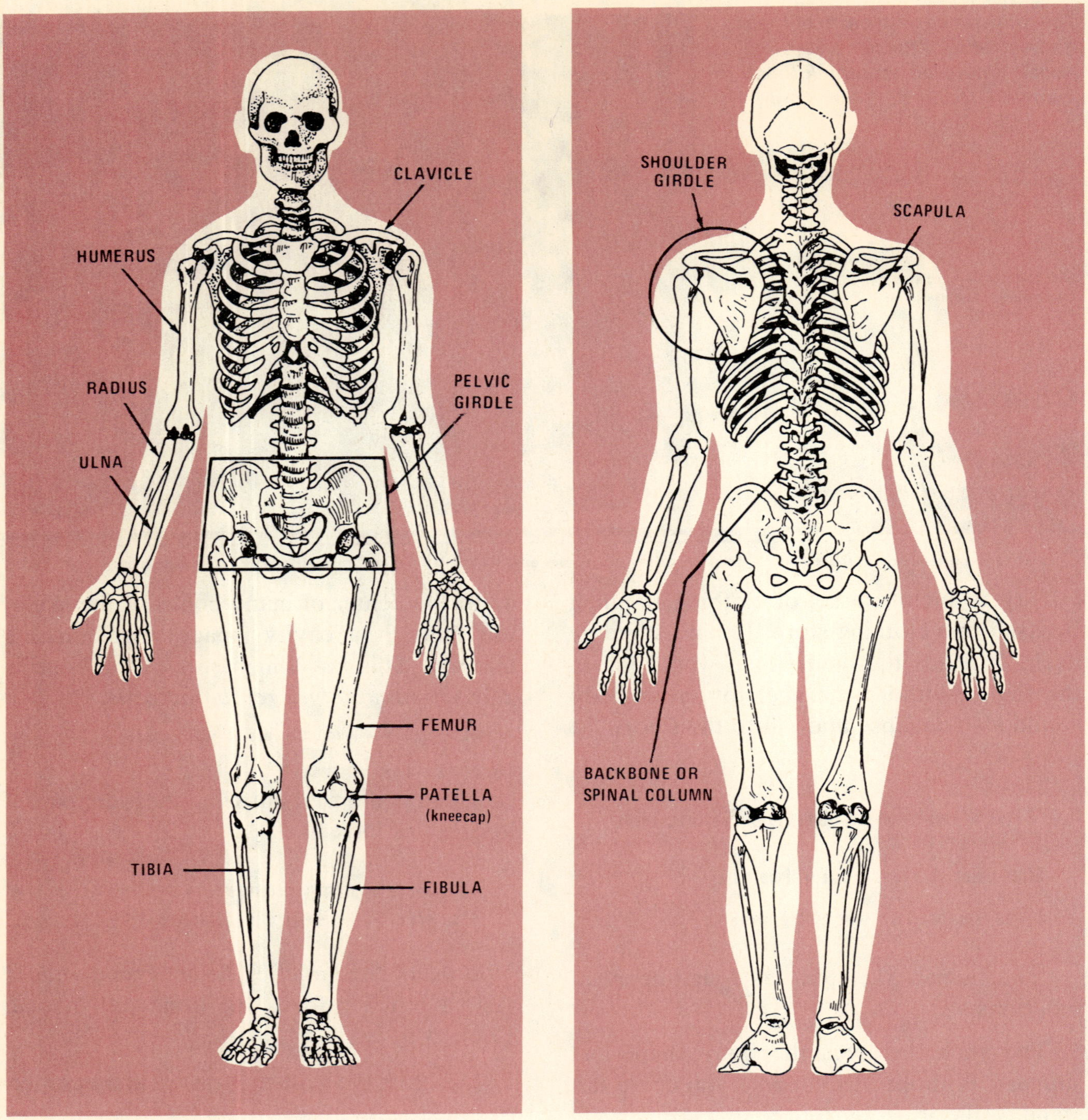

The skeleton.

Check Up A

Ask students to locate the largest and strongest bone in the human skeleton. (Femur)

List four functions of the skeleton.

(1) Gives form and support.

(2) Protects delicate organs.

(3) Makes movement possible.

(4) Marrow manufactures red blood cells.

B. The Bones of Your Body

After doing this lesson you will be able:

B1 • to tell how many bones make up the skeleton

B2 • to give the function of the skull

B3 • to give the names and number of the bones that make up the backbone

B4 • to list the bones that make up the shoulder girdle

B5 • to describe where the humerus, radius and ulna are

B6 • to name the bones that make up the pelvic girdle

B7 • to describe where the femur, patella, tibia and fibula are

The skeleton contains, on the average,
B1 206 bones. This number may vary a bit from person to person.

1. The **skull** is made of eight bones. As a baby grows older, these bones join together to form one solid bony structure. The skull
B2 covers and protects the brain.

The lower jaw bone is attached to the skull. If it were not, you could not open and close your mouth. If you place your hand on your jawbone and pretend to chew, you'll find that only the lower jawbone moves.

The skull.

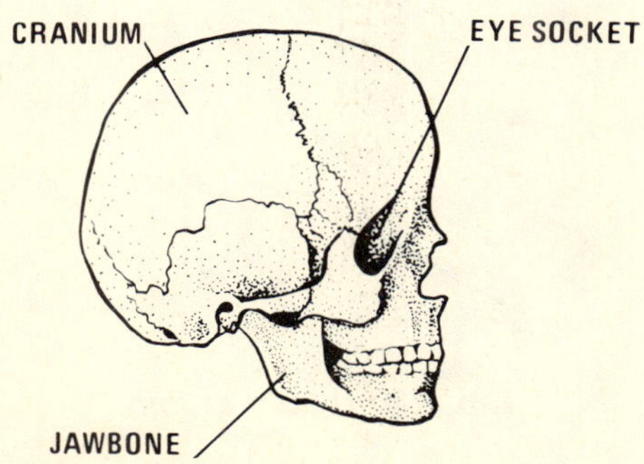

2. The **backbone** (spinal column) is made up
B3 of 32–35 separate bones called **vertebrae.** Some of these are fused (joined together); 24 are separated by a small piece of **cartilage.** Cartilage is a tough, flexible connective tissue. Cartilage gives your nose shape. The cartilage absorbs shock when you walk and keeps the vertebrae from hitting each other. Each vertebra has an opening through which the spinal cord passes. It also has little knobs or spurs to which muscles are attached.

If possible, use a model skeleton to enrich this section.

The backbone is much like a tree trunk that has branches coming out of it. The backbone has two major branches, called **girdles.**

3. The **shoulder girdle** is a group of bones attached to the **sternum** (breastbone). This group of bones includes the four shoulder bones; two collarbones (**clavicle**) in front and B4
two shoulder blades (**scapulae**) in the back.

The arm bones are connected at the outer end of the shoulder bones. The **humerus,** the B5
bone in the upper arm, is attached to the shoulder. Below the elbow there are two more arm bones, the **radius** and the **ulna.** There are several small bones in the wrist and in the hand and fingers.

Attached to the backbone in the chest area are twelve pairs of **ribs.** They bend around the chest cavity and are connected to the breast bone. They form a protective cage around the heart and lungs.

4. The **pelvic girdle** consists of the right and B6
left hip bones and the lower part of the spinal column. Set into each hip bone is a large bone in the upper leg, called the **femur.** At the knee B7
there is a small flat bone called the **patella** or knee cap. Below the knee there are two bones in the lower leg; the **tibia** and the **fibula.** There are several bones in each ankle and foot.

Some people may have 11 pairs of ribs; others as many as 13 pairs.

Check Up B

B1. On the average, about how many bones make up the skeleton? 206 ______

______

______

B2. What is the function of the skull? _____

Covers and protects the brain.

B3. What bones make up the backbone? How many are there? _____

Vertebrae — 32 to 35 bones.

B4. Name the bones that make up the shoulder girdle. Two collarbones and two shoulder blades.

B5. Where are the following bones found?

(a) humerus: Upper arm attached to the shoulder.

(b) radius: Below the elbow.

(c) ulna: Below the elbow.

B6. Name the bones that make up the pelvic girdle. Right and left hip bones and the lower part of the spinal column.

B7. Where are the following bones found?

(a) femur: Large bone in the upper leg.

(b) patella: Small flat bone at the knee.

(c) tibia: Below the knee.

(d) fibula: Below the knee.

C. How Does the Skeleton Move?

After doing this lesson you will be able:

C1 • to explain what a joint is

C2 • to name the three types of joints and give one place where each type is found

You use your skeleton, particularly your hands and arms, for just about everything you do. This means that you must be able to move some bones in almost all directions. Bones are hard and stiff. What makes it possible for them to move? You have special structures called **joints.** A joint is the *place* C1 *where two bones meet.* The joints help make all movement possible. There are several types of joints.

1. **Ball-and-socket joints.** The humerus is connected to the skeleton at the shoulder. The C2 outer end of the shoulder blade is hollowed out, forming a socket. The upper end of the humerus is rounded like a ball which fits right into the socket. The round end of the humerus can move in all directions within the socket. There is a similar ball-and-socket joint at the hip bone. The femur (the bone in the upper leg) fits into this joint.

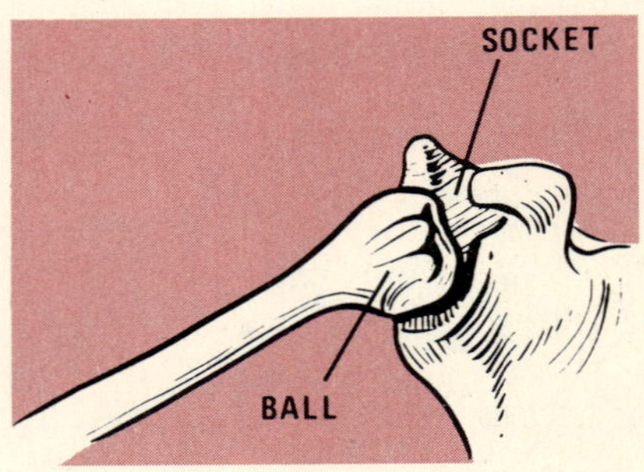

A ball-and-socket joint.

2. **Hinge joint.** You can see, if you try it, that your elbow and your wrist cannot move in all C2 directions. They are like the door in your classroom. The door can only open and close. The kind of joint in the fingers, wrist or elbow is called a **hinge joint.**

This joint is at the knee.

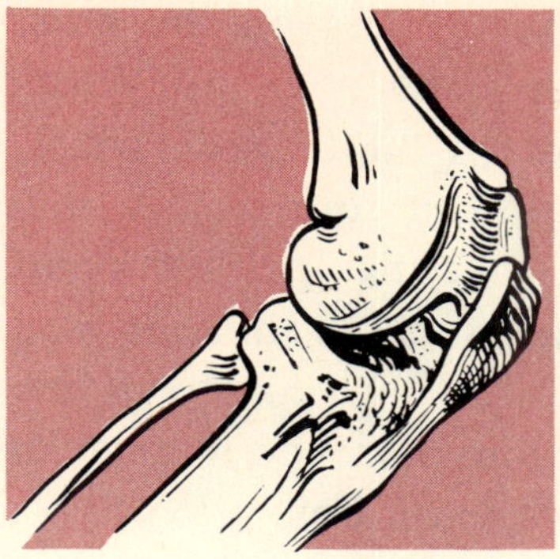

A hinge joint.

An immovable joint is sometimes called a *suture*.

C2 **3. Immovable joints.** Bones in the skull slowly join together much like the pieces in a jigsaw puzzle. And, like pieces in a completed jigsaw puzzle, the bones are held in place where they join. Therefore, the joints are immovable.

In an infant the skull bones are not completely joined. There is a soft spot just above the forehead. Usually by about one and a half years of age, the skull bones are completely joined. Until this happens great care must be taken to protect the infant's head.

Check Up C

C1. What is a joint? An area where two bones meet.

C2. Name the three types of joints and give one place where each is found.

(1) Ball-and-socket — at the shoulder, at the hip.

(2) Hinge — fingers, wrist, elbow.

(3) Immovable — skull.

D. How Connective Tissues Hold the Body Together

After doing this lesson you will be able:

D1 • to explain what a ligament is and what it does

D2 • to explain what a tendon is and what it does

You have a skeleton that gives shape and support to your body, and joints that let you move about. But what holds the entire skeleton together? Why don't the individual bones separate so that the skeleton falls apart?

The skeleton probably would come apart if it weren't for a tough, stretchy material that is called **connective tissue.** Some of this tissue attaches one bone to another bone. D1
This kind of connective tissue is called a **ligament.** When you stretch or twist a ligament and it becomes sore and swollen, it is called a **sprain.**

Students can easily locate the tendon above their heel. (Achilles)

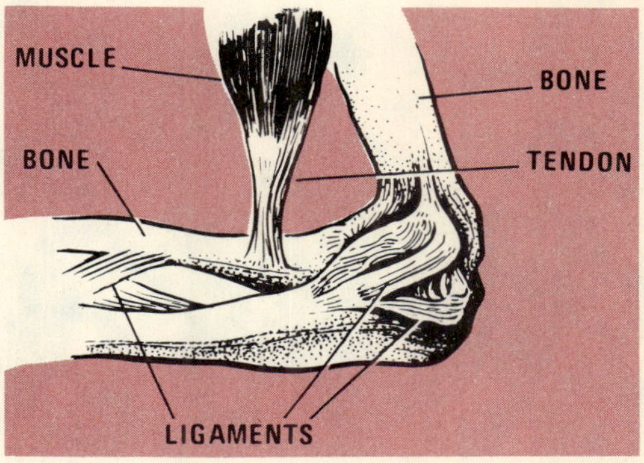

Ligaments and tendons.

A second type of connective tissue is also important. This tissue attaches muscles to D2
bones. At the end of each muscle there is a white, stringy, tough tissue called a **tendon.** This stringy material attaches the muscle close to the bone.

Check Up D

D1. What is a ligament and what does it do?

A type of connective tissue that holds bones together.

D2. What is a tendon and what does it do? A type of connective tissue that connects muscles to bones.

E. What Makes the Bones Move?

After doing this lesson you will be able:

- E1 • to name the tissue that makes it possible for bones to move
- E2 • to explain the difference between voluntary and involuntary muscle

Muscles make the bones move. Each mus-
E1 cle cell can contract but not expand. Because
of this, muscles must always work in pairs.

The voluntary muscles of the upper arm.

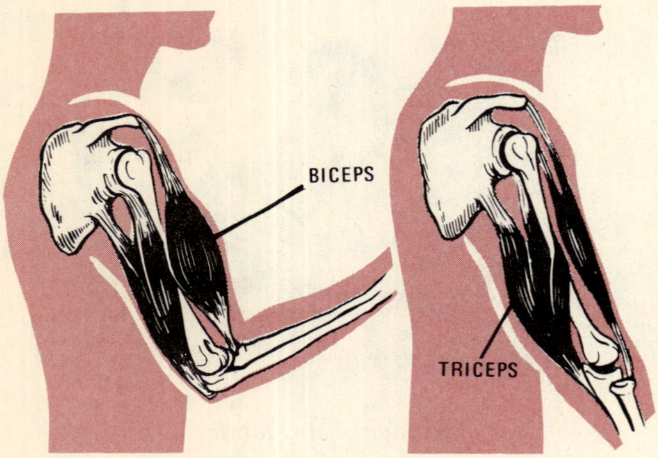

Hold your hand on the muscle in your upper arm. Bend your arm. Notice how the muscle gets a bit bigger as it contracts. If you now lower your arm, a muscle on the backside of your upper arm contracts.

Muscles such as those in your arms and E2
legs are under your control. These are called **voluntary muscles.** All the muscles connected to your skeleton belong to this group. You can move voluntary muscles almost any way you want to.

Your heart, digestive system and circulatory system have muscles that are not connected to your skeleton. You cannot control their movement. Since they cannot be con-
sciously controlled, they are called **invol-** E2
untary muscles.

Check Up E

E1. What kind of tissue makes it possible for bones to move? Muscle

E2. What is the difference between voluntary and involuntary muscle? Voluntary muscles can be controlled. Involuntary muscles cannot be consciously controlled.

Unit 2, Chapter 7
REVIEW

Matching Exercise Match each term in **Column A** with its description in **Column B**. Write the letter of the description in the space beside the correct term.

Column A		Column B
i	**1.** patella	**a.** breastbone
e	**2.** skull	**b.** bones of the backbone
b	**3.** vertebrae	**c.** red blood cells are made here
a	**4.** sternum	**d.** bone in upper arm
j	**5.** scalpulae	**e.** protects the brain
d	**6.** humerus	**f.** bone in upper leg
f	**7.** femur	**g.** spinal column
g	**8.** backbone	**h.** connect bone to bone
c	**9.** marrow	**i.** kneecap
h	**10.** ligaments	**j.** shoulder blades

Multiple Choice Exercise Circle the letter of the word or phrase that best completes each statement.

1. The tissue that makes it possible for bones to move is **(a)** cartilage **(b)** marrow **(c)** muscle **(d)** blood.
2. The shoulder joint is an example of a **(a)** patella **(b)** vertebrae **(c)** hinge joint **(d)** ball-and-socket joint.
3. All of the muscles of the body that are connected to the skeleton are **(a)** voluntary muscles **(b)** involuntary muscles **(c)** cardiac muscles **(d)** ligaments.
4. Muscles that cannot be consciously controlled are **(a)** voluntary muscles **(b)** involuntary muscles **(c)** tendons **(d)** ligaments.
5. Red blood cells are manufactured in the **(a)** tendons **(b)** ligaments **(c)** heart **(d)** bone marrow.
6. The connective tissue that connects bone to bone is called a **(a)** tendon **(b)** ligament **(c)** patella **(d)** involuntary muscle.
7. The kind of joint found in the elbow, wrist and fingers is a **(a)** patella **(b)** ball-and-socket joint **(c)** hinge joint **(d)** vertebrae.
8. The clavicle is also called the **(a)** collarbone **(b)** shoulder blade **(c)** knee cap **(d)** skull.
9. The right and left hip bones and the lower part of the spinal column make up the **(a)** vertebrae **(b)** pelvic girdle **(c)** shoulder girdle **(d)** humerus.
10. The tough, flexible connective tissue that gives the nose shape is **(a)** muscle **(b)** bone **(c)** vertebrae **(d)** cartilage.

Application Exercise Label the parts of the skeletons below. Write your answers next to the arrows that point to each part.

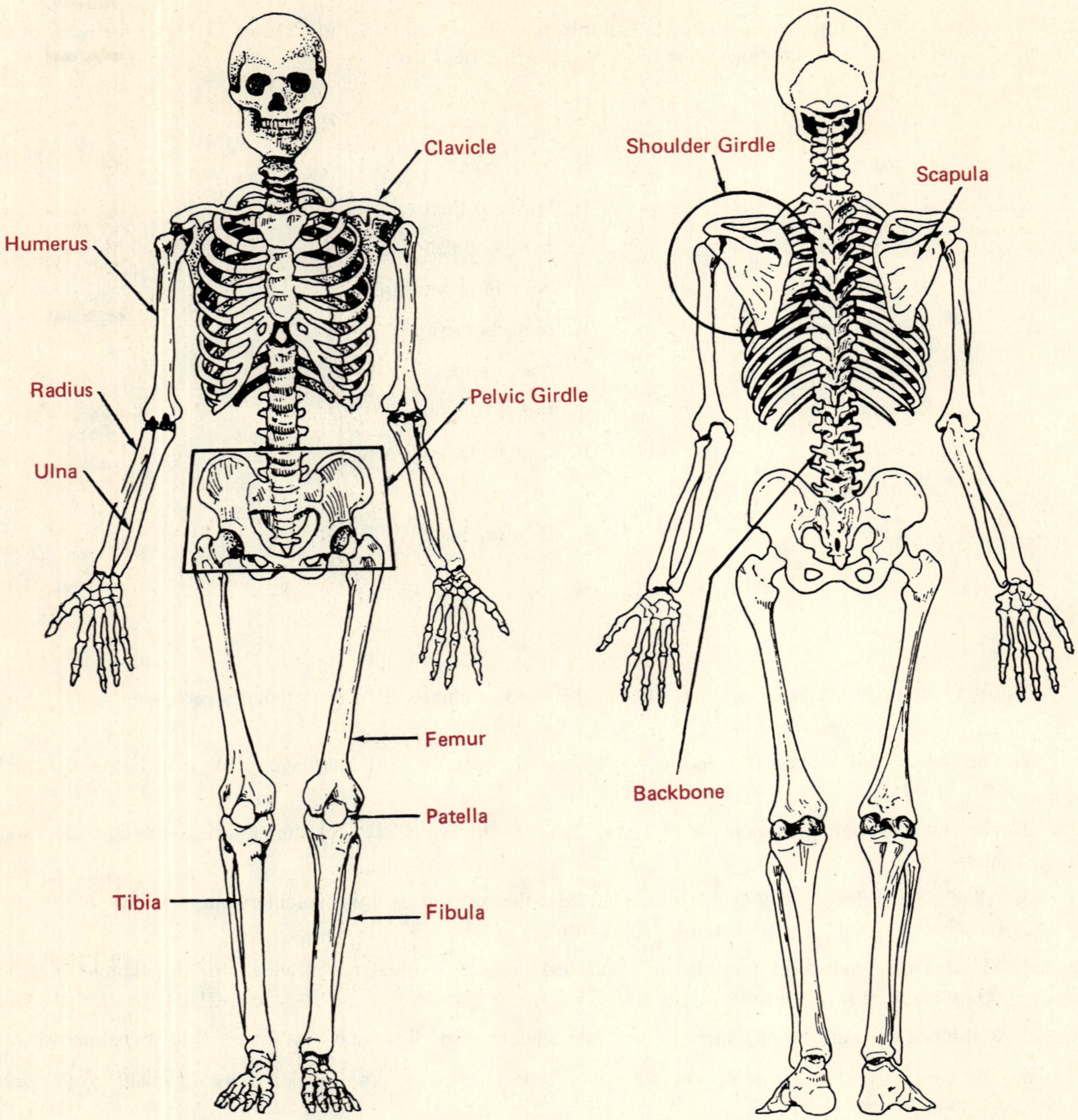

This boy is getting a shot from his doctor. When was the last time you went to your doctor for a check up?

New science words: plaque, oxidation, temperature

Skills developed: observing, identifying, classifying

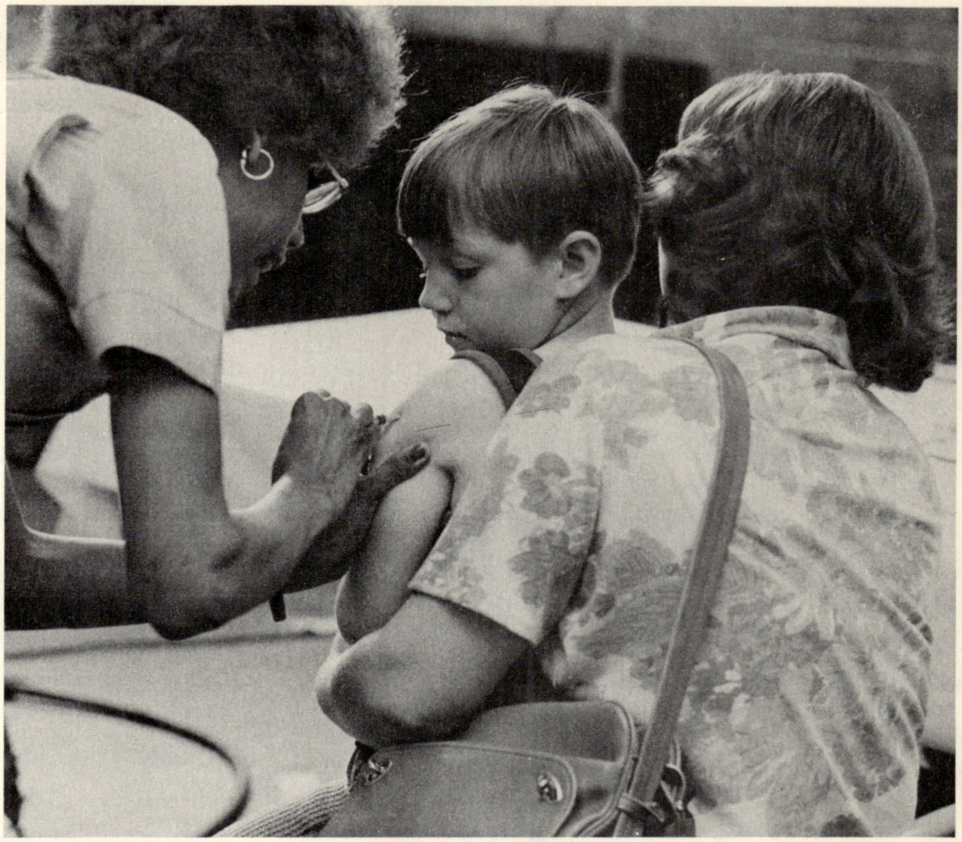

Chapter 8

KEEPING HEALTHY

Plan to have the school nurse or doctor visit your class. Have a conference with them about ways in which they may assist you with this chapter.

If you are in good health you work well, feel well and are happy. Good health means both good physical health and good mental health. Food, rest, sleep and exercise keep you physically and mentally fit.

In order to stay healthy it is important to visit your doctor, and dentist, for a checkup at least once a year. It is not enough to go to a doctor only when you are not feeling well. Let's find out how a doctor can help you.

A. What You Should Know About Physical Health

After doing this lesson you will be able:

A1 • to tell what the following medical instruments do:

(1) thermometer
(2) stethoscope
(3) pressure gauge
(4) otoscope
(5) ophthalmoscope

A2 • to give the normal body temperature of a human being in degrees Celsius and Fahrenheit

A3 • to tell what dental plaque is

A4 • to give five rules for good health

What is a good checkup? Did the doctor you visit do any or all of the following?

Often, doctors check your **temperature** with a **thermometer.** Normal body temperature is about 37°C. However, your normal body temperature may be higher or lower. This does not mean you are ill. It means that your body temperature is a little different from normal. A1(1) A2

Your height and weight are checked and recorded so a comparison can be made next time you visit the doctor.

If possible, have the school doctor or nurse demonstrate some of these instruments to your students.

A1(2) Doctors also listen to your heart. They use an instrument called a **stethoscope.** With the stethoscope they can listen to the action of your heart. Your heart beats about 75–90 times a minute. Your heart makes definite sounds and has a definite rhythm. By listening, doctors can tell if your heart is working well. They do not try to count your heart beats through the stethoscope. Instead, they take your **pulse beat.** As you know, this is the same as the rate your heart beats.

A1(3) Doctors check your **blood pressure** with an instrument called a **pressure gauge.** This instrument is a glass tube filled with mercury, very much like a thermometer. With a **stethoscope** a doctor listens to your lungs. The doctor will ask you to breathe in deeply. A technician might take a chest **x-ray** of your lungs.

At your checkup you will also have an examination of your ears, nose, throat and eyes. These are places where infection can occur easily. If infections are found early, they can be treated so they do not become serious.

A1(4) The doctor uses an **otoscope** to examine your ears. The otoscope has a magnifying lens and a light. The magnifying lens lets the doctor get a better look at your middle ear and remove any hard wax.

Your doctor will also check your eyes with an instrument called an **ophthalmoscope.** This instrument helps him or her to look inside the eye.

If your doctor finds that you do not see properly or that you have an eye disease he
A1(5) or she will send you to see an **ophthalmologist.** An ophthalmologist is an eye doctor who can prescribe (write an order) medicine if you need it and prescribe glasses.

Your dentist can help prevent tooth loss if you visit him or her regularly. But only you can care for your teeth every day. You should visit a dentist regularly. **Dental plaque**
A3 is a sticky, colorless layer of harmful bacteria

Contact lenses to correct vision are becoming popular. A local ophthalmologist may be able to provide your class with additional information about contact lenses.

that is constantly forming on your teeth. The bacteria in plaque cause tooth decay and gum disease. If you do not remove plaque daily, it will collect and turn into a hard coating which can be removed only by your dentist or a dental hygienist.

Toothbrushing removes plaque and other materials from the outer and inner surfaces of your teeth. You can remove plaque and debris from between your teeth with dental floss.

After brushing and flossing your teeth, you should rinse your mouth well with water. This action will remove loosened plaque and debris from your teeth and gums.

There are many things you can do to keep yourself healthy. They take very little of your time. You only have to remember to do them.

1. Eat a well-balanced diet. A4
2. Get enough sleep and exercise.
3. Bathe every day.
4. Wash hands before eating, and brush and floss your teeth after each meal.
5. See a doctor and dentist regularly.

Encourage students to add some health and safety rules of their own to this list.

Check Up A

A1. What do the following medical instruments do?

(1) Thermometer: Measures body temperature.

(2) Stethoscope: Permits the doctor to listen to your heart.

(3) Pressure gauge: Measures blood pressure.

(4) Otoscope: Used to examine the ears.

(5) Ophthalmoscope: Used to examine the eyes.

A2. What is the normal body temperature of a human being in degrees Celsius (C)?

37° C

A3. What is dental plaque? Sticky, colorless layer of harmful bacteria.

A4. Give five rules for good health.

(1) Always eat a well-balanced diet.

(2) Get enough sleep and exercise.

(3) Bathe every day.

(4) Wash hands before eating; brush and floss teeth after each meal.

(5) See a physician and dentist regularly.

B. What You Should Know About Your Diet

After doing this lesson you will be able:

B1 • to name the five nutrients that must be included in your diet

B2 • to tell what each of the following vitamins do:

(1) vitamin A
(2) vitamin B_1
(3) vitamin B_2
(4) niacin
(5) vitamin C
(6) vitamin D

B3 • to tell where the vitamins listed in B2 above are found

B4 • to name the four food groups

What foods does your body need? The materials in food that help the body grow, provide energy and prevent certain diseases are called **nutrients.** The study of these foods and how they are used is called **nutrition.**

There are nutrients that you must include in your diet. These are: (1) **proteins,** (2) **fats,**
B1 (3) **carbohydrates,** (4) **minerals** and (5) **vitamins.** You will see why your body needs each of these nutrients.

1. Proteins are needed for growth and repair of cells. If your body lacks proteins, you will not grow. Nor will your body be able to repair the cells of which it is made up. Proteins are made up of carbon, hydrogen, oxygen and nitrogen. In addition to these elements, proteins may also have sulfur and phosphorus. Proteins are the only nutrients which provide the nitrogen needed for repairing and building new cells. Proteins are found in milk, eggs, fish, lean meat and cheese.

2. Fats are used by the body to release energy. Most of the fats you eat are quickly burned (oxidized) by the body. The body stores extra fat to protect it and help keep it warm. Fats are made up of carbon, hydrogen and oxygen. Much of the fat your body needs may be obtained from peanuts, butter, bacon and oils.

3. Carbohydrates give the body the material it needs for the immediate release of energy. **Sugar** and **starches** are carbohydrates. Candy and fruits contain sugar. Potatoes, bread and spaghetti contain starch.

Carbohydrates, just like fats, are made up of carbon, hydrogen and oxygen. Carbohydrates combine with oxygen in the body's cells to release the energy in the form of heat. This type of chemical reaction is called **oxidation.** Although we always need carbohydrates for energy, too many can be harmful. When the body has too much carbohydrates, it is changed to a kind of animal starch called **glycogen.** Glycogen is stored in your liver to be used by the body when it needs energy.

4. Minerals also help build a strong and healthy body. **Calcium** and **phosphorus** are needed for strong bones and teeth. If you do not have enough calcium and phosphorus, your bones become brittle (easily broken).

If possible, show a motion picture or filmstrip related to foods and proper diet. If your school does not have these materials, they are often available from local libraries and state or county health departments.

Goiter, a swelling of the thyroid gland, is caused by a deficiency of iodine in the diet. It was common among people who lived inland before salt was iodized.

Vitamin D does not come from the sun. When exposed to sunlight, the body produces Vitamin D.

Calcium is also needed for muscles and the blood. Milk is a good source of calcium.

Red blood cells need **iron** in order to form hemoglobin. Not having enough iron can cause anemia. **Anemia** is a disease caused by the lack of hemoglobin in the blood. The thyroid gland found in your neck needs the mineral **iodine.** This gland gives off a substance (material) which controls many important functions of the body. Iodine is found in seafood and is added to common table salt.

5. Vitamins help the body work and grow properly. A lack of vitamins can cause many diseases known as **deficiency diseases.**

There are many vitamins. Each is known by a letter as well as by a chemical name. You will study the more important vitamins.

B2(1) **Vitamin A** helps build up the skin and
helps you grow. This vitamin helps you to
see well in dim light or at night. It also helps
your body to resist colds. Vitamin A is
B3 found in eggs, whole milk, butter and many
vegetables, particularly carrots.

Since there is more than one **vitamin B,**
the group is generally known as the "B com-
plex" vitamins. These vitamins also help pre-
vent disease, help the nerves and the heart
B2(2) and increase growth. **Vitamin B_1** is called
thiamin. It prevents a nerve disease called
beriberi. People with beriberi are weak; they
B3 lose weight and waste away. Vitamin B_1 is
found in cereals, meats and yeast.

B2(3) **Vitamin B_2** is called **riboflavin.** Vitamin
B_2 prevents skin diseases such as sores and
B3 rashes. Eggs, liver and cereal contain vitamin
B_2.

Niacin is another vitamin in the "B com-
B2(4) plex" group. Niacin prevents **pellagra.** Pel-
lagra interferes with the normal activity of
the digestive system and also causes mental
B3 disease. Niacin is found in meat, fish and
dairy products.

Vitamin C, or **ascorbic acid**, prevents B2(5)
scurvy, a disease in which the blood capil-
laries break down. Oranges and grapefruit
and leafy green vegetables have large amounts B3
of vitamin C. Vitamin C is destroyed by
cooking. Vitamin C builds up healthy gums
and strong teeth.

Vitamin D is necessary for the growth of
strong bones and teeth. A lack of vitamin D B2(6)
may make the bones and teeth soft. This dis-
ease is known as **rickets.** This vitamin is B3
found in fish, fish oils and dairy products.

Your body needs water. The living material (protoplasm) found in every cell in your body is made up of a large amount of water. Blood is about 90% water. Digested foods are carried to the cells dissolved in water. Water also carries waste materials away from the cells of your body. Most foods contain water.

All of the nutrients the body needs fall
into four groups. When you eat the right
amounts of the right food, you are eating a
balanced diet. As you read about these groups,
check and see if your diet is a balanced one.
The food groups are: (1) the **milk group,** (2) B4
the **meat group,** (3) the **bread and cereal group**
and (4) the **fruit and vegetable group.**

1. The **milk group** includes milk, milk products, butter and margarine. Children should have at least four glasses of milk every day. Milk products such as cheese and yogurt may be eaten in place of a glass of milk.

2. The **meat group** consists of fish, eggs, meat and poultry. Your diet should have two or three servings of this group daily. This group is high in proteins which the body needs for growth and repair.

3. The **bread and cereal group** has the carbohydrates your body needs for energy. Corn, wheat, rice and oats are grains found in cereals that have carbohydrates. You should eat four servings of this group daily.

Ask students to keep track of their daily diets for a week. Discuss deficiencies and suggest changes. "Anonymous" diets should be submitted so that students are not put on the spot.

4. The **fruit and vegetable group** gives your body the vitamins and minerals it needs. This group includes green and yellow vegetables, tomatoes and potatoes, oranges, grapefruit and other fruits. Everyone should eat at least four servings from this group every day.

Is yours a balanced diet? Do you eat the proper foods? What can you do to be sure that you are on a well-balanced diet? You can start by eating a good breakfast. It will give you the energy you need to do your work. Then, you can check out your other meals. What are you missing? Start including it in your diet.

Older references may still refer to the "7 Food Groups". The newer 4 food group classification includes the same basic foods.

Check Up B

B1. What are the five nutrients that you must include in your diet?

(1) Proteins

(2) Fats

(3) Carbohydrates

(4) Minerals

(5) Vitamins

B2. What do each of the following vitamins do?

(1) Vitamin A: helps build up the skin and promote growth.

(2) Vitamin B_1: prevents beriberi.

(3) Vitamin B_2: prevents skin disorders such as sores and inflammation.

(4) Niacin: prevents pellagra.

(5) Vitamin C: prevents scurvy; builds up healthy gums and teeth; helps blood cells remain healthy.

(6) Vitamin D: essential for growth of strong bones and teeth.

B3. In what foods are the following vitamins found?

(1) Vitamin A: eggs; whole milk; many vegetables, particularly carrots.

(2) Vitamin B_1: cereals; meat; yeast.

(3) Vitamin B_2: eggs; liver; cereal.

(4) Niacin: meat; fish; dairy products.

(5) Vitamin C: citrus fruits; leafy green vegetables.

(6) Vitamin D: fish; fish oils; dairy products.

B4. What are the four food groups?

(1) Milk

(2) Meat

(3) Bread and cereal

(4) Fruit and vegetable

C. Some Things You Should Know About Smoking

After doing this lesson you will be able:

C1 • to name one harmful gas in cigarette smoke

C2 • to list two harmful materials that build up in the lungs of smokers

C3 • to name three lung diseases brought on by tars and nicotine

Tobacco smoke has many chemicals that are harmful and cause serious disease. One of these harmful chemicals is the gas **carbon monoxide.** It acts on the blood in a way that keeps the blood from carrying the oxygen it needs to produce energy by oxidation. Without energy, we cannot live.

The smoke from a burning end of a cigarette has more carbon monoxide in it than the smoke inhaled by the smoker! Even though you are not smoking a cigarette, you breathe in carbon monoxide when you are around smokers. This may make you feel dizzy and sick to your stomach; smoke may burn your eyes and throat. It can be very serious for people who have allergies, asthma, or heart disease. Heart attacks and high blood pressure are common in smokers.

Tobacco smoke also carries harmful materials that build up in the lungs of smokers. C2 One of these is **nicotine,** another is **tar.** Tars and nicotine are now known to cause very serious lung diseases. Three of these dis- C3 eases are lung cancer, bronchitis and emphysema. Tars and nicotine can change the way lung cells divide and make them produce cells that are not normal. Tars and nicotine can also destroy air sacs of the lungs and make the lungs stiff. For these reasons, smoking is very dangerous to health.

Check Up C

C1. What harmful gas is found in cigarette smoke? carbon monoxide

C2. Name two harmful materials that build up in smokers' lungs. nicotine, tar

C3. What are three serious lung diseases that tars and nicotine bring on? ______

cancer, bronchitis, emphysema

D. Some Things You Should Know About Drinking

After doing this lesson you will be able:

- D1 • to describe what alcohol can do to the central nervous system
- D2 • to identify the part of the central nervous system that is first slowed down by alcohol
- D3 • to list some common reactions of the body to alcohol
- D4 • to name an organ that can be seriously damaged by alcohol

Sometimes drinkers feel "pepped up" by alcohol. But alcohol does not really pep you up. Alcohol gets into the blood almost right after it is swallowed. It first goes to the D2 cerebrum of the brain. Remember, that's the part of your brain that controls thinking. Alcohol slows down the cerebrum and D1 drinkers feel very relaxed and even "pepped up." But the alcohol is interfering with the person's normal way of thinking, remembering and understanding.

Not everybody reacts to alcohol in the same way. But there are some reactions that are common to people who drink. Slower thinking and some loss of muscular control D3 are often the first reactions. With increased drinking there is blurred sight, unclear speech and a slowing of reflexes. Heavy drinking causes loss of balance, seeing double and, often, passing out. If someone drinks a very large amount of alcohol all at once, death could result. Now it should be clear why drinking and driving a car just don't mix!

Certain organs of the body may be seriously damaged by alcohol. One of these is the liver. Alcohol is used by the liver cells D4 to make energy very quickly. As a result, other foods, especially fats, go unused. They build up in the liver and are stored there in large amounts. When this happens, the liver is not able to carry out its job for the body.

Check Up D

D1. What can alcohol do to the central nervous system? slows down the brain

D2. What part of the central nervous system is first slowed down by alcohol? ______

the cerebrum of the brain

D3. List some reactions the body may have to heavy drinking.

Answers may be any listed under first reaction.

increased drinking, heavy drinking.

D4. What body organ can be seriously damaged by alcohol? liver

Unit 2, Chapter 8
REVIEW

Matching Exercise Match each term in **Column A** with its description in **Column B**. Write the letter of the description in the space beside the correct term.

	Column A	*Column B*
f	**1.** Vitamin B_1	**a.** prescribes eyeglasses
d	**2.** Vitamin B_2	**b.** used to measure blood pressure
h	**3.** Vitamin C	**c.** measures temperature
c	**4.** thermometer	**d.** riboflavin
g	**5.** otoscope	**e.** sugar and starches
i	**6.** opthalmoscope	**f.** thiamine
a	**7.** opthalmologist	**g.** used to examine the ears
e	**8.** carbohydrates	**h.** ascorbic acid
j	**9.** stethoscope	**i.** used to examine the eyes
b	**10.** pressure gauge	**j.** used to listen to the heart

True/False In the space provided, write **T** if a statement is true. If it is false, change the underlined word or words to make the statement true.

90%	**1.** Water makes up about <u>10%</u> of the blood.
cereals	**2.** Skin diseases and beriberi can be prevented by eating <u>dairy products</u>.
nitrogen	**3.** Proteins are made up of carbon, hydrogen, <u>glycogen</u> and oxygen.
more	**4.** The smoke from the burning end of a cigarette has <u>less</u> carbon monoxide than the smoke that is inhaled.
T	**5.** Driving after drinking is dangerous because alcohol can cause blurred vision, slower thinking, and <u>slowed reflexes</u>.
vitamin A	**6.** Eggs are a good source of vitamin B_2, <u>vitamin C</u> and protein.
glycogen	**7.** The body changes unneeded carbohydrates into <u>starch</u> and stores it in the liver.
liver	**8.** The organ most likely to be damaged by alcohol is the <u>lung</u>.
T	**9.** Doctors use an instrument called an <u>otoscope</u> to examine the ears.
A stethoscope	**10.** <u>An opthalmoscope</u> is used to examine the heart and lungs.
fruits and vegetables	**11.** Scurvy may be caused by not eating enough <u>breads and cereals</u>.
37° C	**12.** Normal body temperature in humans is <u>30° C</u>.
fats	**13.** The body uses <u>proteins</u> to release energy.
T	**14.** Calcium and <u>phosphorus</u> are needed for strong bones and teeth.
Anemia	**15.** <u>Iodine</u> is a disease caused by a lack of hemoglobin in the blood.

Multiple Choice Exercise Circle the letter of the word or phrase that best completes each statement.

1. Scurvy develops from a lack of **(a)** vitamin A **(b)** vitamin B **(c)** vitamin D **(d)** vitamin C.
2. Rickets develops from a lack of **(a)** vitamin A **(b)** vitamin B **(c)** vitamin D **(d)** vitamin C.
3. Two harmful materials that build up in the lungs of smokers are **(a)** carbon monoxide and tar **(b)** tar and nicotine **(c)** carbon monoxide and nicotine **(d)** carbon dioxide and tar.
4. The body temperature of a human is **(a)** 98° C **(b)** 43° C **(c)** 37° C **(d)** 96° C.
5. The body organ that is most likely to be damaged by alcohol is the **(a)** heart **(b)** lungs **(c)** brain **(d)** liver.
6. The disease caused by a lack of hemoglobin in the blood is **(a)** a deficiency disease **(b)** dental plaque **(c)** anemia **(d)** glycogen.
7. A doctor examines the ears with **(a)** a pressure gauge **(b)** an otoscope **(c)** an opthalmoscope **(d)** a stethoscope.
8. A disease caused by a lack of vitamins is **(a)** anemia **(b)** a deficiency disease **(c)** dental plaque **(d)** high blood pressure.
9. Niacin can help to prevent **(a)** beriberi **(b)** anemia **(c)** scurvy **(d)** pellagra.
10. Fish, fish oils and dairy products are good sources of **(a)** vitamin B_1 **(b)** vitamin D **(c)** vitamin C **(d)** iron.

Application Exercise Fill in each space in the vitamin summary chart below.

VITAMIN	WHAT IT DOES	TYPE OF FOOD IT IS FOUND IN
Vitamin A	helps build up the skin and promotes growth	eggs; whole milk; many vegetables, particularly carrots
Vitamin B_1	prevents beriberi	cereals; meat; yeast
Vitamin B_2	prevents skin disorders such as sores and inflammation	eggs; liver; cereal
Niacin	prevents pellagra	meat; fish; dairy products
Vitamin C	prevents scurvy	citrus fruits; leafy green vegetables
Vitamin D	essential for strong bones and teeth	fish; fish oils; dairy products

Get Ready To Read UNIT 3

USING DIAGRAMS TO REINFORCE WHAT YOU READ

In Unit 2, you learned how graphic aids help to present new information on a subject. Graphic aids also help you understand information in the text. Now you will find out how you can test yourself on what you learn in this Science Work-A-Text by making diagrams of what you read.

One use of diagrams is to show directions. People often ask for directions on how to get from one place to another. It is helpful to sketch a simple diagram, or map, to illustrate the directions.

Read the directions that tell how to get from the Lakeville Library to the Post Office. Then make your own diagram according to the directions you read. Use arrows to show the direction of movement.

> Turn left out the front door of the library and walk along Ridge Street until you reach the corner of Ridge and King Streets. Turn right on King Street and walk until you reach the first traffic light. The Post Office will be on the right, facing King Street.

Does your diagram look like one of these?

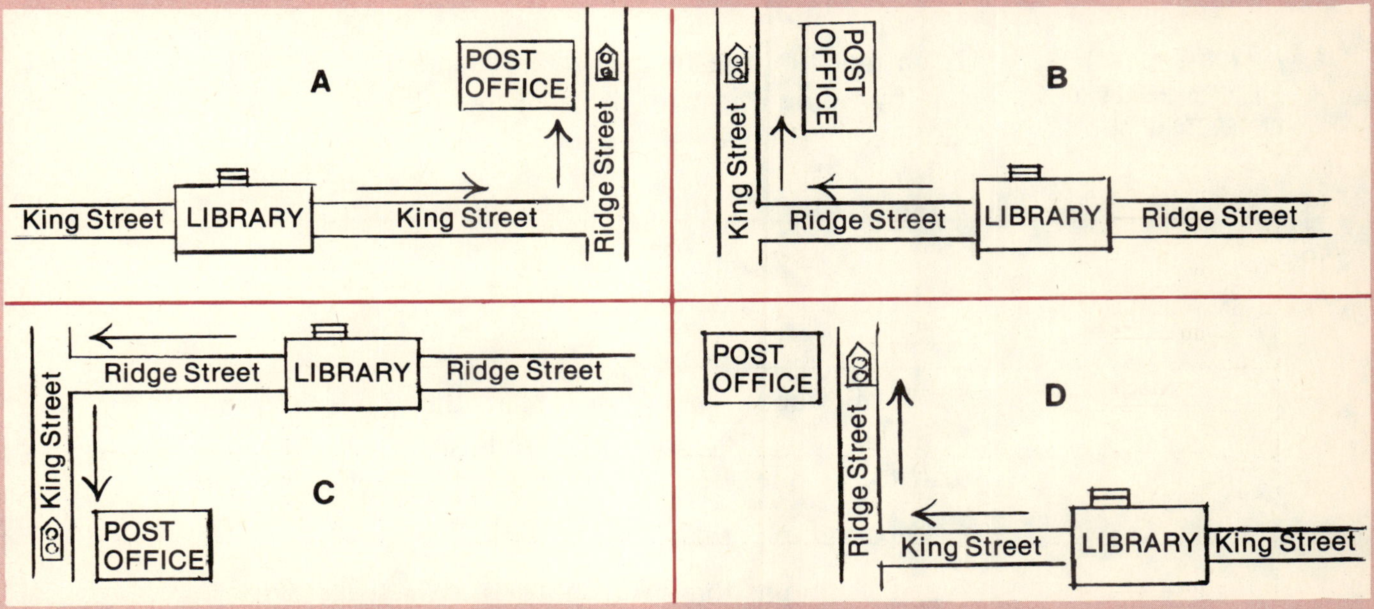

If your diagram looks like B, you understood the directions. If your diagram looks more like one of the others, read the paragraph again. Then compare your diagram with diagram B to see how yours is different. Being able to draw a diagram of something you read is one way to show that you understand what you have learned.

Read the next paragraph about the architectural styles that were popular at different times in history. Then finish the time-line that follows the paragraph.

Throughout the ages, styles in building, or architecture, have changed to show the times in which people lived and their needs during those times. The earliest people to have had a great influence on architecture may have been the Egyptians around 2500 B.C. They built pyramid-shaped vaults in which they buried their dead, along with certain valuable objects. The objects buried in these pyramids have survived to the present time. Somewhat later, about 500 B.C., the Greeks introduced a way of building with tall marble columns. This style had a great effect on later architectural styles. The ancient Mayan civilization of Central and South America had yet a different architectural style that was popular around 750 A.D. This was centuries after the Romans had done large-scale building about 100 A.D. Much later, during the 1700's, European colonists left their mark on American architecture by building simple "saltbox" homes made to withstand rough New England winters. Since the 1950's, modern architecture, with its tall, sharp-looking steel and glass towers, has replaced many of the older styles. Tall buildings show the need to expand upward in a crowded world.

DO YOU SEE THAT THE DIAGRAM REINFORCES UNDERSTANDING?

Each mark on the time line represents 100 years. Finish the time line by filling it in with the missing numbers that show when each architectural style was popular.

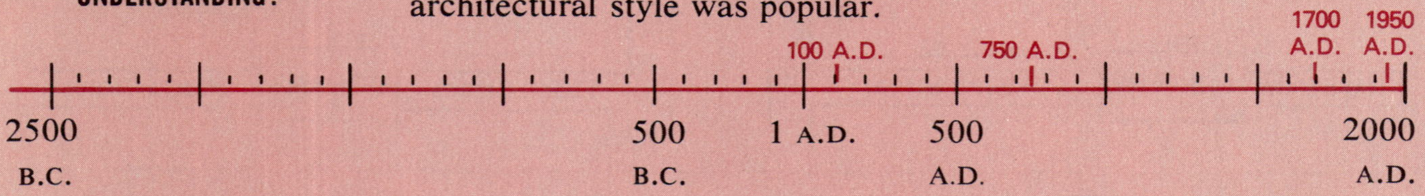

Think about which types of reading material are best suited to be made into a diagram or simple drawing. Put an *X* on the line next to the ones below that can best be made into a diagram.

_X_ paragraphs that describe processes

___ paragraphs that give opinions

___ paragraphs that describe people

_X_ paragraphs that tell order of events

_X_ paragraphs that compare things

___ paragraphs that present arguments

USING HEADINGS AS CLUES TO WHAT YOU WILL LEARN

As you read through this Science Work-A-Text, notice that there are headings in heavy black type. This kind of type is called **boldface** type and it signals that something important is about to follow. Headings give you a clue to the material you are going to read. A good heading, or title, may sum up the material.

Read the following paragraphs. Then from the list at the end choose the correct heading that matches each paragraph. Write the heading on the line above each paragraph. Remember to capitalize the headings correctly.

Tuning-up Your Body

Exercise is good for everyone. It allows you to release energy while keeping you physically fit. Exercising will help tone your muscles and strengthen your heart. For many fitness "nuts", exercise provides a much needed break from their daily routines.

The Importance of Exercise

If you don't exercise regularly, you probably should consult your doctor before starting an exercise program. It is very important not to get too much exercise the first time out. Before starting any strenuous, or fast-paced, activity such as jogging, biking or swimming, warm up your body. Do this with exercises that stretch your muscles. And remember, exercise regularly. Regular exercise will improve your ability to keep going.

Exercise and Digestion

It is not a good idea to exercise right before or right after you eat. A greater amount of blood is needed in your digestive tract an hour or two after you eat. Exercising too soon after you eat interferes with the digestive process. It takes the blood away from the digestive tract to fill the needs of the muscles used during exercise. The result of mixing exercise and eating is usually stomach pains. This is something you want to avoid.

tuning up your body
avoiding exercise
the importance of exercise
exercise and digestion
muscular control

UNIT 3 Introduction to Physical Science

This is a photograph of Irene Curie-Joliot, daughter of Pierre and Marie Curie, and her husband, Frederic. They continued her parents' work in radioactivity for which they won the Nobel Prize for chemistry in 1935.

New science words: conservation, decomposition, element, equation

Skills developed: observing, inferring, stating a conclusion

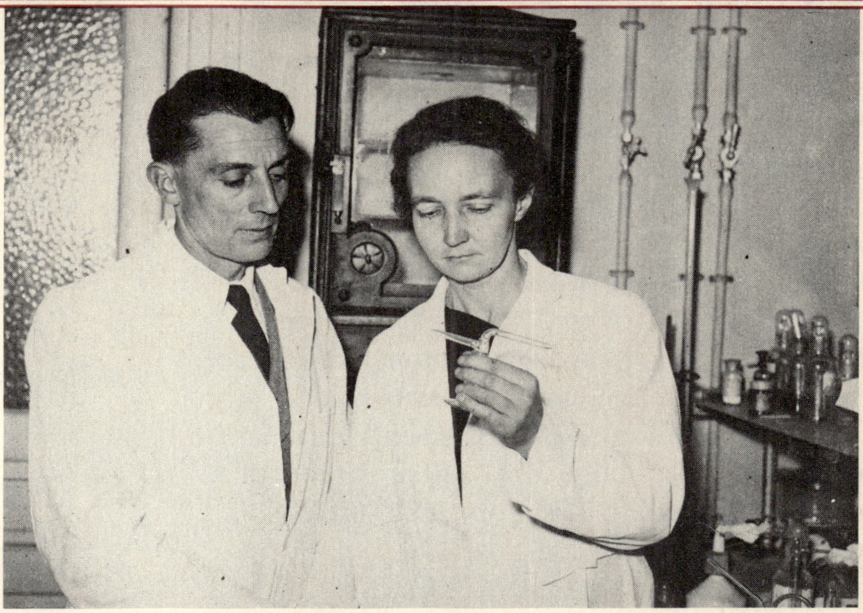

Chapter 1

THE STRUCTURE OF MATTER

Look around you. What do you see—a book, a glass of water, another person? Open your mouth and take a deep breath. What rushes into your mouth—air? The book, the glass of water, the person and the air are all **matter.**

The book, the glass, the water and the person do not look alike. And you can't even see air. This tells you that one kind of matter can be very different from another. Matter can be a **solid,** a **liquid** or a **gas.**

Whatever shape matter takes, all matter has two basic common properties (characteristics). All matter takes up space: that is, it has **volume.** And, all matter has **mass.** The mass of an object depends on how much and what kind of matter the object is made of.

A. What Is An Atom?

After doing this lesson you will be able:

A1 • to explain what an atom is

A2 • to name the three parts of an atom and tell what charges they have and where they are found in an atom

A3 • to explain what an ion is

The word "atom" may not be new to many of you. But, do you know what an atom is? An **atom** is made up of one or more negatively charged particles which circle around a positively charged central part. The negatively charged particles are called **electrons.** The positively charged center is called the **nucleus.** Most kinds of atoms do not exist alone. An **atom** is the smallest part A1
of a particular type of matter called an **element** that has all the properties of that element.

An electron has only about 1/2000 the mass of a proton.

A model of an atom. The orbits of electrons around the nucleus are not as fixed as shown in this model. They move freely about the nucleus in swarms.

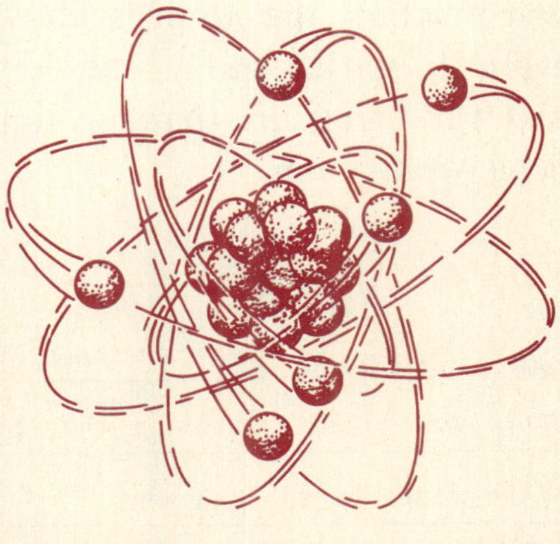

Other physical properties include color, odor, solubility, density, hardness, melting and boiling points, crystalline or amorphous form.

A neutron has about the same mass as a proton.

In 1897, the British physicist, Joseph Thomson, discovered the electrons in an
A2 atom. **Electrons** are tiny particles with a negative (−) electrical charge. Electrons circle (go around) the **nucleus** of the atom. They spin around the nucleus in paths called **energy levels.**

In 1919, the British scientist, Sir Ernest Rutherford, discovered small particles inside
A2 the nucleus that he called "protons." **Protons** have a positive (+) charge of electricity.

In 1932, another British scientist, James C. Chadwick, found that there were still other particles inside the nucleus. He called
A2 them **neutrons** (n) because they have no electrical charge and are neutral.

Since the neutrons have no electrical charge, the nucleus of an atom is positively charged because it has protons. Since protons and neutrons are heavy particles, most of the mass of an atom is in the nucleus.

In a normal state, an atom is electrically neutral or uncharged. The number of positive protons in its nucleus equals the number of negative electrons in its energy levels. When an atom has more electrons than protons, it has a **negative** charge. When an atom has more protons than electrons, it has a **positive** charge.

When an atom loses or gains electrons, it is no longer neutral. The atom is electrically
A3 charged and is called an **ion.** When an atom gains electrons and, therefore, has more electrons than protons, the atom is known as a **negative ion.** When an atom loses electrons and has fewer electrons than protons, it is known as a **positive ion.**

Check Up A

A1. What is an atom? The smallest part of a particular type of matter called an element that has all the properties of that element.

A2. Name three parts of an atom. Give the charge each has and tell where each is located.

(1) Electrons; negative charge; around the nucleus in paths called energy levels.

(2) Protons; positive charge; inside the nucleus.

(3) Neutrons; no charge, inside the nucleus.

A3. What is an ion? An electrically charged atom.

B. Elements

After doing this lesson you will be able:

- B1 • to describe what an element is and tell how many are known
- B2 • to explain what the atomic number of an element is and what it tells you about an element
- B3 • to explain what the atomic mass of an element is based on
- B4 • to name the atom to which other atoms are compared and give its atomic mass

Elements are pure materials that contain B1
only one kind of atom. Their makeup never changes. All the atoms in a piece of iron are iron atoms. Elements cannot be broken down into more simple materials by ordinary chemical means.

There are **92** different kinds of elements that occur naturally on Earth. All of the different kinds of matter that make up planet Earth and the air around it are made of these 92 natural elements.

Borrow periodic tables (i.e., from a nearby high school). Help students interpret its symbols to find out what information it holds.

Up to now scientists have been able to make **14** other elements in the laboratory. These elements are called synthetic elements. The number of natural elements and the num-
B1 ber of synthetic elements make a total of **106** elements.

What makes one element different from another element? Each element has a different number of protons in the nucleus of its atom. For example, carbon has 6 protons, oxygen has 8 protons, and silver has 47 protons.

B2 Every atom has an **atomic number.** The atomic number of an element is *the number of protons in the nucleus of an atom of that element*. Since an atom has the same number of electrons as protons, the atomic number is also equal to the number of electrons in the energy levels of an atom.

Atomic numbers allow scientists to classify the elements and to predict (foretell) their properties. Scientists have arranged all known elements in a table. The table begins with hydrogen which has the atomic number of 1. It ends with the synthetic element 106. This arrangement is called the **Periodic Table of Elements.**

The mass of an atom is called its **atomic mass.** Since the greatest mass of an atom is found in the nucleus, the atomic mass is based on *the number of neutrons and protons in the nucleus*. Thus, the atomic mass of an element is about equal to the total number of protons and neutrons in the nucleus. The unit that is usually used when talking about atomic mass is the **atomic mass unit (amu).**

B3 The atomic mass of an atom of an element is a **relative mass.** This means that the mass of that atom is *compared to the*
B4 *mass of another atom*. Scientists have chosen the carbon atom, with an atomic mass of 12, as the atom to which all others are compared.

The atomic mass of hydrogen, the lightest atom, is 1 because it has 1 proton in its nucleus. The atomic mass of oxygen is 16 because it has 8 neutrons and 8 protons in its nucleus.

Scientists use symbols for each element. Sometimes the symbol is the *first* letter of the name of the element: H stands for hydrogen. Sometimes the symbol is the first *two* letters: Al stands for aluminum. Some of the symbols are abbreviations of foreign (Greek, German, etc.) names for the element: Au stands for the Latin word for gold, *aurum*.

Check Up B

B1. What is an element? How many elements are known to exist? Elements are pure substances that contain only one kind of atom. 106 elements now exist.

B2. What is the atomic number and what does it tell about an element? The atomic number of an element is the number of protons in the nucleus of an atom of the element. The atomic number is also equal to the number of electrons in the energy levels.

B3. What is the atomic mass of an element based on? The atomic mass of an atom of an element is a relative mass in which the mass of that atom is compared with the mass of an atom of a chosen standard.

B4. What is the atom to which all others are compared? What is its atomic mass? carbon; atomic mass of 12

Using an encyclopedia, have students list the names and some common facts about the 106 known elements.

Word to review: maximum

C. Chemical Behavior Of Elements

After doing this lesson you will be able:

C1 • to explain what the chemical behavior of an element depends on

C2 • to tell the difference between inert elements and active elements

C3 • to list the three ways active elements unite with other elements

When scientists refer to the energy levels of an atom, they count from the nucleus outward. The energy level closest to the nucleus is the first energy level, the next is the second energy level, and so on.

The energy level farthest from the nucleus is the outer energy level. This energy level tries to become filled so that it holds as many electrons as it can. The number
C1 of electrons in the outer energy level determines how one element reacts with another element. Thus, **the number of electrons in the outer energy level determines the chemical activity of an element.**

The outermost energy level contains 1 electron. It has a capacity of 18; a certain stability is reached at 8.

PRINCIPAL ENERGY LEVEL 1
PRINCIPAL ENERGY LEVEL 2
PRINCIPAL ENERGY LEVEL 3

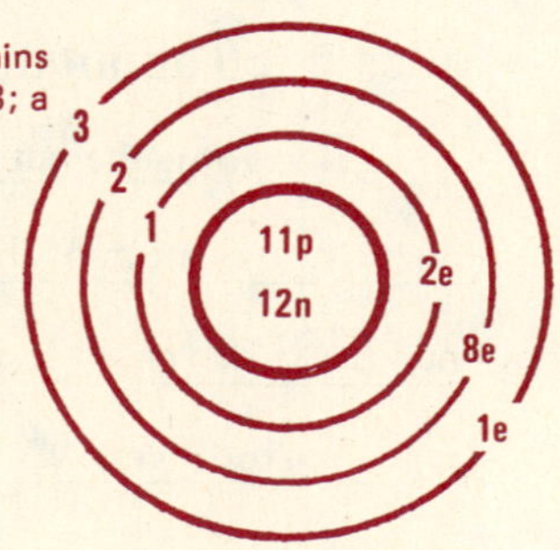

Principal energy levels of the sodium atom. How many electrons are in the outermost level?

C2 **Inert elements** do *not* normally react with other elements. The reason is that the outermost energy levels of their atoms already have as many electrons as they can hold. All of these elements are gases: helium, neon, argon, krypton, xenon, and radon.

There are some elements whose outermost energy levels are incomplete, so they either try to "fill" them, or "give away"

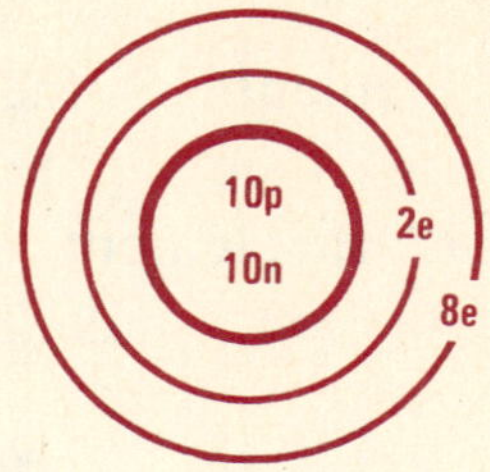

Principal energy levels of the neon atom. What letter represents the outermost energy level? Is the outermost energy level filled?

Neon's outermost energy level is L. It is filled.

the electrons they have. The next energy level to the outermost one then becomes the outermost energy level. And, this energy level is already "filled."

Sometimes atoms do not give away or borrow electrons when they unite (join). They share electrons as the outermost energy level of one atom "overlaps" the outermost energy level of the other atom. Thus, C2
atoms of **active elements** unite with the atoms of other active elements by **lending,** C3
borrowing or **sharing** electrons.

Check Up C

C1. What does the chemical behavior of an element depend on? ______

The number of electrons in the outermost energy level.

______

C2. How are inert elements different from active elements? Inert elements do not normally react with other elements. Active elements unite with atoms of other active elements.

______

C3. Name the three ways that active elements unite with other elements.

(1) Lending electrons

(2) Borrowing electrons

(3) Sharing electrons

D. Changes in Matter

After doing this lesson you will be able:

D1 • to tell the difference between a physical change and a chemical change

D2 • to describe how a compound is formed

D3 • to tell what a chemical equation is

D4 • to explain what decomposition means

D5 • to be able to state the law of conservation of matter and energy

Matter may go through two types of
D1 change: physical and chemical. A **physical change** is a change in the size or shape of a substance. No new substance forms when a physical change occurs. In a **chemical change,** a substance breaks down (loses its properties) and a new substance is formed.

D2 **1. Elements join together to form compounds (synthesis).** A **compound** is a substance formed by the chemical joining of two or more elements.

For example, when hydrogen is burned in air, it combines with the oxygen in air to form water. To show what takes place during this chemical change or reaction,
D3 chemists write an **equation.** An equation shows what substances react and what substances are formed by their reaction. In a chemical equation, a symbol stands for an element. For example, H is the symbol for the element hydrogen. A combination (more than one) of symbols, called a **formula,** shows what a substance is made of. A **formula** is used to represent one **molecule** (smallest amount) of a compound or sometimes of an element.

In an equation the two sides are separated by an arrow (→) which stands for "yields" or "produces." The materials that react with each other are put on the left side of the arrow. The resulting new substances, products, are put on the right side. Study this equation for the reaction between hydrogen and oxygen that produces water.

(Reacting substances)				(Product)
$2H_2$	+	O_2	→	$2H_2O$
hydrogen		oxygen		water

Read the equation again. The equation states that 2 molecules of hydrogen ($2H_2$) have joined with 1 molecule of oxygen (O_2) and produced 2 molecules of water ($2H_2O$).

2. Compounds break down into elements. D4
The second kind of chemical change is the **decomposition** (breaking down) of a compound into the two or more elements of which it is made up. When an electric current is passed through water (H_2O), it is **decomposed,** or broken down, into hydrogen (H_2) and oxygen (O_2). These are the elements of which it is made up. The equation for the chemical change is shown below.

$2H_2O$ →	$2H_2$	+	O_2
water	hydrogen		oxygen

Discover for yourself
how you could show that oxygen is present in the compound hydrogen peroxide.

PURPOSE: To show a decomposition reaction • To develop skills of observation, inference and stating a conclusion

You need: 2 test tubes, test tube rack, water, 3% solution hydrogen peroxide (H_2O_2), wood splints, matches

a. Pour some hydrogen peroxide solution into a test tube. Place the test tube in a test tube rack.

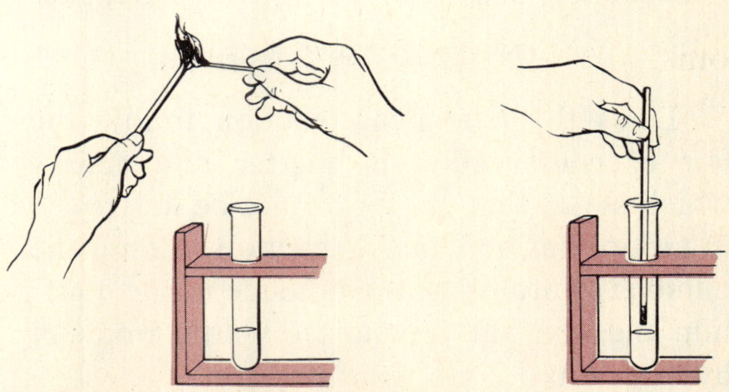

A small amount of manganese dioxide, if available, is an inorganic catalyst which can speed up this reaction.

b. Light the top of a wood splint and let it burn for a few seconds. Blow it out so that it has only a faint glow.

c. Insert the glowing splint inside the test tube close to but not in the hydrogen peroxide. See diagram.

1. What happened to the splint? It burst into flame.

2. What element causes the burning? Oxygen

d. Repeat procedures **a** and **b**, but use water in the test tube.

1. Did the splint burn? No

2. What have you shown about the compound hydrogen peroxide? One of the elements in this compound is oxygen.

In the 18th century, Antoine Lavoisier, a French chemist, set down the **law of conservation of matter.** This law states that "matter can neither be made nor destroyed but may only be changed in form."

Later the law was changed to the **law of**
D5 **conservation of matter and energy.** This law states that "in *ordinary* chemical reactions, matter *and* energy cannot be made or destroyed."

Look again at the equation that shows the reaction between hydrogen and oxygen

$$2H_2 + O_2 \rightarrow 2H_2O$$

Does this chemical reaction follow the **law of conservation of matter and energy** which states that matter cannot be destroyed in an ordinary chemical reaction? Count the number of atoms on the left side of the equation and see that you started with 4 atoms of hydrogen and 2 atoms of oxygen.

On the right side of the equation you still have 4 atoms of hydrogen and 2 atoms of oxygen in the 2 molecules of water. Thus, you see that matter has not been destroyed in the chemical reaction. It has only changed its form.

Check Up D

D1. Explain the difference between a physical change and a chemical change. In a physical change, there is a change in the size or shape of a substance, but no change in its chemical makeup. In a chemical change, a new substance or substances are formed.

D2. How is a compound formed? By the chemical combination of two or more elements.

D3. What does a chemical equation show? An equation shows what takes place during a chemical reaction.

D4. What happens in a decomposition reaction? A compound is broken down into the two or more elements of which it is made up.

D5. State the law of conservation of matter and energy. In ordinary chemical reactions, matter and energy cannot be created or destroyed.

Word to review: nuclei

E. Nuclear Changes

After doing this lesson you will be able:

- **E1 • to explain what a radioactive element is**
- **E2 • to name three radioactive elements**
- **E3 • to tell what nuclear fission is and name the "bullet" used in this process**
- **E4 • to explain what a chain reaction is**
- **E5 • to name and describe two types of nuclear reactions**
- **E6 • to list three ways that nuclear energy is being used**

At the end of the 19th century the French scientist, Henri Becquerel, experimented with, or tested, the element uranium. He found that uranium gave off energy in the form of rays. This did not fit in with the idea that an atom was the smallest part of matter.

Today, thanks to the discoveries of the Polish and French scientists, Marie Curie and Pierre Curie, we know that an atom can be broken down.

The Curies discovered that the uranium
E2 atom could be divided into two new elements, polonium and radium. They found that atoms of these two elements gave off rays of en-
E1 ergy. Because of this, they called these elements **radioactive elements.**

A British scientist, Ernest Rutherford, later found that some so-called "rays" were really tiny particles, now called **alpha** and **beta particles.** The rest are rays called **gamma rays.** These particles were much smaller than the radium atom. Could it be possible, scientists asked, that some atoms or all atoms might be made up of smaller particles?

In the 20th century Ernest Rutherford and Niels Bohr showed that atoms of elements *not* radioactive are also made up of particles. In fact, **all atoms are made up of particles.** Although atoms are *different* from each other, they all have the *same* kinds of particles.

"Splitting an atom" really means *splitting the nucleus of an atom.* The splitting or breaking apart of the nucleus of an atom is called **nuclear fission.** How can the nucleus of an E3
atom be split?

Neutrons, the tiny neutral particles in the nucleus of an atom, are used as "bullets" to E3
split the nucleus of an atom. Certain atoms give up neutrons from their nuclei. These are called **unstable** atoms.

Scientists have found that the nuclei of the heavier atoms (those having greater mass) are easier to split. Uranium is an example of such an element. When the uranium nucleus goes through fission (breaks apart), it splits into two or more smaller nuclei. The result is a different element. Some elements that may be formed through fission of uranium atoms are barium, krypton, cesium and rubidium.

Energy, usually in the form of heat energy, is released when the nucleus of an atom is split. To gain speed, neutrons are shot from a machine called a **cyclotron.** The cyclotron is also called a **nuclear accelerator.** Dr. E. O. Lawrence, and American scientist, invented the cyclotron.

There are two types of nuclear reactions: E5
(1) fission and (2) fusion.

1. Fission. The breaking down of a heavy nucleus of an atom into two lighter nuclei of other elements is called **nuclear fission.** When the nucleus of an atom splits, particles shoot out in all directions. A great deal of heat is produced. In order to use this energy, the reaction must happen *over and over again.*

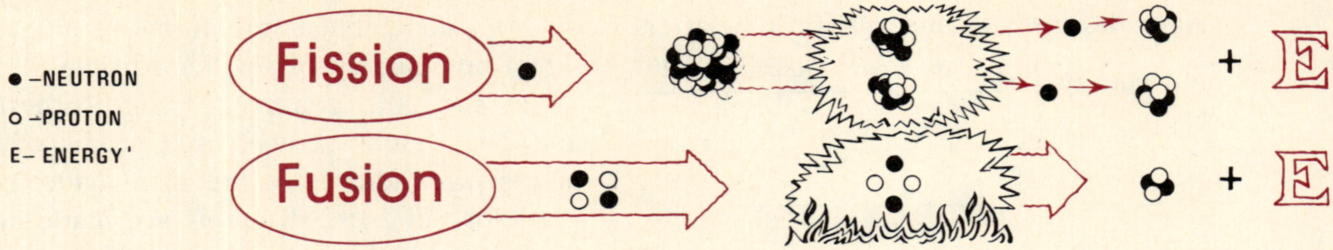

Nuclear fission and fusion.

E4 This continuous splitting, or fission, is called a **chain reaction.** To keep a nuclear chain reaction going, at least one neutron from each fission splits another nucleus. In a chain reaction the amount of energy keeps on increasing. This is how nuclear energy is used to produce electricity in the **nuclear reactor** of a nuclear power plant. Neutrons continuously split atoms of uranium. This is also how the first atomic bomb was exploded.

In a nuclear reactor great amounts of energy, heat and nuclear radiation are given off. Nuclear power plants are built of thick concrete to prevent radiation from escaping. The people who work in the plant wear clothing that protects them. They work behind lead shields and they are checked daily for radioactivity. But even with this protection, accidents happen.

Nuclear fission chain reaction.

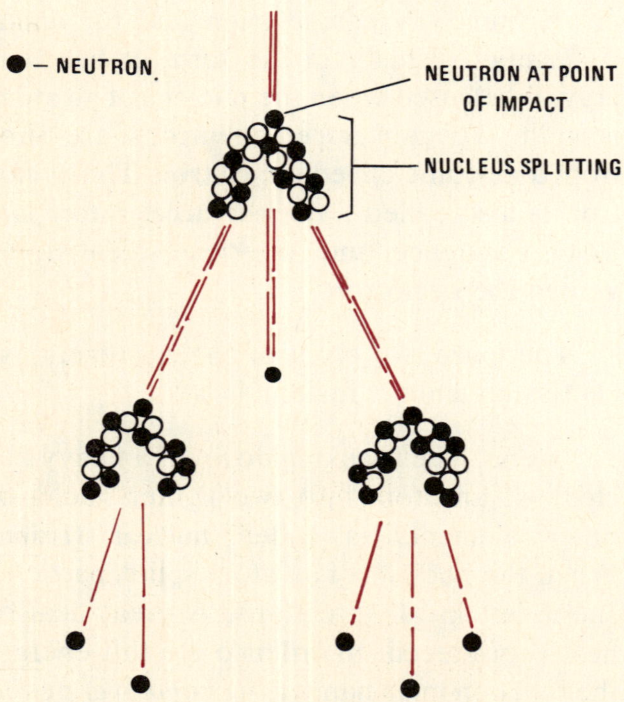

To deal with the great amounts of heat released, many nuclear power plants have been built beside a river, lake or inlet. Water is taken from the river (lake or inlet) and used to cool the reactor. But when this water is returned to its source, it is hot. It kills fish and other living things in the water. You will read more about **thermal (heat) pollution** in the last chapter of this unit.

2. Fusion. Fusion takes place when the nuclei E5
of light elements join together to form a heavier nucleus of another element. An example of such a reaction is the fusion of four nuclei of hydrogen atoms to produce one nucleus of helium. This reaction can take place only at very, very high temperatures. Fusion is always taking place in the sun, and huge amounts of energy are always being given off. A hydrogen bomb is exploded by fusion.

Nuclear energy is being used to produce E6
electricity for homes and factories. Nuclear energy is used to make some submarines work. Radioactive materials, such as radium, are used in the treatment of cancer. Other peaceful uses of atomic energy have yet to be explored.

Have students do some research on the Three Mile Island station. Find out if the station is still in operation.

Cooling towers cool down heated water before releasing it back into the nearby river. Why is this done?

to prevent killing of plants and animals

Check Up E

E1. What is a radioactive element? Elements which give off rays of energy are called radioactive. Uranium, polonium, and radium are radioactive elements.

E2. List three radioactive elements.

(1) Uranium

(2) Polonium

(3) Radium

E3. What is nuclear fission? What acts as the "bullet" in this process? The splitting or breaking apart of the nucleus of an atom. Neutrons are used as "bullets."

E4. What is a chain reaction? A chain reaction is one that continues over and over again. In a nuclear chain reaction, at least one neutron from each fission must cause another nucleus to split.

Energy is a major issue today. Have students clip articles on energy from newspapers for a week. Discuss the problems and possible solutions for the conservation of energy.

E5. Name and describe two types of nuclear reactions.

(1) Fission: The breaking down of a heavy nucleus of an atom to produce two lighter nuclei of other elements.

(2) Fusion: Nuclei of light elements fuse to form a heavier nucleus of another element.

Words to review: conversion, absorb

E6. List three ways that nuclear energy is being used today.

(1) Generate electricity

(2) Power submarines

(3) Treat cancer

THINK ABOUT THIS

A More Efficient Breeder Reactor

The idea of a nuclear breeder reactor that can "breed" more fuel than is burned goes back to the earliest days of nuclear power. Uranium 235 is the usual fuel in a nuclear reactor, but very little of it occurs naturally. Most uranium occurs as U-238 and must be changed to U-235 before it can be used in the reactor. This process is expensive and wastes energy.

The purpose of a breeder reactor is to change U-238 to a more useful form. If an atom of U-238 absorbs (takes in) a neutron, it changes to plutonium 239, which can be used as reactor fuel. This change happens to some extent in all reactors, but happens much faster in a breeder reactor.

So you see, the breeder reactor makes better use of fuel. Even more important, uranium 235 will be rare by the year 2000. So the cost of nuclear power will be almost impossible to afford. Oil and natural gas probably will be used up by then. And the use of solar energy may move ahead too slowly to meet energy needs. What are the pros (arguments for) and cons (arguments against) of building breeder reactors to supply energy for planet Earth?

Unit 3, Chapter 1
REVIEW

Matching Exercise Match each term in **Column A** with its description in **Column B**. Write the letter of the description in the space beside the correct term.

	Column A	*Column B*
d	**1.** atom	**a.** smallest part of a compound
e	**2.** proton	**b.** unit for atomic mass
j	**3.** neutron	**c.** number of protons in the nucleus of an atom
h	**4.** electron	**d.** smallest part of an element
i	**5.** ion	**e.** positively charged particle
g	**6.** atomic mass	**f.** pure material that contains only one kind of atom
c	**7.** atomic number	**g.** number of protons and neutrons in the nucleus of an atom
b	**8.** amu	**h.** negatively charged particle
a	**9.** molecule	**i.** charged atom
f	**10.** element	**j.** neutral particle

True/False In the space provided, write **T** if a statement is true. If it is false, change the underlined word or words to make the statement true.

volume **1.** The amount of space an object takes up is its mass.

T **2.** The nucleus of an atom is positively charged.

negative ion **3.** When an atom has more electrons than protons, it is a positive ion.

fusion **4.** The energy from the sun comes from nuclear fission.

gases **5.** All of the inert elements are liquids.

T **6.** The chemical activity of an element is determined by the number of electrons farthest from its nucleus.

T **7.** A hydrogen bomb explodes because of nuclear fusion.

compound **8.** Water is a common element.

cannot **9.** The law of conservation of matter states that matter can be created and destroyed in an ordinary chemical reaction.

An atom **10.** A molecule is the smallest part of an element that has all of the properties of that element.

Neutrons **11.** Protons have no electrical charge.

electrons **12.** In most atoms, the number of protons is equal to the number of neutrons.

protons **13.** The atomic number of an element is the number of neutrons in the nucleus of one atom of that element.

neutrons **14.** The atomic mass of an element is equal to the number of electrons and protons in the nucleus of one atom of the element.

T **15.** There are 92 different kinds of elements that occur naturally on Earth.

Multiple Choice Exercise Circle the letter of the word or phrase that best completes each statement.

1. The amount of space an object takes up is its **(a)** atom **(b)** mass **(c)** volume **(d)** density.
2. All matter has **(a)** mass **(b)** mass and volume **(c)** only volume **(d)** neutrons.
3. The smallest part of an element that has all the properties of that element is **(a)** a molecule **(b)** an atom **(c)** a compound **(d)** a proton.
4. The nucleus of an atom is **(a)** positively charged **(b)** electrically neutral **(c)** negatively charged **(d)** where the electrons are located.
5. The atomic mass of an element is determined by the **(a)** number of protons in its nucleus **(b)** number of electrons in its nucleus **(c)** number of protons and number of neutrons in its nucleus **(d)** number of electrons orbiting its nucleus.
6. A substance formed by the chemical joining of two or more elements is **(a)** an atom **(b)** a molecule **(c)** a formula **(d)** a compound.
7. The joining of two atomic nuclei to form a heavier nucleus of another element is **(a)** fusion **(b)** fission **(c)** a compound **(d)** a formula.
8. Elements that do not normally react with other elements are said to be **(a)** molecules **(b)** valence **(c)** inert **(d)** active.
9. The number of electrons in the outermost energy level of an atom determines its **(a)** chemical activity **(b)** mass **(c)** volume **(d)** atomic number.
10. An electrically charged atom is **(a)** inert **(b)** an ion **(c)** a molecule **(d)** radioactive.

Application Exercise Use the drawing of the neon atom below to answer the following questions.

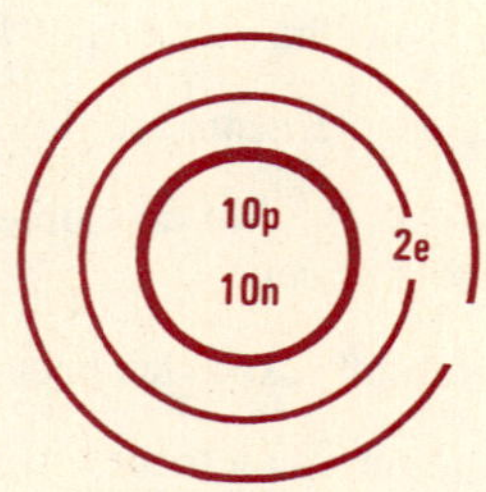

1. The atomic mass of this atom is 20 amu.
2. The atomic number of this atom is 10.
3. This atom has a total of 10 electrons.
4. There are 8 electrons in the outermost energy level of this atom.
5. The outermost energy level of this atom cannot hold any more electrons.

New science words: circuit, conductor, insulator, resistance

This is a solar energy collector system, in Minneapolis, used to heat and cool the new building next to it. This collector is located on a parking ramp adjacent to the building.

Skills developed: observing, comparing, inferring, stating a conclusion, constructing a model

Chapter 2

ELECTRICITY IS ENERGY

How many kinds of energy can you think of? In the first chapter of this unit, you found out about nuclear energy. In this chapter, you will look into electrical energy. In the chapters that follow, light, heat and sound will be considered. All of these are kinds of energy that make it possible for you to do work and enjoy life.

This discovery was made by the ancient Greeks.

A. Static Electricity

After doing this lesson you will be able:

A1 • to state the law of electrical charges

A2 • to tell when an atom is electrically charged

A3 • to tell what causes lightning

A4 • to tell how lightning rods protect buildings

As long ago as 600 B.C. it was known that when a material called amber was rubbed with wool, it pulled small pieces of material toward itself. At the end of the 16th century, the British scientist, Sir William Gilbert, discovered that other substances behaved like amber. He found that when glass was rubbed with silk, or when hard rubber was rubbed with fur, the same thing happened. Gilbert called this attraction (pull) **electricity** (from the Greek word *elektron,* meaning "amber").

Later, in the 18th century, the Frenchman, Charles du Fay and the American, Benjamin Franklin, carried on separate experiments with electricity. To both there seemed to be two kinds of electricity: *positive electricity* and *negative electricity.* Scientists now know that the tests made by du Fay and Franklin were experiments with *static electricity.*

What is **static electricity**? The crackling sound you hear when you rub the fur of a cat or dog is caused by static electricity. When you pull a comb through your hair, static electricity makes your hair stand up. When you drag your feet across a rug and then touch a doorknob, the shock you get is static electricity. The fur, your comb and your shoes picked up static electric charges because of friction (rubbing).

Word to review: repel

All matter has electricity. As you have learned, matter is made up of atoms. Atoms, in turn, are made up of **electrons,** which are negative charges, and **protons,** which are positive charges.

A1 The **law of electrical charges** states that: *Unlike charges attract* (draw closer to) *each other, and like charges repel* (move away from) *each other.* Thus, electrons are attracted to protons. Sometimes however, a force may be stronger than the attraction between the unlike charges. When this happens, electrons may be pulled from one atom and added to another.

When a neutral, or uncharged, atom *loses* some of its electrons, it has fewer electrons than protons. Then the atom is no longer neutral but has a positive charge. When a neutral atom gains electrons, it has more electrons than protons and has a negative charge. An
A2 atom is **electrically charged** when its number of electrons *does not* equal its number of protons.

When you pull your comb through your hair, friction between the comb and your hair causes electrons to be drawn from your hair to the comb. It is **negatively charged.** The comb now has a static electric charge.

Since your hair has lost some of its electrons, it has fewer electrons than protons, so it is positively charged. Any object can be charged either positively or negatively by friction. Such a charged, or electrified, object can attract other materials which are oppositely charged.

Discover for yourself

what would happen it you rubbed different materials on paper.

PURPOSE: To demonstrate static electricity • To develop the skills of observation, comparison, and inference

You need: a few newspaper sheets, pieces of fur, rubber, wool, cotton, silk, plastic, coin

Place a sheet of newspaper on a flat wooden surface. Rub the paper with a piece of fur, then rubber, wool, cotton and silk, plastic and a coin. After each rubbing, try to lift the newspaper from the wooden surface. Choose a different spot on the newspaper sheet to rub each material. Write your findings on the chart below.

	Attracted	Not Attracted
Fur	✓	
Rubber	✓	
Wool	✓	
Cotton		✓
Silk	✓	
Plastic	✓	
Coin	✓	

1. What did you do to the materials by rubbing them? You charged some of them with static electricity.

2. Which materials took on the static charge?

fur, rubber, wool, silk, plastic, coin

A lightning bolt, if it could be harnessed, contains enough electricity to light a city.

What causes lightning? In 1752, Benjamin Franklin flew a kite during a thunderstorm. He held the kite by a silk thread to which he had tied a metal key. When he put his hand near the key an electric spark jumped to his fingers. This was dangerous because he could have been killed by lightning. **(Don't try this experiment!)** By this experiment, however, he showed that lightning is a form of electricity.

Scientists now know that **lightning** is static
A3 electricity. Lightning happens when electrons pass quickly between oppositely charged clouds or between charged clouds and the earth. The charges are caused by friction between particles in the air. This friction takes away electrons from atoms in these particles and causes lightning.

If a negatively charged cloud moves near a positively charged cloud, electrons jump from the negative cloud to the positive cloud. This movement causes a giant spark, which is lightning.

Lightning can also happen when a charged cloud moves over an object on earth. Electrons arc, or jump, between the object and the charged cloud.

A clap of **thunder** always follows a flash of lightning. This is because lightning heats the air and causes the air to expand (spread out) quickly. This expansion of air causes thunder.

Is static electricity dangerous? Lightning can be very dangerous. It can kill people. It can set a building or a forest on fire. Lightning rods are put on rooftops to protect buildings. These rods set up a pathway for the flow of electrons between the clouds and the earth. The electrons move through a wire leading from the lightning rod to the ground.

A lightning bolt may strike a distance of 3 kilometers.

Never stand under a tree during an electrical storm. Lightning could hit the tree and move through your body on the way to the ground. Your body can be a path for the flow of electrons. The ground is usually electrically neutral. So the word "ground" is used to mean any electrically neutral point. The chain that drags along the ground under a gasoline truck "grounds" static electricity caused by the friction between the road and the tires, and by the movement of the gasoline inside the truck.

Check Up A

A1. What is the law of electrical charges?

Unlike charges attract each other; like charges repel each other.

A2. When is an atom electrically charged?

An atom is electrically charged when its number of electrons does not equal its number of protons.

A3. When does lightning happen?

when a negatively charged cloud moves near a positively charged cloud; or when a charged cloud moves over an object on earth

A4. How do lightning rods protect buildings?

They provide a pathway for the flow of electrons from the clouds into the earth.

B. Electron Flow

After doing this lesson you will be able:

- **B1 • to tell what electricity is and what a circuit is**
- **B2 • to list three common sources of electricity**
- **B3 • to name the parts that make up a wet cell**
- **B4 • to tell what electromotive force is and give its common name**
- **B5 • to describe the difference between electrical conductors and insulators**
- **B6 • to tell what electrical resistance is**
- **B7 • to explain the difference between series wiring and parallel wiring**
- **B8 • to tell the reason for fuses and circuit breakers and how they work**

In about 1800, Alessandro Volta, an Italian scientist, found that electrons flow when two different metals are connected by a wire and then dipped in a liquid that can carry electrons. Volta went on to invent an electric cell by using strips of copper and zinc separated by cardboard made moist by salt water. This device is a wet cell. It was the first means of producing a steady electron flow.

B1 **Electricity** is *electrons in motion.* The closed path along which the electrons flow is called a **circuit.** In order for electrons to flow in a circuit, it is necessary to have a **source** (supply) of electrons and a **force** to move the electrons through a circuit. The
B4 scientific name for this force is **electromotive force,** commonly known as **voltage.** Voltage is measured in units called **volts,** named for Alessandro Volta. Electron flow is measured in **amperes,** named for André Ampère, a French physicist who lived in the 19th century.

B2 Some common sources of electrons and voltage are wet cells, dry cells and electric generators.

1. A wet cell is a container filled with a liquid, which is made up of one or more acids.

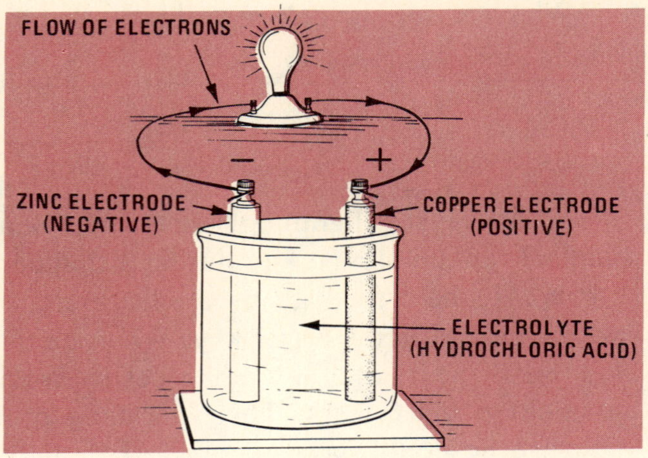

A wet cell connected to an electric light bulb.

Two metal strips are placed in the liquid; one is copper, the other is zinc. These materials have a large supply of electrons that can be made to flow.

However, there must be a complete circuit or path through which electrons can flow. To make this path, the two metal strips are connected by a wire. Electrons can then flow from the cell through the circuit. A very small electric light bulb will usually light up if connected to the circuit.

In a wet cell the two metals which become electrically charged are called **electrodes,** or terminals. The liquid through which the electricity flows is called the **electrolyte.**
Two electrodes and an electrolyte make up B3
an **electric cell.** The chemical reaction between the electrodes and the electrolyte causes the zinc electrode to gain electrons and the copper electrode to lose them. Thus, the zinc electrode becomes negatively charged, while the copper electrode becomes positively charged.

An electromotive force, or voltage, builds up between the two electrodes. This force, in turn, causes electrons to flow from the negative electrode to the positive one *through the wire connecting them.* This electron flow is **electricity.** It can be used to make light bulbs light up or to run machines.

An electric cell stores chemical energy *not* electrical energy. The chemical energy is

 Words to review: invent, alternately

If possible, obtain a dry cell and hand generator for students to examine.

changed to electrical energy when the cell is in use. An electric cell is "dead" when the chemical reaction that causes electrons to flow stops.

Before scientists knew about electrons, they thought that an "electric current" flowed from the positive electrode to the negative electrode. However, since the electrons are negative, they must flow from a negative point to a positive point. Engineers still refer to electron flow as "current." The term **electron flow** more exactly explains what takes place.

A **battery** is made up of several electric cells joined together by wires so that the negative terminal of one cell is connected to the positive terminal of the next cell. In a storage battery wet cells are joined together so that each connection runs between positive and negative terminals. A terminal is a place where a wire is connected. Each cell has two sets of lead plates (the electrodes). The electrolyte is usually sulfuric acid.

Chemical action causes extra electrons to collect on the negative electrode. When a complete circuit is made, electrons flow through the wire from the negative to the positive terminal.

A "dead" storage battery can be **recharged** so that it will work again. This is done by causing the electrons to flow through the electrolyte in the opposite direction. This will bring a new supply of electrons to the negative electrode.

2. A **dry cell** is another source of electricity. The container of a dry cell is usually made of zinc. It is filled with chemicals in the form of paste. The zinc container serves as the negative electrode. A carbon rod in the center is the positive electrode. A "dead" dry cell cannot be recharged. Dry cells have very few uses. They are often used to run small appliances, such as flashlights and portable radios, that need only small voltages to work.

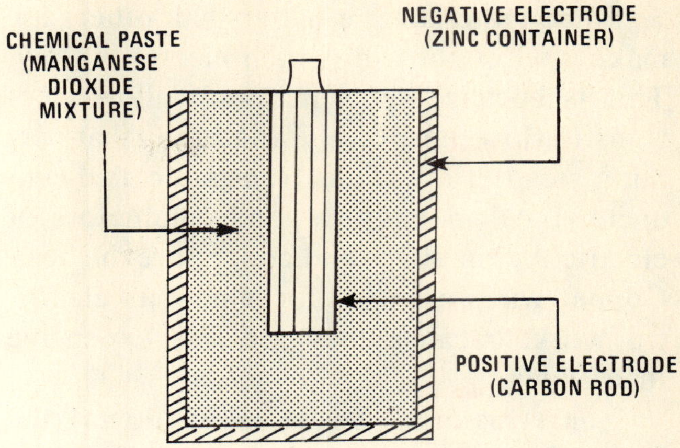

A dry cell.

3. A **generator** is a machine that can produce a large amount of electricity. In a generator a large coil of wire called an **armature** turns between magnets. This causes electrons to flow in the coils of the armature.

Almost all electrical energy today is produced in very large generators. These huge machines use steam power, atomic power or falling water (natural waterfalls) to move the huge coils past the magnets. Electricity from generating plants comes into most buildings through wires.

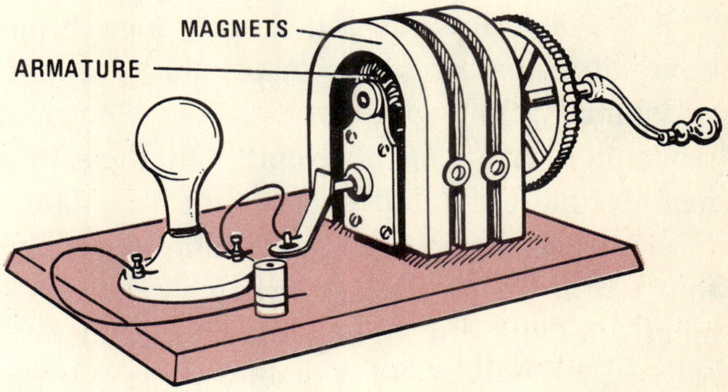

A hand generator.

Electrons will flow from one place to another only when there is an electromotive force, or voltage, between the two places. They flow from a place with extra electrons to a place with fewer electrons. The electrons will keep on flowing until both places are **electrically neutral.** This means one place is neither negative nor positive with respect to the other.

There was electric lighting before Edison's light bulb, but these were arc lights which were undependable and expensive to operate.

Some materials are better than others for
making wires through which electrons flow.
This is because some materials allow elec-
trons to flow more freely. Materials that per-
B5 mit a steady flow of electrons are called **con-
ductors.** All metals are good conductors of
electricity, but silver and copper are the best.
Copper wires are used for almost all electri-
cal work, because they are less expensive
than silver.

Some materials resist (slow down) the
flow of electrons and they are, therefore,
poor conductors of electricity. Some do not
conduct electricity at all. Such materials are
B5 called **insulators** or **nonconductors.** Glass,
plastic and rubber are good insulators. Be-
cause of this, copper wires are insulated by
being covered with plastic or rubber.

The flow of electrons can be directed by using a circuit. Conductors lead electricity to where it is wanted. Insulators keep it from going where it is not wanted. Insulators keep you from getting electric shocks, and also prevent the loss of electrical energy.

All materials slow down electron flow
to some extent, even copper and silver wire.
B6 This slowing down is called **resistance.** Non-
conductors or insulating materials have high
resistances. The greatest possible electrical
resistance is an **open circuit.** On the other
hand, a material that offers little resistance
to electron flow causes something to happen
that is known as a **short circuit.** Very quick
electron flow through a short circuit will
cause a lot of heat. That is why short circuits
are dangerous. They can cause fires. You
can think of electric appliances as resistances
or **resistors.**

Notice that an electric light bulb becomes hot if it is kept on for a time. Light bulbs offer resistance to a circuit so they are resistors. But, all resistors are not light bulbs. Resistance is measured in **ohms** (named for the German scientist, Georg Simon Ohm, who lived in the 19th century).

An electric light bulb contains a very fine wire called a **filament.** It is made of a rare metal called tungsten. When electrons flow through the filament it becomes hotter and hotter until it glows. The **incandescence,** or glow, given off by the filament in the bulb is the light.

Thomas Edison, an American inventor, made the first practical incandescent lamp in 1879. He also invented the sockets, switches, fuses and wiring systems that make electric light useful to people.

A **switch** is used to open or close a circuit. In the "off" position, a switch is open, or opens the circuit. In the "on" position, a switch is closed and allows electrons to flow through the circuit.

A **fuse** acts as a switch and opens a cir- B8
cuit when there is danger of a short circuit.
That is, when electron flow becomes too
fast for a circuit to carry. When there is too
much electron flow in a circuit, the circuit is
overloaded. A fuse has a thin wire called a
filament. This wire is very much like the
filament in an electric light bulb. When a
circuit is overloaded, this filament melts and
opens the circuit, thus stopping electron flow.

An electron flow is measured in *amperes.* Thus, a fuse opens when there are a certain number of amperes of electron flow in a circuit. Different-size fuses will open at different amperes. Therefore, it is important to get a fuse for your home that the circuit calls for. The size of the fuse in amperes is written on the fuse. Make sure you put the right fuse in your fuse box.

When a circuit containing a fuse is overloaded, the fuse will "blow," or open. A blown fuse must be replaced; it cannot be repaired. When the cause of the overload has been found, a new fuse must be placed in the fuse box in the socket from which the blown fuse was removed.

A **circuit breaker** is like a fuse except that B8
it has a switch which opens when the circuit

is overloaded. Circuit breakers do not have to be replaced. They can be reset when the problem has been taken care of.

A circuit can be pictured in a *diagram* by means of symbols. This is called a **schematic** of a circuit. Some of these symbols are:

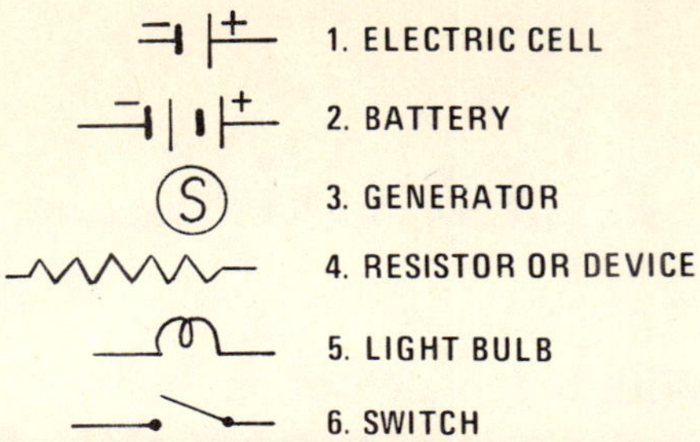

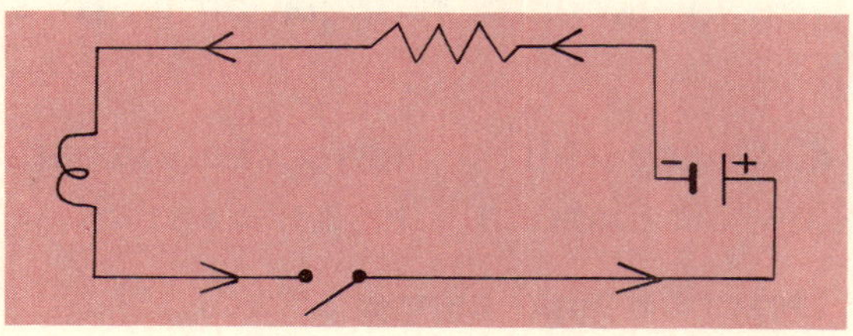

A schematic with some common symbols.

The two principal kinds of electric circuits are the **series circuit** and the **parallel circuit.** The parallel circuit is the circuit most commonly used in buildings.

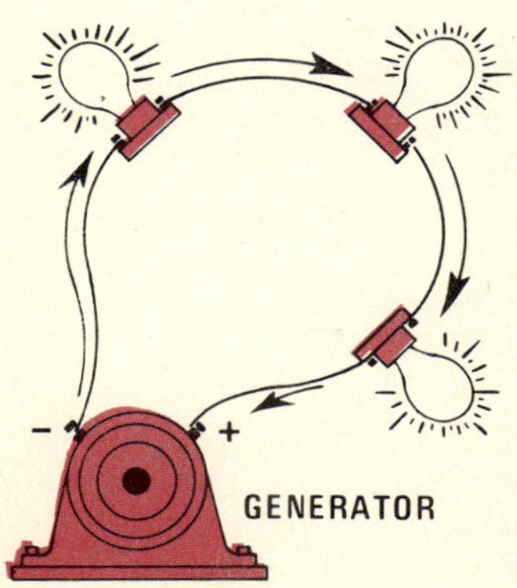

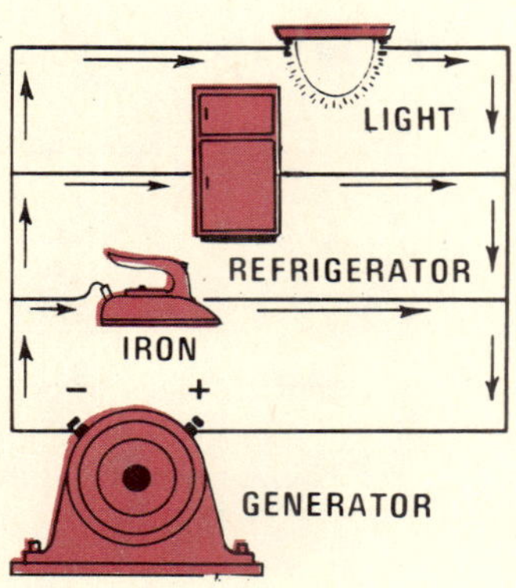

B7 1. A **series circuit** has only *one* path for electron flow. Therefore, all resistors are connected into this path. If one device is taken away, the circuit will open. If, for example, electric light bulbs are connected in series and one bulb is removed, all of the other light bulbs will go out.

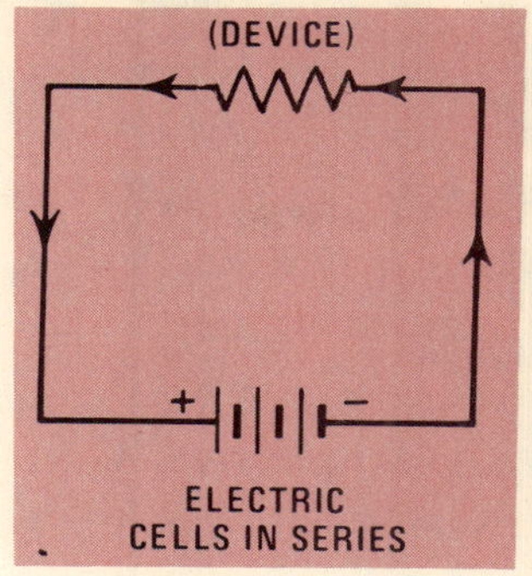

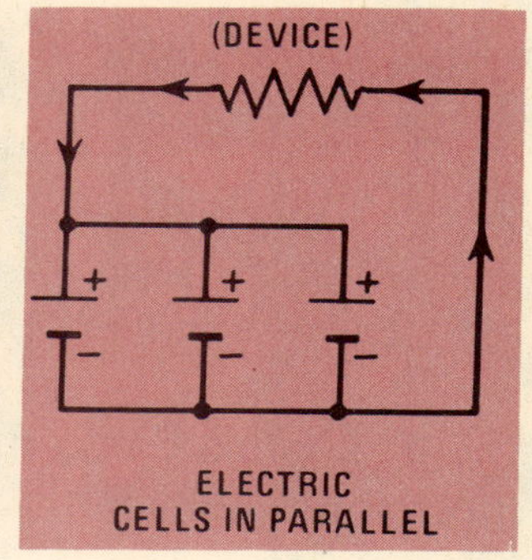

2. A **parallel circuit** has more than one path for electron flow. Each resistor in a parallel circuit is connected into a separate path. Therefore, if one resistor is removed, the circuit will *not* open because electrons can still flow through the other resistors which are in separate paths. A good example of a parallel circuit is a string of Christmas tree lights. If one bulb goes out, nothing will happen to the other bulbs. This is why buildings are wired with parallel circuits. B7

If dry cells are connected so that the negative electrode of one cell is wired to the positive electrode of the next cell, they are wired in series. When cells are wired in series, the voltages of each cell are added to each other. Therefore, the voltage of cells in series is the total voltage of all the cells in the battery. For example, if you place three cells of $1\frac{1}{2}$ volts in series, the total voltage is $4\frac{1}{2}$ volts.

If dry cells are connected so that the positive electrode of one cell is wired to the positive electrode of the next cell, and the negative electrodes are joined in the same way, they are wired in **parallel.**

Discover for yourself

how you could show how a series circuit works.

PURPOSE: To build a series circuit • To develop the skills of observation, demonstration, stating a conclusion and constructing a model

You need: insulated copper wire, switch, 2 dry cells, 3 light bulbs in porcelain sockets

Extension: Make up several diagrams on cards showing different kinds of circuits. Have students identify each symbol used in these schematics.

Make certain all dry cells are disconnected after activities to avoid damage to the cells.

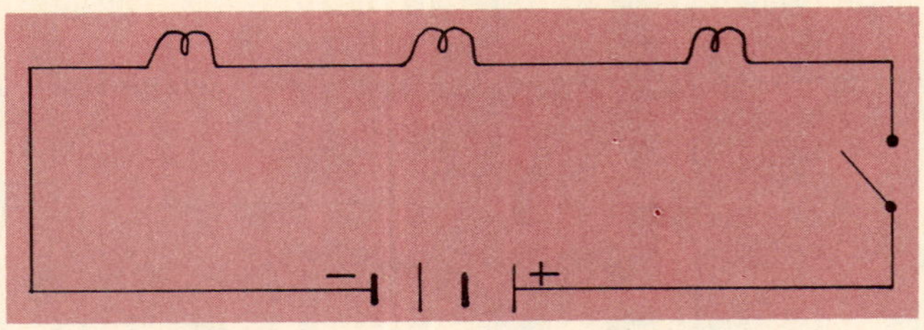

a. Connect two dry cells in series. Connect three light bulbs in series.

b. Connect the switch to the center terminal of one dry cell and to the terminal of the bulb socket.

c. Connect the last bulb socket to the outer terminal of the second dry cell.

d. Close the switch.

1. Tell what happens. The bulbs light up.

2. Tell what happens when you remove one of the bulbs. All bulbs go out.

3. Why? A break in a series circuit opens the circuit and the flow of current stops.

Discover for yourself
how you could show how a parallel circuit works.

PURPOSE: To build a parallel circuit • To develop the skills of demonstration, observation, stating a conclusion and constructing a model

You need: the same materials you used in the last activity

a. Connect the center terminal of one dry cell to the outer terminal of the second dry cell.

b. Using a short piece of wire, connect the switch to the center terminal of the second dry cell.

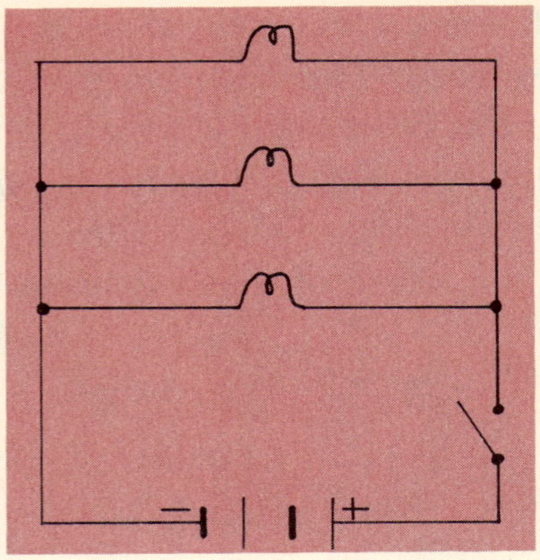

c. With the switch open, connect a piece of wire to the outer terminal of the first dry cell.

d. Connect all the bulb sockets in parallel. Follow the diagram. Close the switch.

1. Describe what happens. The bulbs light up.

2. Describe what happens when you remove one of the bulbs. The other bulbs stay lighted.

3. Describe what happens when you remove two of the bulbs. The other bulb is still lit.

4. Why? In parallel circuits, each branch is independent. A break in one branch does not interrupt the flow of current in any other complete branch.

Check Up B

B1. What is electricity? What is a circuit? Electricity is electrons in motion. The closed path along which electrons flow is called a circuit.

B2. What are three common sources of electricity?

(1) wet cells

(2) dry cells

(3) generators

B3. What parts make up a wet cell? Two electrodes and an electrolyte make up a wet cell.

B4. What is electromotive force? What is it commonly called? Electromotive force is the force that moves electrons through a circuit. It is commonly called voltage.

B5. How are electrical conductors different from electrical insulators? Conductors permit a steady flow of electrons; insulators do not conduct electricity well.

B6. What is electrical resistance? Slowing down of electron flow.

B7. How is a series circuit different from a parallel circuit? A series circuit contains only one path for electron flow. A parallel circuit contains more than one path.

B8. What is the reason for fuses and circuit breakers? How do they work? Fuses and circuit breakers open a circuit when it is overloaded. A fuse contains a filament which melts and opens a circuit. A circuit breaker contains a switch which opens when the circuit is overloaded.

C. Magnetism

After doing this lesson you will be able:

- C1 • to explain the difference between magnetic and nonmagnetic materials and give one example of each
- C2 • to explain the difference between artificial and natural magnets
- C3 • to tell what the poles of a magnet are and name the two poles every magnet has
- C4 • to state the law of magnetic poles
- C5 • to give two ways in which electricity and magnetism are alike
- C6 • to explain what an electromagnet is
- C7 • to describe the difference between alternating current and direct current

Magnetism is very important in today's world. A magnetic compass is used on ships and boats to find direction. Many appliances use electromagnets: radios, television sets, electric motors and many others.

What is a magnet? A **magnet** is *a material that can attract, or pull to itself, other materials*. Common magnetic substances include iron, steel, nickel and cobalt. Materials that are not attracted to a magnet are **nonmagnetic.** Tin, brass and aluminum are nonmagnetic. There are two types of magnets: (1) natural and (2) artificial. C1 C1

1. **Natural magnets** are found in nature. Early peoples discovered that certain stones, called **lodestones,** had magnetic properties. These stones contain an iron ore called **magnetite.** C2

2. **Artificial magnets** are manufactured. That is, they are made by people. Artificial magnets made of soft iron lose their magnetism easily. Therefore, they are **temporary magnets.** Magnets that keep their magnetism for a long time are **permanent magnets.** Very strong artificial magnets have been made of iron, steel and alnico. Alnico is made up of aluminum, nickel and cobalt. C2

Pocket compass pointer attracted to the N (north) pole of a bar magnet.

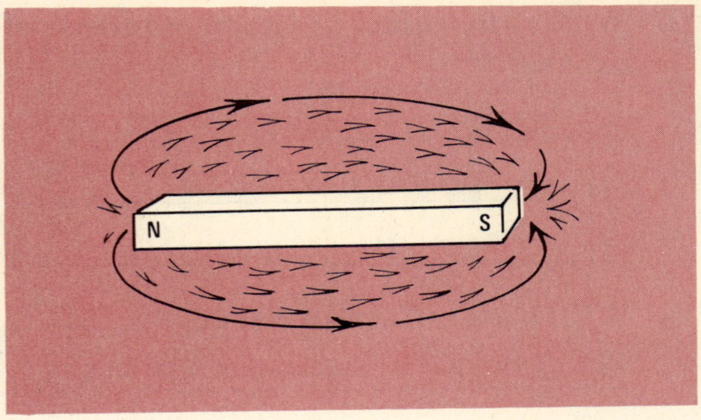

The magnetic field around a magnet showing lines of force.

Students may wonder why this is so, since like poles are supposed to repel and unlike poles to attract. The *N* and *S* poles of magnets were so named because they are *N* and *S seeking* poles, respectively. They are attracted to the *N* and *S* geographical poles of the earth.

C3 The parts of a magnet where its force is strongest are called the **poles** of the magnet. Every magnet has two poles: a **north pole** and a **south pole.** These are at the opposite ends of the magnet.

Magnets behave toward other magnets according to the **law of magnetic poles.** It
C4 states that: *Like magnetic poles repel each other and unlike magnetic poles attract each other.* If you bring the north poles of two magnets together, they will repel (drive away) each other. If you bring the north pole of one magnet close to the south pole of another magnet, they will attract each other. If a bar magnet is hung so that it swings freely, it will stop with its north pole pointing to the earth's north pole. The other end of the magnet, the south pole, will point to the earth's south pole.

Magnetism is a force that you cannot see, touch, smell or hear. Yet around every magnet there is an area called the **magnetic field** which has invisible magnetic **lines of force.** You can see these lines of force by covering a bar magnet with a piece of glass and sprinkling iron filings (tiny bits of iron) on the glass above the magnet. The iron filings form a design of lines around the entire magnet with more lines at the ends of the magnet (its poles).

Magnetic lines of force are curved and reach from one of the magnet's poles to the other. Since there are more magnetic lines at the poles of the magnet, the force of a magnet is greatest at its poles.

If you compare magnetism with electricity, you can see certain likenesses. The law of magnetic poles is very much like the C5
law of electrical charges (see page 122). Electricity has an **electric** field and magnetism has a **magnetic** field. But magnetism and electricity are *not* the same type of force. If you put your charged comb close to iron filings, it will not attract the filings because the comb is not a magnet.

In 1820, Hans Christian Oersted, a Danish scientist, was the first to notice that when electrons flowed through a wire, a nearby compass needle moved. This showed that there is a magnetic field around a wire carrying electricity.

Soon after this, André Ampère, a French scientist, took Oersted's discovery one step further. He found that a coil of wire acted like a magnet when electricity flowed through it. Two such coils placed close together moved toward each other or away from each other just as two bar magnets do.

The magnetic field of force can be made C6
stronger by sliding a piece of soft iron into the coil of wire. This soft iron builds up lines of force and produces an electron flow through the coil of wire. This magnet is called an **electromagnet.**

This can be demonstrated with a piece of heavy paper, a magnet and some iron filings. Sprinkle the filings on the paper. Hold the magnet underneath against the paper. Vibrate slightly. A pattern showing the lines of force will form.

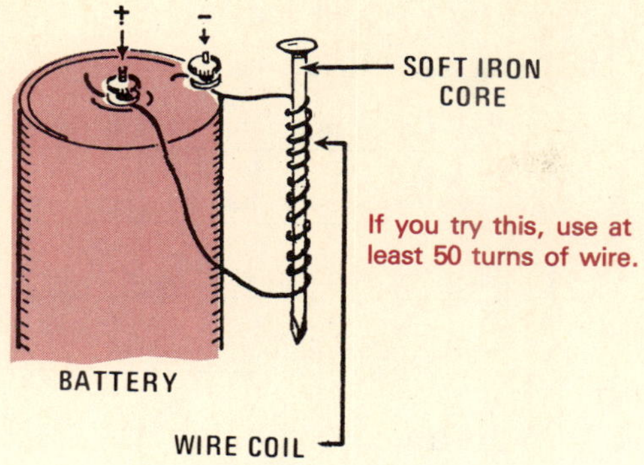

A simple electromagnet.

In the 19th century, Joseph Henry, an American physicist, found that he could make an even stronger electromagnet if he used insulated wire instead of bare wire. Also, by increasing the number of turns in the coil and increasing the amount of electron flow, the force of the magnetic field was made stronger.

Electromagnets are more useful than ordinary magnets because their strength can be controlled. They can be turned on and off at will. Electromagnets are used in telephones, radios, doorbells, telegraphs, electric motors and many other electrical appliances.

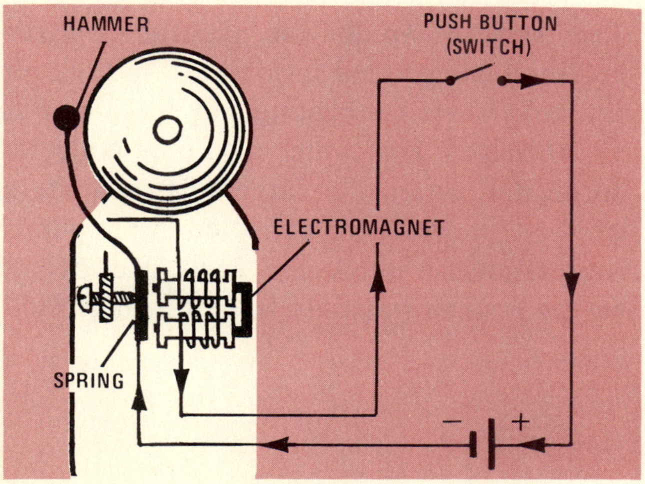

An electric bell circuit.

This diagram shows how an electric doorbell works. When the push button is pushed, the circuit is closed and the electricity magnetizes the electromagnet. The metal hammer is attracted by this magnetic force and hits

Encourage students to do library research and report on Samuel Morse's invention of the telegraph.

the gong. The hammer opens the circuit, the flow of electricity is stopped, and the electromagnet is no longer attracted to the hammer. A spring pulls the hammer back to its original position.

Samuel Morse, an American, built the first **telegraph** in 1844. It used electromagnets to send messages over wires. He worked out a code of short and long pauses tapped out with a key that acted as a switch.

An **electric motor** is made up of two bar magnets and a spinning coil of wire called an **armature.** The coil becomes an electromagnet when it is charged with electricity. One end becomes the north pole, and the other end becomes the south pole. The magnetized electromagnet turns in the magnetic field of the bar magnets.

Because like poles of magnets repel each other and unlike poles attract each other, the coil spins on its axis. This causes the electron flow in the wires of the motor to change direction during one complete spin of the coil. An electron flow that goes in one direction and then in another is called **alternating current** (AC). To stop this change of direction in electron flow a **commutator** is used. This C7
switch keeps the electron flow moving in the same direction. A flow of electrons in the same direction is called **direct current** (DC). Dry cells supply direct current. Here current means "electron flow."

Today, most buildings use alternating current. Alternating current changes direction sixty times a second.

Check Up C

C1. Explain the difference between magnetic and nonmagnetic materials and give one example of each. A magnet is a substance that attracts other substances — iron, steel. Materials that are not attracted to a magnet are nonmagnetic — tin.

C2. How is a natural magnet different from an artificial magnet? Natural magnets are found in nature; artificial magnets are manufactured.

C3. What are the poles of a magnet? Name the two poles every magnet has. Poles are the parts of a magnet where its force is strongest; every magnet has two poles — north and south.

C4. What does the law of magnetic poles state? Like magnetic poles repel each other and unlike magnetic poles attract each other.

C5. Give two ways that electricity and magnetism are alike.

(1) The law of magnetic poles is similar to the law of electrical charges.

(2) Electricity has an electric field; magnetism has a magnetic field.

C6. What is an electromagnet? An electromagnet is a coil of wire wrapped around a soft iron core. When electricity flows through the wire, it acts like a magnet.

C7. What is alternating current? Direct current? AC is electron flow in one direction and then in another. DC is electron flow in the same direction.

THINK ABOUT THIS

Aquifiers to Store Solar Energy

Everyone seems to agree that solar energy is the best alternate energy source for many areas. But how do you store electricity and hot water formed by solar energy cells for a community?

Some chemical engineers in Texas have designed a plan for a large-scale energy storage system that could serve a community. The system would use **aquifiers,** water-holding underground formations, to hold hot and cold water for heating and cooling up to 3,000 homes. Hot water would be collected during the summer in large solar collectors. Cold water would be collected in the winter from a cooling pond. The earth itself would be the insulator against loss of heat or cold.

Unit 3, Chapter 2
REVIEW

Matching Exercise Match each term in **Column A** with its description in **Column B**. Write the letter of the description in the space beside the correct term.

	Column A	*Column B*
e	**1.** conductors	**a.** voltage
j	**2.** insulators	**b.** negative charges
a	**3.** electromotive force	**c.** electrons in motion
g	**4.** amperes	**d.** a liquid through which electrons can flow
c	**5.** electricity	**e.** materials through which electrons can flow
i	**6.** current	**f.** positive charges
b	**7.** electrons	**g.** units for measuring electron flow
f	**8.** protons	**h.** path along which electrons flow
d	**9.** electrolyte	**i.** electron flow
h	**10.** circuit	**j.** materials that do not conduct electricity

True/False In the space provided, write **T** if a statement is true. If it is false, change the underlined word or words to make the statement true.

repel	**1.** The law of electric charges states that like charges attract each other.
static	**2.** Lightning is caused by current electricity.
expands	**3.** When lightning heats the air, the air contracts quickly, producing thunder.
different from	**4.** An atom is electrically charged when its number of protons is equal to its number of electrons.
T	**5.** Protons are the positively charged particles in an atom.
amperes	**6.** Electron flow is measured in volts.
electrolyte	**7.** A liquid through which electricity can flow is called an electrode.
more	**8.** A negatively charged atom has fewer electrons than protons.
T	**9.** Voltage is the force that pushes electrons through a circuit.
parallel	**10.** A series circuit has more than one path for electrons to follow.
ohms	**11.** Resistance is measured in amperes.
insulators or nonconductors	**12.** Materials that do not conduct electricity are called resistances.
T	**13.** Switches are used to open or close circuits.
electromagnet	**14.** An electrical resistance is a coil of wire wrapped around a soft iron core that has electricity flowing through it.
T	**15.** Materials that have their own magnetism are natural magnets.

Multiple Choice Exercise Circle the letter of the word or phrase that best completes each statement.

1. The force that pushes electrons through a conductor is **(a)** ampere force **(b)** electromotive force **(c)** resistance **(d)** armature.
2. An atom that has fewer electrons than protons is **(a)** negatively charged **(b)** electrically neutral **(c)** positively charged **(d)** an armature.
3. A liquid through which electricity can flow is called **(a)** an electron **(b)** an electrode **(c)** an electrolyte **(d)** a terminal.
4. A circuit with only one path for electrons to follow is **(a)** a series circuit **(b)** a parallel circuit **(c)** a schematic **(d)** an electrolyte.
5. A circuit with more than one path for electrons to follow is **(a)** a series circuit **(b)** a parallel circuit **(c)** a schematic **(d)** an electrolyte.
6. Magnetite and lodestone are examples of **(a)** electrolytes **(b)** artificial magnets **(c)** temporary magnets **(d)** natural magnets.
7. The law of magnetic poles states that like magnetic poles repel each other and unlike magnetic poles **(a)** attract each other **(b)** have no effect on each other **(c)** repel each other **(d)** produce an electric current.
8. The force of a magnet is greatest **(a)** in the center **(b)** at the poles **(c)** at the south pole **(d)** at the negative pole.
9. An electron flow that goes in one direction and then in another direction is called **(a)** direct current **(b)** a series current **(c)** a parallel current **(d)** alternating current.
10. A coil of wire wrapped around a soft iron core that produces a magnetic force when electricity flows through the wire is **(a)** an armature **(b)** an electromagnet **(c)** a natural magnet **(d)** a schematic.

Application Exercise Use the diagram below to complete each of the statements.

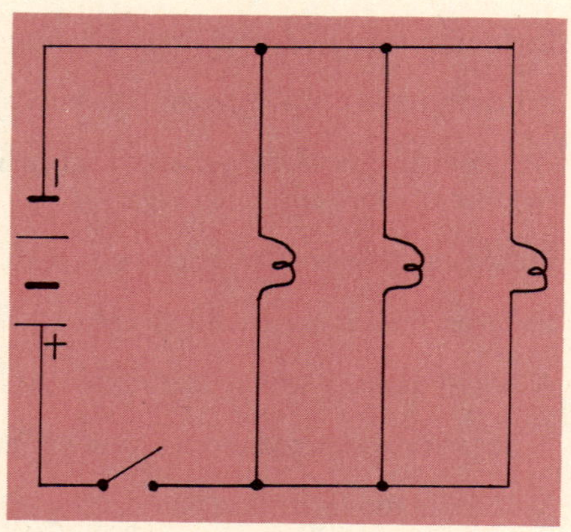

1. This diagram is called a ______________________.
2. The type of circuit shown in the diagram is a ______________________ circuit.
3. If one of the bulbs in this circuit is removed, the other bulbs will ______________________.
4. The source of power for this circuit is a ______________________.
5. This circuit has __________ switch(es) and __________ light bulb(s).

This photograph of Chicago was taken at night. The lights you see are not from the sun. How are the lights in and outside the buildings produced?

from artificial sources

New science words: focus, prism, spectrum

Skills developed: observing, describing, comparing, classifying, stating a conclusion

Chapter 3

LIGHT IS ENERGY

Instruct students to carry out a simple activity such as writing a sentence with their eyes shut. Discuss the results.

Have you ever wondered what the earth would be like without light? It would be in complete darkness. Its temperature would be so low that it would be a frozen mass in space. Plants could not make food. There would be no life on this planet without sunlight.

A. What Is Light?

After doing this lesson you will be able:

- A1 • to explain what kind of energy light is
- A2 • to explain what electromagnetic radiation is and list two kinds of electromagnetic waves
- A3 • to list the names of four theories that try to explain what light is and tell which one is the most recent

Since the 17th century, scientists have had many different theories to explain what light is and the way it behaves. Today, however, they agree that light is a form of **energy** to which your eyes react.

The light and heat from the sun or from a fire are radiant energy. This form of energy is A1
called **electromagnetic radiation.** All forms of electromagnetic radiation are thought to be A2
in part electrical and in part magnetic. Light is one of a large group of waves that are electromagnetic. This group also includes radio waves, heat (infrared) waves, ultraviolet light waves and X rays.

After the invention of lenses and telescopes, scientists had new ways to study light. Some of their theories are:

1. The wave theory. In 1678, Christian Huygens, a Dutch scientist, noticed that the A3
edges of the shadows of objects were not sharp and clear. He compared this to the way that waves in the water spread out and flow into each other. His theory was that *light behaves like waves in water.*

Word to review: theory

A3 **2. The particle theory.** In 1704, Sir Isaac Newton, a British scientist, said that *light was made up of tiny particles* (bundles). His theory was that a shining object gives off its own light by sending these particles out into space.

A3 **3. The electromagnetic theory.** In 1864, James Clark Maxwell, another British scientist, worked on a theory which states that *light waves are partly electrical and partly magnetic in nature.*

A3 **4. The quantum theory.** In 1900, the German scientist, Max Planck, stated that light is made up of particles of electromagnetic waves. These particles are called **quanta** (plural of quantum). Albert Einstein did further work on this theory. He called the quanta **photons.**

Check Up A

A1. What kind of energy is light? ____________

Radiant energy

____

A2. What is electromagnetic radiation? Give two examples of electromagnetic waves.

Electromagnetic radiation is partly electrical and partly magnetic. Two examples are light and radio waves.

A3. What are four theories that try to explain what light is? Which one is the most recent?

(1) The wave theory

(2) The particle theory

(3) The electromagnetic theory

(4) The quantum theory — most recent

B. Where Does Light Come From?

After doing this lesson you will be able:

B • to name two types of light sources and give one example of each type

There are two sources of light: (1) natural sources, and (2) artificial sources. B

1. Natural sources. The sun is the main source of light on earth. The gases on the sun's surface are heated until they are white-hot. The hot gases glow and give off light. B

Some living things make their own light. Fireflies, for example, make a chemical called **luciferin.** When it reacts with oxygen, luciferin gives off light. The angler fish, certain shrimps, mollusks and worms also produce their own light. This process of "living light" has puzzled scientists because no heat is given off with the light.

2. Artificial sources. In the modern electric lamp there is a bulb which gives off light when electrons flow through it. The metal wire in the bulb, called the **filament,** is heated until it glows. This type of light is **incandescent.** B

Some modern electric lights are **fluorescent lights.** In a fluorescent light tube electricity passing through a gas, such as mercury vapor or neon, gives off light. This kind of light is brighter and much cooler than incandescent light.

A 40-watt fluorescent lamp produces about three times more light than a 40-watt incandescent bulb.

Check Up B

B. What are the two kinds of light sources? Give one example of each type.

(1) Natural — sun.

____

____

____

(2) Artificial — electric light bulb.

C. How Does Light Travel?

After doing this lesson you will be able:

- C1 • to explain how light travels
- C2 • to tell how much light passes through transparent, translucent and opaque materials
- C3 • to describe absorption, reflection and refraction
- C4 • to tell what causes refraction
- C5 • to tell where the speed of light is the fastest and where it is the slowest

Light travels outward in all directions in straight lines from its source. It can travel through a vacuum, a liquid, a gas, and some solids, but *not* around a corner.

Discover for yourself
how you could show that light travels in a straight line.

PURPOSE: To show how light travels • To develop skills of observation, description and stating a conclusion

You need: 4 pieces of cardboard, hole puncher, 4 blocks of wood, clay, candle, matches

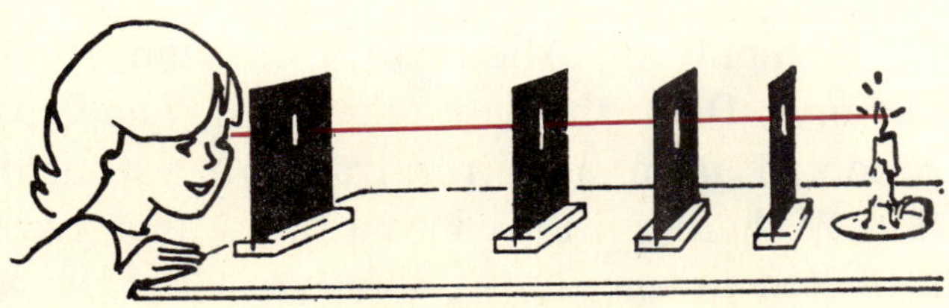

a. Punch a hole in each of the pieces of cardboard. Make sure that the hole is in the same place on each piece of cardboard.

Point out the effects that transparent, translucent and opaque materials have on light.

b. Put each cardboard on a block of wood, using clay to hold up the cardboard.

c. Line up the four cardboard stands about 12 centimeters apart in front of a lighted candle. Adjust the wooden stands until the light can be seen through all the holes.
How did you arrange the cardboard stands?
In a straight line.

d. Now move the cardboard nearest the candle about 2 centimeters to the right. Look into the hole in the cardboard nearest you.

1. What do you see? Darkness — no light.

2. Explain what happened. Light travels in straight lines; it cannot turn. The cardboards must be in a straight line.

You can see an object only if the light coming from the object reaches your eyes. Materials through which light can pass are **transparent.** Objects can be seen clearly through transparent materials such as glass or air. C2

Some light can pass through frosted glass or a sheet of white paper. But most of the light is blocked so that you cannot see an object clearly. Such materials are **translucent.** C2

No light can pass through some materials, such as a book or a wall. Such materials are called **opaque.** C2

Discover for yourself
what would happen if you flashed a light on different materials

PURPOSE: To investigate the effects of transparent, translucent and opaque materials on light • To develop skills of observation, comparison and classification

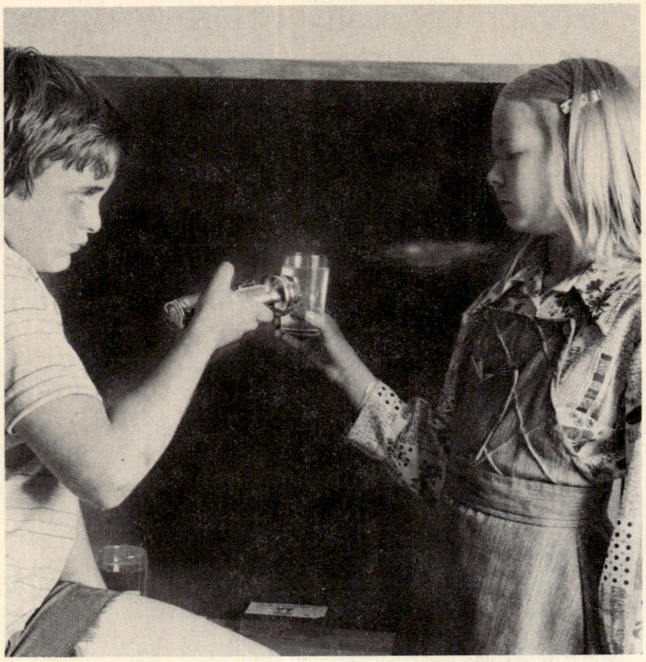

You need: a flashlight, clear plastic, tissue paper, wax paper, wood, aluminum foil, glass of water, pane of glass, cardboard, glass of milk, frosted glass

Make the classroom dark. Hold up an object in front of the chalkboard. Flash a beam of light on the object and look at the chalkboard to see if light passes through the object. Repeat with a number of different materials.

1. Which materials are transparent? ______
   Clear plastic, water, pane of glass.

2. Which materials are translucent? ______
   Tissue paper, wax paper, frosted glass.

3. Which materials are opaque? ______
   Wood, aluminum foil, cardboard, milk.

Light, as it travels outward in all directions from its source, comes into contact with materials or objects. What happens to the light depends on what the object is made of.

Some objects absorb (take into themselves) light energy. This taking in of light C3
is called **absorption.** Dark materials absorb a great deal of sunlight. Light-colored materials absorb very little. Light also contains heat energy. Because they can absorb light energy, dark clothes are warmer than light-colored clothes.

You can see some objects because they are **luminous.** That means that they give off their own light. Most objects are not luminous. You can see these objects only because they receive light from some other source. Such objects are said to be **illuminated.** Some of the light that illuminates (lights up) the object bounces off so that you can see the object. This bouncing off of C3
light is called **reflection.**

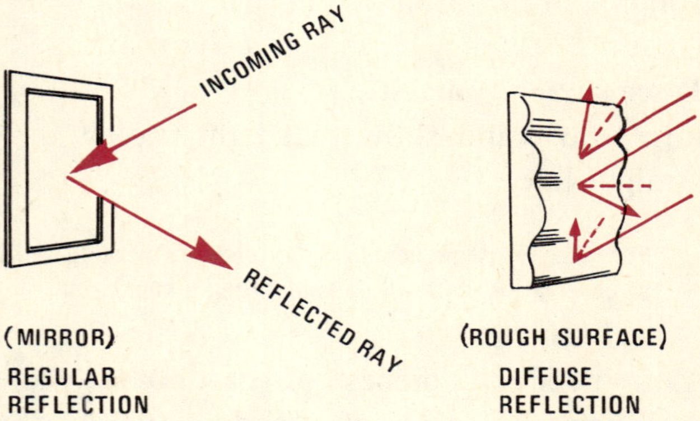

Light rays reflecting from a smooth surface (mirror) and a rough surface.

A highly polished surface, such as a mirror, reflects the light that hits it. The light reflected from a mirror travels in straight lines. When you look into a flat mirror your reflected image looks like you, but it is *reversed;* that is, your real right side is the left side of the image.

When light strikes a smooth, flat surface, such as a mirror, the reflected light rays always travel in the *same direction* from which

Light travels: 297,600 km/sec through space (slightly slower through air); 224,000 km/sec through water; 196,800 km/sec through glass.

they came. Reflection of this kind is called **regular reflection.** When light is reflected in *all directions,* it is called **diffuse reflection.**

Materials such as white paper, the walls of a room and snow reflect light in all directions. There are very few surfaces that are perfectly smooth. So most of the reflected light you see is diffused.

Have you ever looked down into a fish tank or glass of water and thought that the bottom was closer than it really was? Put a pencil into a partly filled glass of water. Does the pencil look broken at the point where it enters the water? What causes these "optical illusions"?

Light travels in a straight line as long as it travels through the same material. However, if light passes at an angle other than 90° from one material to another as, for example, from air into water, it bends and changes its direction of travel. This bending of light as it
C3 passes from one material into another is called **refraction.** Refraction happens because light
C4 travels at different speeds through different
C5 materials. Light travels fastest in a *vacuum* and slowest in a *solid.*

Check Up C

C1. How does light travel? Outward from a source in all directions in straight lines.

C2. What are the effects of each of the following materials on light?

(1) Transparent: Allow light to pass through.

(2) Translucent: Allow light to pass through but scatter it.

(3) Opaque: Do not allow light to pass through.

C3. Explain each of the following in regard to light rays.

(1) Absorption: Taking in light rays.

(2) Reflection: Bouncing off of light rays.

(3) Refraction: Bending of light rays.

C4. What causes light to be refracted? Refraction results because light travels at different speeds through different materials.

C5. Through what does light travel the fastest? The slowest? Light travels fastest in a vacuum and slowest in a solid.

D. Is White Light Really White?

After doing this lesson you will be able:

D1 • to describe what a prism does to white light

D2 • to list the seven colors of the visible spectrum

D3 • to name one kind of invisible light

Have you ever seen a rainbow in the sky? Do you know why this happens? Learning about a prism will help you to answer this question.

Light bends toward the base of the prism because the waves are passing through more glass and are slowed down more. As soon as the light emerges from the glass, it resumes its original speed.

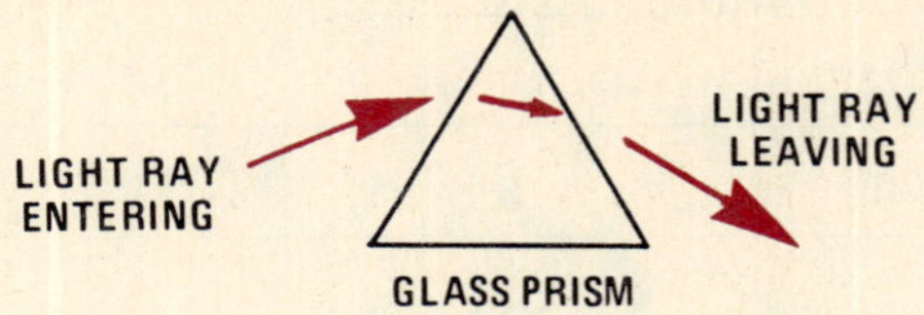

Light rays refract, or bend, when passing through a prism.

What does a prism do? A prism is a solid piece of glass or other transparent material shaped like a wedge (a five-sided object with two parallel triangular sides). It is used to bend, or refract, light. Light enters a prism
D1 from the air. The speed of light is slowed down in the prism. The prism always bends light toward its base. When light leaves a prism, it is refracted again.

When white light from the sun enters a glass prism it is separated into bands of dif-
D2 ferent colors. These colors are red, orange, yellow, green, blue, indigo and violet. The band of colors formed when sunlight passes through a prism is called the **visible spectrum.**

Drops of water are tiny prisms. They break the sunlight into a band of colors called a **rainbow.** This shows you that white light is not really white, but made of many different colors.

A prism can also separate light that you cannot see. In 1800, the British scientist, Sir William Herschel, was studying the colored bands of the visible light spectrum. To find out which color had the most heat he placed the bulb of a thermometer in each band of color.

He read the temperature of each band and found that the red band had the highest temperature. He also found that, although he could see no light, a thermometer placed just outside the red band showed an even higher temperature than the red band. He reasoned that although he could not see any light, light must be there. This kind of in-
D3 visible light is called **infrared light.** Infrared light is outside the red band and is part of an **invisible spectrum.**

Check Up D

D1. What does a prism do to white light?

It bends, or refracts, light.

D2. List the seven colors of the visible spectrum.

Red, orange, yellow, green, blue, indigo, violet

D3. Name one kind of invisible light.

Infrared light

E. Lenses

After doing this lesson you will be able:

- E1 • to explain what a lens does
- E2 • to name two types of lenses
- E3 • to describe a convex lens and explain what it does to light rays
- E4 • to explain what the focus of a convex lens is
- E5 • to describe a concave lens and explain what it does to light rays
- E6 • to explain what myopia is and tell the kind of lens that is used to correct this defect
- E7 • to name two kinds of farsightedness and tell the kind of lens that is used to correct these defects

You have learned that light is a form of energy that can be absorbed, reflected or refracted. How is this knowledge used?

Do you wear glasses? Look around you. How many classmates wear glasses? People wear glasses to help them see better.

Eyeglasses are made up of two *lenses* in a frame. A **lens** is a transparent material, usu- E1
ally glass or plastic, that can refract, or bend, light. Lenses do not have the same thickness

 Challenge students to find out what ultraviolet light is.

throughout. Lenses are divided into two
E2 types depending on their shape: (1) convex or (2) concave.

E3 1. A **convex lens** is thicker in the center than at the edges. All the light rays that pass through it bend so that they *come together*
E4 at one point. This point is called the **focus.**

Take a magnifying glass and a sheet of paper outdoors on a sunny day. Hold it so that light passes through it and falls on the sheet of paper. This dot shows you where the focus is. Be careful! The paper may begin to burn because you have focused the sun's energy in one spot.

E5 2. A **concave lens** is thicker at the edges than at the center. This lens is curved in such a way that the light rays are refracted and *spread apart*. This is just the opposite of what a convex lens does.

How do lenses help your eyes? To answer this question you must first understand how light enters the human eye.

Follow the path of light as it enters your eyes and until you see an image.

Word to review: image

1. Light rays from some luminous object enter the eye through the cornea.

Sunlight is refracted when it passes through a convex and a concave lens.

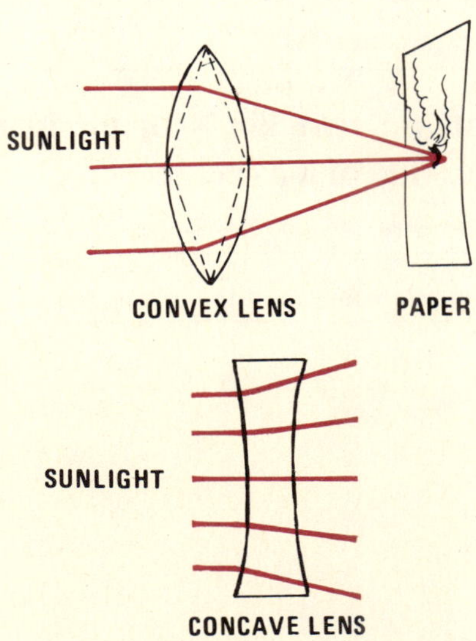

Obtain samples of various kinds of lenses for students to examine.

2. The light passes through the pupil and then through the lens.

3. The lens refracts the light rays to form an inverted (upside down) image on the retina.

4. As soon as the image is formed, nerve cells in the retina send a message to the brain.

5. The brain interprets the message from the retina and also turns the image right side up.

In the eyes of some people, the image of an object is not focused right on the retina. The image may be formed either behind the retina or in front of it. This results in two eye defects: (1) **myopia** or **nearsightedness** and (2) **hyperopia** or **farsightedness.**

1. A **myopic,** or nearsighted, person can see E6
close objects clearly but cannot see far off objects clearly. The images of far off objects come to a focus *in front of* the retina and look blurred.

To correct myopia, concave lenses are E6
used in glasses. The concave lenses cause the light rays to be spread out a little before they enter the eye. This causes the image to be focused right on the retina.

Myopia usually begins between the ages 13 to 19 but it can start earlier and sometimes later. If you cannot see the chalkboard clearly from a seat at the back of the classroom, you should have your eyes examined.

2. A **hyperopic,** or farsighted, person can E7
see far off objects better than nearby objects. The images of nearby objects come to a focus behind the retina.

To correct this eye defect, glasses with E7
convex lenses are used. The convex lenses bend the light rays together before they enter the eye, so that images come to a focus right on the retina.

Hyperopia usually begins at about 13 to 19 years. People with hyperopia usually can-

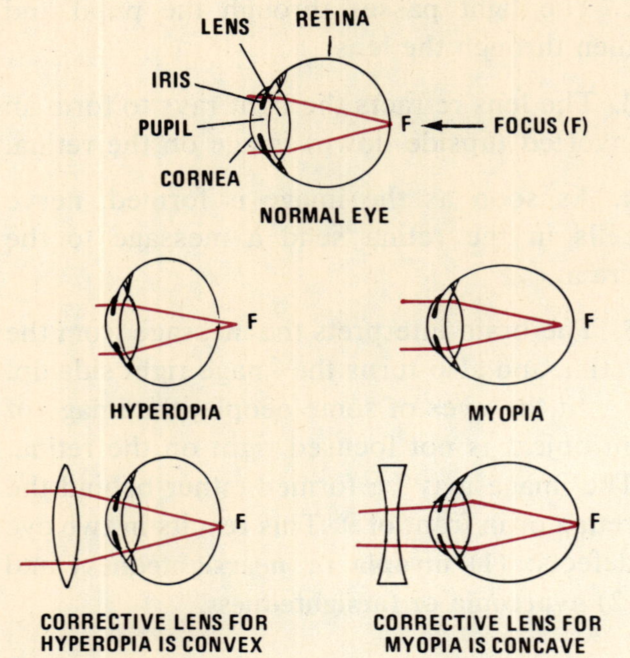

In myopia the distance between the lens and the retina is a little longer than normal; in hyperopia this distance is a little shorter than normal. A convex lens corrects hyperopia and a concave lens corrects myopia.

not see well either near or far, especially as this grows worse. It begins with eye strain from reading which gives headaches. If you have this problem, you should have your eyes examined.

E7 Farsightedness often occurs in adults in their 40s. This is called **presbyopia** (old eyes). The lens in the eye fails to adjust (thicken) for close work. Headaches occur and there is blurred vision in reading. The size of the eyeball is not changed, only the lens. Therefore, glasses are needed only for reading, unless a person has another disorder causing poor distance vision. Nearly everyone gets this condition; it is part of growing older.

Check Up E

E1. What is a lens? A lens is made of transparent material, usually glass or plastic, that can refract light.

E2. What are the two types of lenses? (1) Convex (2) Concave

E3. How is a convex lens shaped? What does it do to light rays? A convex lens is thicker in the center than at the edges. It bends light rays so that they come together at one point.

E4. What is meant by the focus of a convex lens? The point at which light rays all come together.

E5. How is a concave lens shaped? What does it do to light rays? A concave lens is thicker at the edges than at the center. It bends light rays so that they are spread apart.

E6. What is myopia? What kind of lens corrects this defect? Myopia is nearsightedness. Concave lenses are used to correct it.

E7. What are two kinds of farsightedness? What kind of lens corrects these defects? (1) Hyperopia (2) Presbyopia. Convex lens are used to correct these defects.

Unit 3, Chapter 3
REVIEW

Matching Exercise Match each term in **Column A** with its description in **Column B**. Write the letter of the description in the space beside the correct term.

	Column A	*Column B*
a	**1.** convex	**a.** thicker in the center than at the edges
e	**2.** concave	**b.** farsighted
h	**3.** myopic	**c.** light is relfected in all directions
b	**4.** hyperopic	**d.** materials through which light can pass
f	**5.** refraction	**e.** thicker at the edges than in the center
c	**6.** diffuse reflection	**f.** bending of light
j	**7.** luminous	**g.** material through which no light can pass
i	**8.** absorption	**h.** nearsighted
d	**9.** transparent	**i.** the taking in of light
g	**10.** opaque	**j.** object that gives off light

True/False In the space provided, write **T** if a statement is true. If it is false, change the underlined word or words to make the statement true.

concave	**1.** A convex lens is thicker at the edges than at the center.
T	**2.** When light rays are passed through a concave lens, they spread apart.
nearsighted	**3.** A person who is myopic is farsighted.
nearsighted or myopic	**4.** A hyperopic person can see close objects clearly, but cannot see far off objects clearly.
T	**5.** A convex lens is thicker at the center than at the edges.
refraction	**6.** The bending of light as it passes from one material into another is called reflection.
Luminous	**7.** Illuminated objects give off their own light.
opaque	**8.** Materials that do not allow light to pass through them are transparent.
Fluorescent	**9.** Incandescent light is given off by electricity passing through gas in a tube.
T	**10.** The sun is a natural source of light.
an artificial	**11.** A light bulb is a natural source of light.
radiant	**12.** Light is photon energy.
T	**13.** The quantum theory of light states that light is made up of particles of electromagnetic waves.

electromagnetic 14. The particle theory of light states that light waves are partly electrical and partly magnetic in nature.

cannot 15. Light can travel around corners.

Multiple Choice Exercise Circle the letter of the word or phrase that best completes each statement.

1. Light travels in **(a)** curved paths **(b)** straight lines and curved paths **(c)** around corners **(d)** only in straight lines.
2. A material through which light passes is **(a)** opaque **(b)** transparent **(c)** photon **(d)** quanta.
3. A material through which some light passes is **(a)** opaque **(b)** translucent **(c)** transparent **(d)** quanta.
4. A person who has a shorter than average distance between the lens of the eye and the retina is **(a)** myopic **(b)** presbyopic **(c)** hyperopic **(d)** opaque.
5. When light passes through a concave lens, the rays of light **(a)** spread apart **(b)** come together **(c)** are absorbed **(d)** are reflected.
6. Light rays travel fastest in a **(a)** gas **(b)** liquid **(c)** solid **(d)** vacuum.
7. When white light passes through a glass prism, it is **(a)** absorbed **(b)** reflected **(c)** separated into bands of different colors **(d)** diffused.
8. A diffused reflection occurs when light is **(a)** absorbed **(b)** reflected in all directions **(c)** separated into bands of different colors **(d)** passed through a prism.
9. An object that gives off its own light is **(a)** illuminated **(b)** opaque **(c)** luminous **(d)** transparent.
10. Infrared light is **(a)** a type of invisible light **(b)** a type of visible light **(c)** a type of x-ray **(d)** a type of prism.

Application Exercise Label the parts of the eye shown in the diagram below. Write your answers in the provided spaces.

1. retina
2. lens
3. iris
4. pupil
5. cornea

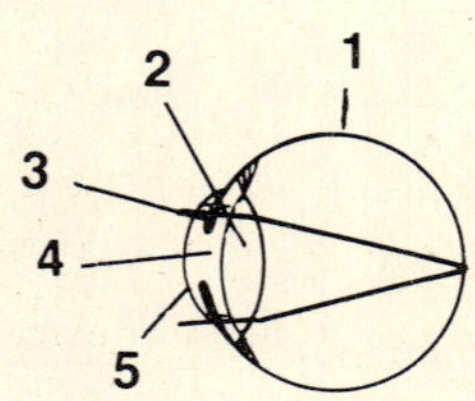

The forest fire shown on this photograph took place in Ochoco National Forest, Oregon. One hundred acres of trees were destroyed as a result of this fire. How could this fire have been prevented? What kind of energy does fire give off? Why is it difficult to put out a fire as large as this one?

By not littering the forest with flammatory materials; fire gives off light and heat energies; heat from burning trees is so great firefighters cannot get close.

New science words: calorie, contraction, expansion

Skills developed: observing, comparing, inferring, stating a conclusion

Chapter 4

HEAT IS ENERGY

The word *energy* comes from the Greek word for *work*.

If you were asked to list the most important things that human beings need to stay alive, what would be on your list? If you think about it, heat probably would be very high on your list.

Human beings use heat to cook food; to heat houses, factories and schools; to produce power; to remove metals from their ores; to run machinery; and to carry out many other industrial and scientific activities.

It seems safe to say that human beings have been able to control much of their environment by learning how to produce and control heat.

A. What Is Heat?

After doing this lesson you will be able:

- A1 • to explain what heat is
- A2 • to list two things that the amount of heat a substance has depends on
- A3 • to explain what a calorie is
- A4 • to explain the difference between the small calorie and the large Calorie

You have learned that all matter is made up of atoms. Most of these atoms combine to form molecules. These molecules are always moving—they have kinetic energy. **Heat** is

The *translational* movement of the molecules of liquids and gases leads to many collisions; in solids, this movement is *vibratory*.

A1 the energy of motion (kinetic energy) of the particles that make up any piece of matter.

A2 The amount of heat a material has depends on how many molecules it has and how fast the molecules are moving. The greater the number of molecules and the faster they move, the greater the number of collisions among them. These collisions produce a large amount of heat.

How is heat measured? Scientists measure heat by using a unit called a **calorie.** A
A3 calorie is the amount of heat needed to raise the temperature of 1 gram of water 1 degree centigrade (Celsius).

A **gram** is a unit used for measuring mass. There are about 454 grams in 1 pound.

Scientists use a *small calorie* and a *large Calorie*. The unit used to measure the amount of heat needed to raise the temperature of 1
A4 gram of water 1 degree centigrade is the *small calorie*. There are 1,000 small calories in one large Calorie. The **large Calorie** is used to measure units of heat in food. For example, a glass of milk when burned in your body produces about 125 Calories.

Check Up A

A1. What is heat? Heat is the energy of motion (kinetic energy) of the particles that make up any piece of matter.

A2. What two things does the amount of heat a substance has depend on? (1) How many molecules it contains (2) How fast the molecules are moving.

A3. What is a calorie? The amount of heat needed to raise 1 gram of water 1 degree Celsius.

A4. Explain the following:

(1) small calorie Amount of heat needed to raise 1 gram of water 1 degree Celsius.

(2) large Calorie Equals 1000 small calories.

B. What Is Temperature?

After doing this lesson you will be able:

- B1 • to explain what temperature is
- B2 • to tell what expansion and contraction are
- B3 • to list the two materials that are used in thermometers
- B4 • to name the person who invented the first thermometer
- B5 • to name the two temperature scales
- B6 • to explain the fixed points on a thermometer and name them on the Celsius scale
- B7 • to explain what the boiling and freezing points of a liquid are

The amount of hotness in an object is called its **temperature.** A **thermometer** is B1
used to measure temperature in units called **degrees.** Most thermometers contain a liquid.

When the thermometer measures an increase in temperature, the molecules of the liquid move more rapidly. This causes the liquid to take up a larger space, or volume, so it rises. This increase in space is called B2
expansion.

When the temperature of a liquid in the thermometer is lowered, the molecules move more slowly; the liquid **contracts** (takes up B2
less space), and it falls. This expansion and **contraction** of liquids makes a thermometer work.

Most liquid thermometers have mercury, B3
a silvery-colored metal. Alcohol is used in some thermometers but it cannot be used to measure high temperatures because it evaporates (changes to a gas) easily.

Collect and examine various types of thermometers.

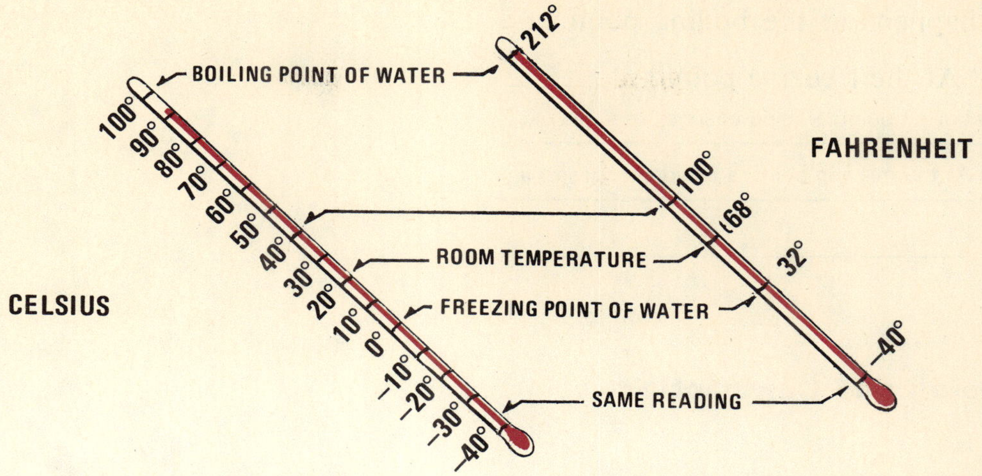

The Celsius (centigrade) temperature scale was until recently used only by scientists. Now it is being used by the general public. Some weather forecasters are using both the Celsius and Fahrenheit readings. The Celsius scale will become more important as the metric system comes into use.

The first thermometer was made in about
B4 1593 by the Italian astronomer, Galileo
Galilei. However, the first *accurate* ther-
mometer contained alcohol and was made in
the early 1600s. In 1714, Gabriel Fahrenheit,
a German physicist, made a mercury ther-
B5 mometer and introduced the Fahrenheit scale.
In 1742, Anders Celsius, a Swedish astrono-
mer, introduced the centigrade scale, which
is now called the Celsius scale.

On the Celsius scale, water freezes at 0°
and boils at 100°. The freezing point and the
B6 boiling point of water are called the **fixed
points** of a thermometer. The Celsius scale
has 100 divisions between the freezing point
of water and the boiling point of water. The
Celsius scale is used by scientists and in
most countries of the world.

The **boiling point** is the temperature at
which a liquid changes into a gas or vapor.
B7 Different liquids have different boiling points.
The **freezing point** is the temperature at
which a liquid changes into a solid. Different
liquids have different freezing points.

Check Up B

B1. What is temperature? ________________
the amount of hotness an object contains

B2. What is expansion and contraction? ____
Expansion is the increase in volume of a substance. Contraction is the decrease in volume of a substance.

B3. What two substances are used in thermometers? mercury and alcohol

B4. Who invented the first thermometer?
Galileo Galilei

B5. Name two temperature scales. ________
Fahrenheit and Celsius scales

B6. What are fixed points on a thermometer? What are they on the Celsius scale?
Fixed points on a thermometer are the freezing and boiling points of water. On the Celsius scale, water freezes at 0° and boils at 100°.

B7. What happens at the boiling point of a liquid? At the freezing point? ________

At the boiling point, a liquid changes to a gas or vapor. At the freezing point, a liquid changes to a solid.

C. Expansion and Contraction

After doing this lesson you will be able:

C1 • to describe how most solids, liquids and gases react to being heated and cooled

C2 • to explain what causes substances to expand and contract

Water is an exception. On cooling, it expands to form ice.

C1 Most solids, liquids and gases *expand* when heated and *contract* when cooled. When matter is heated, its molecules move faster. As they move, they collide with their neighbors very rapidly. The collisions force
C2 the molecules to spread farther apart. The farther apart they spread, the more the matter expands.

Air, which is a mixture of gases, expands and becomes lighter when its temperature rises. Warm air rises because the cold, heavier air sinks and pushes up the lighter warm air.

What happens when solids or gases are cooled? The molecules slow down and collide
C2 less. The molecules move closer together, causing the material to contract.

Do you ever wonder why there are spaces between the concrete blocks of sidewalks? The spaces are there to allow the concrete to expand in hot weather. Telephone and electric wires are hung loosely. Do you know why? In the summer they become loose; in the winter they become tight. If they were stretched tight in the summer they would snap when they contract from the winter cold. Bridges are built to allow for the expansion of materials in the summer and the contraction of these materials in the winter.

This bridge was built to allow for contraction and expansion. Otherwise it would snap in the winter and become too loose in summer.

Discover for yourself
how you could prove that metal expands when heated.

PURPOSE: To show how heat affects contraction and expansion of an object • To develop the skills of observation, comparison and stating a conclusion

You need: a ring with a long handle, steel ball with a long handle, Bunsen burner, matches

a. Be sure the ring and the ball are about the same size in diameter. The ball should be able to fit through the ring.

b. Heat the ball over the Bunsen burner for a time. Try to pull the ball through the unheated ring.

Have students look for other examples of this: sagging power lines; spaces between concrete road sections; meshed grates on bridge surfaces.

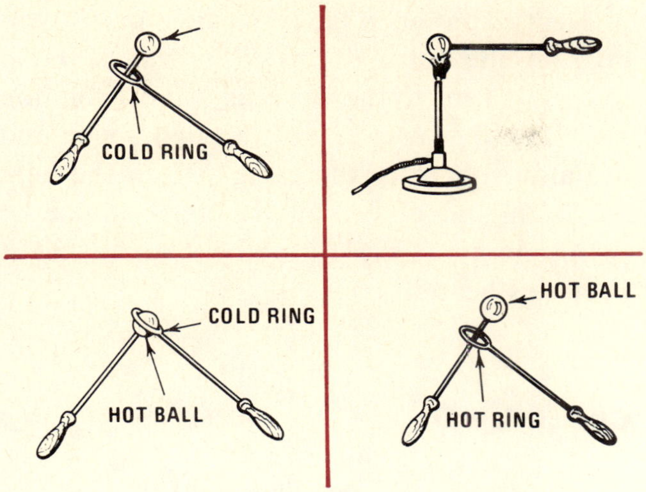

1. What happened? ______________________
The ball did not fit through the ring.

2. Why? The ball expanded when it was heated.

c. Heat both the ball and the ring over the Bunsen burner. Try to put the ball through the ring.

1. What happened? The ball went through the ring.

2. Why? Since both ball and ring were heated, both expanded.

Check Up C

C1. How do most solids, liquids and gases react when they are heated? When they are cooled? Most solids, liquids and gases expand when heated and contract when cooled.

C2. What causes a material to expand and contract? ______________________
A substance expands when its molecules move farther apart. It contracts when its molecules move closer together.

D. How Does Heat Travel?

After doing this lesson you will be able:

- D1 • to list three ways heat can be transferred from one object to another
- D2 • to explain what heat conduction is
- D3 • to name one good conductor and two poor conductors of heat
- D4 • to explain what heat convection is
- D5 • to explain what heat radiation is

Have you ever gone to the mountains or seashore in the summer to escape the heat in the city? Do you know why such places have lower temperatures than the city? The answer depends on the ways in which different objects gain and lose heat. Heat is carried from one object to another in three ways:
(1) by *conduction,* (2) by *convection* and (3) D1
by *radiation.*

1. Conduction. The movement of heat
through a solid object as a result of the col- D2
lisions of its atoms or molecules is **conduction.** You know better than to put one end of a needle into a flame and hold on to the other end. You will soon drop the needle or get burned! Why does heat flow from the hot end of the needle to the cooler end?

The molecules of the metal that is in the flame become hot and move about faster.

Point out how closely packed the molecules of solids are; how there is more space between the molecules of a liquid and even more between gas molecules. This space between the molecules of liquids and gases makes them poor conductors.

They begin to hit the molecules near them and set them into faster motion. This goes on until all the molecules in the needle are moving around faster and the entire needle becomes hot. The heat travels from one molecule to another; that is, from one end of the needle to the other end. This is conduction.

D3 Solids, especially metals, are good conductors of heat. Liquids and gases are poor conductors of heat. If you go on to study physics, you will find out why this is so.

D4 2. **Convection.** In **convection** heat travels through liquids or gases by the movement of atoms and molecules from one place to another. In a room heated by a radiator, the air next to the radiator is warmed and it expands. Since the heated air is lighter than the cold air around it, it rises. The heavier cold air sinks, and pushes up the lighter warm air. This movement of air is called a **convection current.**

Water is also heated by convection. When water is heated in a pot, the water near the burner is heated first. The heat causes the water to expand, become lighter and rise. The heavier cold water near the top sinks down pushing up the hot water. This continues until all the water is heated evenly. Both hot air and hot water heating systems produce heat by convection.

This beaker contains a coloring material that shows the direction in which the convection current is moving.

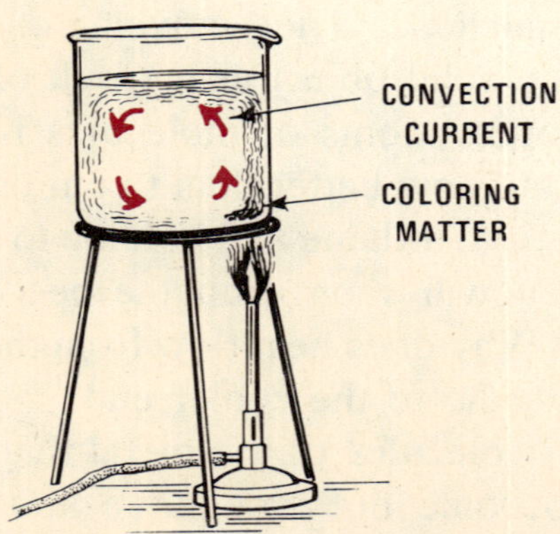

3. **Heat radiation** is the movement of heat energy in the form of invisible waves. Heat energy is sent to the earth as waves from the sun. These waves travel through space and the atmosphere of the earth. When the heat waves hit an object on the earth, some of their energy passes into the object, and it becomes heated. D5

Check Up D

D1. What are the three ways that heat can be carried from one object to another?

(1) Conduction

(2) Convection

(3) Radiation

D2. What is heat conduction? The transfer of heat through a solid as a result of the collisions of its atoms or molecules.

D3. Name one good conductor and two poor conductors of heat. Solids, especially metals are good conductors; liquids and gases are poor conductors.

D4. What is heat convection? The transfer of heat through liquids and gases by the movement of heated atoms and molecules from one place to another.

D5. What is heat radiation? The movement of heat energy in the form of invisible waves.

 Heat (infrared) waves are a type of radiant energy transmitted by electromagnetic radiation. This radiant energy generates the convection currents which transfer heat throughout the room.

E. Heat Insulators and Conductors

After doing this lesson you will be able:

E1 • to describe what a heat insulator is and name two good heat insulators

E2 • to explain how heat loss by convection can be prevented

E3 • to explain how heat loss by radiation can be prevented

Some materials are better conductors of heat than others. Those that are poor con-
E1 ductors of heat are the best **insulators** against heat when heat travels by conduction.

Most cooking utensils—pots, pans, etc.—are made of stainless steel, aluminum or copper. Metals are used for cooking utensils because they are good conductors of heat. Look at the handles. Most of them are wood
E1 or plastic. Why? The wood or plastic will not conduct heat as well as the metal does and so will not get as hot.

Discover for yourself
how you could show that metals conduct heat.

PURPOSE: To compare the rates at which different metals conduct heat • To develop the skills of observation, comparison, inference and stating a conclusion

You need: a ring stand; Bunsen burner; matches; iron, aluminum, brass and copper rods (all the same size); wax; thin wire; scissors

a. Make a ring of wax on the outer end of each metal rod.

b. Set the rods on a ring stand over a Bunsen burner so that the inner tip of each rod can be heated. Arrange the rods evenly around the ring stand. Fasten them to the ring stand with thin wire.

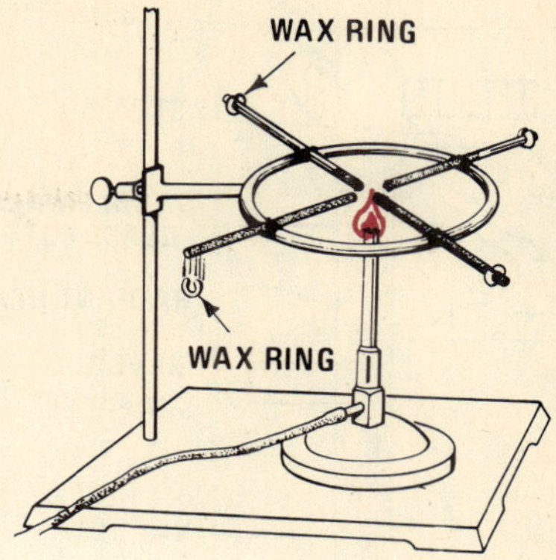

c. Light the Bunsen burner and heat the inner tips of the metal rods.

1. Which rod lost its wax ring first? ______
Copper ______

2. Which rod lost its wax ring last? ______
Iron ______

3. Since the wax rings were not directly over the flame, why did the wax melt? ______
The heat was conducted through the rods to the wax. ______

4. Why didn't the rods lose their wax rings at the same time? Metals conduct heat at different rates. ______

Air is a poor conductor of heat, but it is a good insulator. All materials that have a lot of air spaces are good insulators. Such materials are used in building houses to keep the heat out during the summer and to keep the heat in during winter.

Convection takes place mainly in air. To prevent heat loss by convection currents, the E2
air must be removed. In a thermos bottle, there is a vacuum between the two walls so that heat cannot be lost by convection. In houses with storm windows only a few con-

Obtain a thermos bottle for students to examine. Caution students to handle it carefully to avoid breaking the vacuum bottle.

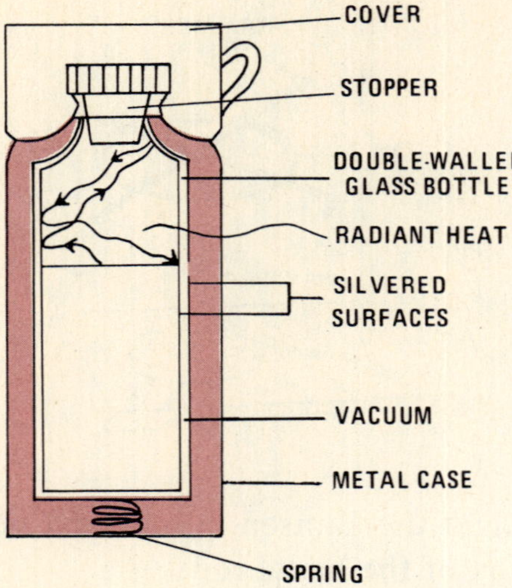

This cross section of a thermos bottle shows materials in the bottle that prevent heat radiation—the smooth silvered surfaces, the vacuum area around the inside of the bottle and the double wall.

Air Force fire fighters use a special foam to control a fire of blazing aircraft fuel. Why are their suits made of shiny material? Why are the suits loose-fitting?

vection currents can develop, and heat loss by convection is less.

Heat loss by radiation can be prevented by using surfaces that either absorb or reflect heat waves. Shiny, smooth and light-colored surfaces reflect heat and do not allow much heat to pass in or out by radiation. Have you seen builders putting aluminum foil between the walls and under the roof of a house? They do this so that radiant heat cannot escape. E3

Check Up E

E1. What is an insulator? What are two good heat insulators? An insulator is a poor conductor of heat. Wood and plastic are good insulators.

E2. How can heat loss by convection be prevented? By removing air.

E3. How can heat loss by radiation be prevented? By using surfaces that either absorb or reflect heat.

Shiny materials are good insulators because they reflect heat waves. Loose-fitting clothing permits air to act as additional insulation.

Unit 3, Chapter 4
REVIEW

Matching Exercise Match each term in **Column A** with its description in **Column B**. Write the letter of the description in the space beside the correct term.

	Column A	*Column B*
h	**1.** convection	**a.** temperature at which a liquid becomes a gas
d	**2.** conduction	**b.** unit of temperature
f	**3.** calorie	**c.** amount of heat in an object
b	**4.** degree	**d.** movement of heat through a solid
g	**5.** thermometer	**e.** substance that does not allow heat to pass through it
c	**6.** temperature	**f.** unit of heat
i	**7.** conductor	**g.** instrument used to measure temperature
e	**8.** insulator	**h.** movement of heat through a liquid or gas
a	**9.** boiling point	**i.** substance that allows heat to pass through it
j	**10.** freezing point	**j.** temperature at which a liquid becomes a solid

True/False In the space provided, write **T** if a statement is true. If it is false, change the underlined word or words to make the statement true.

contracts **1.** When the temperature of the liquid in a thermometer is lowered, the liquid expands.

temperature **2.** Scientists measure heat in units called degrees.

boiling **3.** The freezing point is the temperature at which a liquid changes to a gas or vapor.

large Calorie **4.** The small calorie is used to measure units of heat in food.

T **5.** The unit of mass is the gram.

calorie **6.** The unit of heat is the degree.

0° **7.** On the Celsius scale, water freezes at 32°.

alcohol **8.** Mercury and water are the two most commonly used substances in thermometers.

T **9.** Heat is the energy of motion of the particles that make up any piece of matter.

conduction **10.** Heat moves through solids by convection.

heat radiation **11.** Heat travels from the sun to the earth by conduction.

T **12.** Substances that are poor conductors of heat are called insulators.

T **13.** The transfer of heat through a solid is the result of collisions between the molecules or atoms.

poor **14.** Liquids and gases are good conductors of heat.

Celsius **15.** Most scientists use the Fahrenheit scale when measuring temperature.

Multiple Choice Exercise Circle the letter of the word or phrase that best completes each statement.

1. The temperature at which a liquid changes to a solid is the liquid's **(a)** boiling point **(b)** expansion point **(c)** contraction point **(d)** freezing point.
2. The amount of heat needed to raise the temperature of one gram of water one degree Celsius is **(a)** one Calorie **(b)** one calorie **(c)** one gram **(d)** ten degrees.
3. The liquids most likely to be found in a thermometer are **(a)** mercury and water **(b)** water and alcohol **(c)** mercury and alcohol **(d)** water and iodine.
4. The fixed points of a thermometer are the **(a)** freezing point and boiling point of mercury **(b)** freezing point and boiling point of alcohol **(c)** freezing point of mercury and boiling point of alcohol **(d)** freezing point and boiling point of water.
5. Water freezes at **(a)** 0° Celsius **(b)** 40° Celsius **(c)** 32° Celsius **(d)** 100° Celsius.
6. The amount of hotness in an object is its **(a)** heat **(b)** temperature **(c)** calorie **(d)** boiling point.
7. Heat moves through a solid by **(a)** heat radiation **(b)** convection **(c)** conduction **(d)** expansion.
8. Heat moves through a liquid by **(a)** heat radiation **(b)** convection **(c)** conduction **(d)** contraction.
9. Most solids, liquids and gases **(a)** expand when cooled **(b)** contract when heated **(c)** expand when heated **(d)** take up the same amount of space at all temperatures.
10. Most metals are **(a)** good insulators **(b)** poor conductors **(c)** good sources of heat **(d)** good conductors.

Short Answer Answer each of the following in complete sentences.

1. What is an insulator? Give three examples of materials that are good insulators. An insulator is a material that is a poor conductor of heat. Three examples of heat insulators are: plastic, wood and air. (Answers will vary. Accept all reasonable answers.)

2. What happens to the molecules of most solids, liquids and gases when they are heated? What happens when they are cooled? The molecules of most solids, liquids and gases expand, or spread apart, when they are heated. In most substances, molecules move closer together, or contract, when the substance is cooled. Water is an exception to this. The molecules in water expand when they are cooled.

These children are making sound. What part of their bodies is the sound coming from?

New science words: pitch, tone, vibration

Skills developed: comparing, inferring, stating a conclusion

Chapter 5

SOUND IS ENERGY

Can you imagine what it would be like if you lived in a world in which there were no sounds? You would never know the voices of your family, friends or teachers. You would never know the sound of the wind in the trees, the songs of the birds, the noise of trains and airplanes, or the honking of car horns. Without sound there would be no radio, TV or talking motion pictures. A world without sound would be rather dull.

A. What Is Sound?

After doing this lesson you will be able:

A1 • to tell what sound is

A2 • to tell how sound spreads out

A3 • to define the term frequency

A4 • to describe the motion of a sound wave

Sound is a form of energy caused when matter moves back and forth in a motion called **vibration.**

The vibration of matter can be caused in many different ways. For example, you can make sound by hitting a drum, blowing into a horn or plucking the strings of a guitar. When you hit a drum with your fingers or a drumstick, the drum cover is pushed out of its normal position. It then springs back and vibrates until it comes to rest in its original position.

Where does the sound come from? The vibrating drum cover is the source of the sound. When you blow into a bugle or trumpet, the air inside the horn vibrates and makes sound. When you pluck the string of a guitar, the string vibrates and makes sound. When you speak, your vocal chords vibrate and make sound.

To demonstrate that sound is caused by movement, immerse a large tuning fork into a glass of water immediately after striking it. The water will splash out of the glass.

Suggested Activity: Have students watch their favorite television program without the sound turned on. This might be an impressive way to begin this chapter and emphasize the importance of sound.

Emphasize the parts of a wave: crest, trough, wavelength, amplitude.

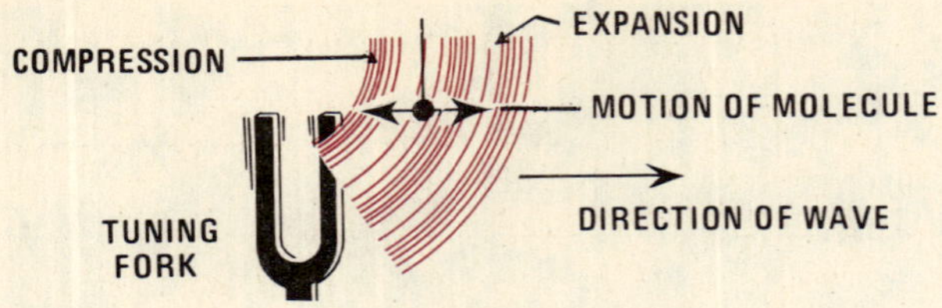

Sound waves from tuning fork—molecules move parallel to the direction of the wave.

A vibrating object makes the air around
A2 it vibrate too. These vibrations spread out in all directions as **sound waves.**

Strike tuning forks of different sizes with a rubber hammer. If you listen carefully, you will discover that each tuning fork makes a different sound. The sounds differ because each tuning fork vibrates at a different speed. The number of vibrations made by a vibrat-
A3 ing object in one second is called its **frequency.**

As the vibrating arm of a tuning fork moves away from its original position, it pushes against the molecules of air that touch it. This causes these air molecules to push against other air molecules and a wave motion spreads through the air.

As the vibrating arm of a tuning fork moves back to its original position, the molecules of air lying next to it rush in to take up the space left by the arm. Other molecules of air rush in to take their place. This causes another wave motion.

A sound wave is formed by **compression**
A4 and **expansion.** When molecules of air are pushed together, they are *compressed.* When the molecules of air move back, they are spread out or *expanded.*

Sound wave motion can be compared to a coiled spring that is stretched and then suddenly let go. The compression and the expansion of the air molecules near the source of the sound compress and expand the air molecules farther away. Because of this, vibrations travel in air in all directions.

Check Up A

A1. What is sound? Sound is a form of energy caused when matter vibrates.

A2. How does sound spread out? Sound spreads out in all directions as waves.

A3. What does the term *frequency* mean? the number of vibrations made by a vibrating object in one second.

A4. Describe the motion of a sound wave:

(1) Compression molecules of air pushed together

(2) Expansion molecules of air spread out

B. How Does Sound Travel?

After doing this lesson you will be able:

B1 • to list three materials that sound can travel through

B2 • to explain why sound will not travel through a vacuum

B3 • to tell what the speed of sound depends on

Long ago native Americans put their ears to the ground in order to hear the coming of the enemy. These people knew that sound traveled through matter like the ground or the air. But they didn't know how or why.

For sound waves to travel, molecules of matter must be present. Sound waves can
travel through solids, liquids or gases but not B1
in a vacuum. In a vacuum there are no mole-
cules to carry the energy of the vibrating B2
object.

A solid carries sound waves best. If you remember, molecules in a solid are close together. Air is the most common medium (material) for carrying sound waves. But water is better than air because water molecules are closer together than air molecules.

Does sound always travel at the same speed? **The speed of sound depends on the**
B3 **medium through which it travels.** Sound travels about 1,410 meters per second in water and about 330 meters per second in air.

Sound travels faster when the temperature of the air is higher. This is because the molecules of a gas move faster when the temperature of a gas is high.

The speed of sound is much slower than the speed of light. That is why you always see a flash of lightning before you hear the sound of thunder. If you can figure out the amount of time between the flash of lightning and the sound of thunder, you can tell how far away the lightning is. Usually the lightning is about 1.6 kilometers away for every five seconds between the flash and the sound of thunder.

In order for you to hear sound, the vibrations must reach your ears. You need (1) a source of sound, that is, a vibrating object, (2) a medium, such as air or water to conduct the sound waves, (3) the ear to receive the sound waves and (4) the brain to interpret (explain) the sound.

You can hear sounds with frequencies between 20 and 20,000 vibrations per second. Different animals can hear different ranges of sound frequencies. Dogs and bats, for example, can hear sounds with higher frequencies than human beings can.

Check Up B

B1. Which materials does sound travel through?

(1) Solids

(2) Liquids

(3) Gases

B2. Why can't sound travel through a vacuum? There are no molecules to transfer the energy of the vibrating object.

B3. What does the speed of sound depend on? The speed of sound depends on the medium through which it travels.

Energy, in this case, refers to the amplitude of a sound wave.

C. Some Characteristics of Sound

After doing this lesson you will be able:

- C1 • to list three characteristics of sound
- C2 • to tell what pitch is and what determines it
- C3 • to tell what the loudness of a sound depends on
- C4 • to explain what an echo is
- C5 • to tell what the tone of a sound depends on

Your ears can pick up all kinds of sounds and you can tell one sound from another. Sounds have three characteristics which allow you to tell them apart: (1) pitch, (2) loud- C1
ness and (3) tone or quality.

1. The **pitch** of a sound refers to how low or how high the sound is. Pitch depends on how C2
fast or how *slow* an object is vibrating; that is, on its frequency. The faster an object vibrates, the greater is its frequency and the higher its pitch.

2. The **loudness** of a sound depends on the C3
energy of the sound wave. What is the "energy of the sound wave"? This is the greatest distance a molecule in a sound wave moves from its normal position. Usually, the greater the energy of the sound wave, the louder the sound. If you pluck a guitar or violin string very hard, it will make a loud sound. If you pluck the same string gently, it will make a soft sound.

The loudness of a sound changes with the distance of the listener from the sound wave's

Have students "feel" the energy of loud sounds by holding a piece of cardboard next to the speaker of a television set or radio. Turn the sound control up and down. The increased energy of the sound can be felt in the cardboard.

source. The loudness becomes less as the distance between the listener and the source of the sound becomes greater. This is because sound waves travel from their source in all directions.

Have you ever cupped your hands around your mouth to make sure that someone far away will hear you? By cupping your hands, you keep the sound waves from spreading out in all directions right away.

C4 An **echo** is a sound wave that is bounced off a reflecting surface such as a cliff or a wall. The walls of a **soundproof** room are built of materials that *absorb* sound waves instead of reflecting them.

3. The **tone** or **quality of sound** allows you to tell the difference between a note played on a piano and a note played on a guitar even though both may have the same pitch and
C5 loudness. Tone depends on how an object producing the sound is vibrating. A violin and a guitar do not sound alike because of the difference in quality of the tone.

Check Up C

C1. What are three characteristics of sound?

(1) Pitch

(2) Loudness

(3) Tone

C2. What is pitch? What does it depend on? Pitch refers to how low or how high a sound is. It depends on the frequency of an object.

C3. What does loudness of sound depend on? Loudness depends on the energy of the sound waves.

C4. What is an echo? An echo is a sound wave that is bounced off a reflecting surface.

C5. What does the tone of a sound depend on? Tone depends on how an object producing the sound is vibrating.

Students may have musical instruments at home. Encourge them to bring them to class.

D. What Is Music?

After doing this lesson you will be able:

- D1 • to tell what makes the difference between noise and music
- D2 • to tell what the pitch of a stringed instrument depends on
- D3 • to tell what the pitch of a wind instrument depends on
- D4 • to explain how percussion instruments make sound

What is the difference between musical sound and noise? **Noise** results when *irregu-* D1
lar vibrations are made at irregular times. **Musical sound** results when *regular* vibrations of the voice or a musical instrument are produced.

There are three main kinds of musical instruments: (1) string instruments, (2) wind instruments and (3) percussion instruments.

1. String instruments have strings stretched over a box or board. The violin, guitar and banjo are string instruments. Sound is made by plucking the strings. (The piano is a percussion instrument because its strings are struck.)

The pitch of these instruments depends
on the length, thickness and tightness of the D2
strings. The longer the string, the fewer the vibrations and, therefore, the lower the pitch. Tight, thin strings make sounds with high tones. Loose, thick strings make lower tones.

2. You have blown a whistle or have blown across the top of an open bottle to make sounds. In **wind instruments,** such as the trumpet or flute, air is blown through small openings into a tube or pipe. The stream of air (air column) in the instrument vibrates.

Music results when regular vibrations of the voice or musical instrument are produced at regular times.

A section of the guitar strings produces low sounds. Is it the section under the right hands or left hands of these girls? Will the air the boys are blowing into the horns produce high or low sounds?
Under the right hands: loose, thick strings make lower sounds. Sound from the horns will be low pitched.

D3 The pitch that is produced depends upon length of the air column and how fast it is vibrating. A long air column produces low-pitched sounds. A short column produces high-pitched sounds.

In the human voice, speaking and singing sounds are produced by air columns made to vibrate by the vibrating vocal cords.

Discover for yourself
how you could make a musical instrument.

PURPOSE: To show how the pitch of a wind instrument depends on the length of a vibrating air column • To develop the skills of comparison, inference and stating a conclusion

You need: 3 empty soda bottles (same size), water

a. Line up three empty soda bottles and add water in different amounts to each one.

b. Blow across the tops of the bottles.

1. Describe what happens when you blow across the tops of the bottles. ________
Sounds with different pitches are produced.

2. What causes these sounds? ________
The vibrating air columns in the bottles.

3. Are all the sounds the same? ________
No, the pitches are different.

4. What does the pitch of the sounds have to do with the amount of water in the bottles.
The more water, the higher the pitch.

5. What does the amount of water in the bottles have to do with the length of the vibrating air column. ______

The more water, the shorter the vibrating air column.

3. Percussion instruments include drums, tambourines, cymbals, bells and gongs. If
D4 you stretch a material tightly across a tube and hit it, it will vibrate. This is the basis of the drum or tambourine. When you hit cymbals or gongs and ring bells, they also vibrate and make sounds.

Check Up D

D1. What makes music different from noise?

A noise results when irregular vibrations are produced at irregular times.

D2. What does the pitch of a stringed instrument depend on? ______

Length, thickness and tightness of the strings.

D3. What does the pitch of a wind instrument depend on? ______

Length of an air column and how fast it vibrates.

D4. How do percussion instruments make sound? ______

(1) Drum If you hit material tightly stretched across a tube, it will vibrate.

(2) Gong If you hit a gong, it will vibrate.

E. How Sound Is Used In Communication

After doing this lesson you will be able:

E1 • to list three things used to reproduce sound

E2 • to name the two parts of a telephone and tell what each part does

E3 • to tell what kind of waves radio waves are

E4 • to explain how a microphone works

E5 • to explain the purpose of a radio speaker

The telephone, tape recorder, television, E1
radio and record player are all used to reproduce (make again) sound. We will talk about the telephone and radio.

1. A **telephone** is made up of a **transmitter** (sender) and a **receiver** connected by wires. E2
The transmitter picks up the spoken sounds and changes them into electrical impulses called the **message.** The connecting wires carry the electrical impulses to the receiver which then changes them back into sounds that can be heard.

How is the sound energy changed into electric energy and then back to sound again? Sound waves cause a disc (the diaphragm) in the transmitter, or sender, to vibrate. The transmitter also contains a small box of carbon particles that vibrate in the same way as the vibrations caused by the voice. As the carbon particles move, they are pushed together or separated. The distance between the particles controls the amount of electric current that flows through the wire from the transmitter to the receiver.

An electromagnet is in the receiver. It becomes weaker or stronger depending on the amount of electricity brought to it by the wire. The change in the strength of the magnet causes a thin metal disc in the receiver to vibrate back and forth in the same way as the vibrations of the sound entering the trans-

Your local telephone company may have battery-powered telephones and teaching materials related to communication for loan to your class.

Did you know that the two special buttons marked "" and "#" on push-button telephones will someday allow you to talk to five people at once?*

mitter. So, what you say in the mouthpiece (transmitter) is what someone hears in the earphone (receiver) of the telephone.

2. Radio waves are produced when an electric current moves back and forth in a wire
called an **antenna.** Radio waves are electro- E3
magnetic; that is, they have both *electrical* and *magnetic* properties. Radio waves are sent and received very much like sounds in a telephone.

In radio, you speak into a **microphone.** Vi-
brations are set up in the microphone which, E4
in turn, produce electrical impulses (electrical waves). The microphone is like the transmitter in a telephone. In the radio, the vibration is carried by electromagnetic waves. A broadcasting **antenna** sends these waves out into the air. A radio receiver in your home picks up the waves with the help of another
antenna. A **speaker** in the radio changes the E5
electromagnetic waves into sound waves. Now you can hear your favorite program.

When you hear a voice or music on your radio coming from a faraway place, the sound is being "carried" from the radio station to your home by an electromagnetic radio-frequency *carrier wave. A radio-frequency wave cannot be heard because it is above the sound range of the human ear. You can think of it as having very many vibrations per second—much higher than the very highest note that can be played on the violin.*

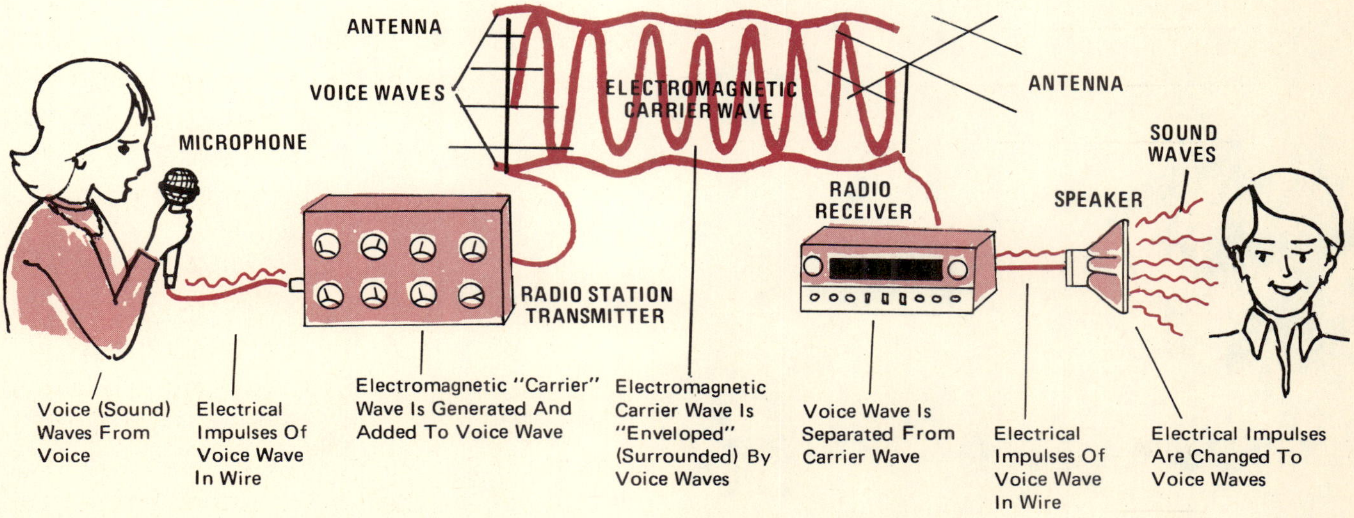

Check Up E

E1. Name three things used to reproduce sound.

(1) Telephone

(2) Television

(3) Radio

E2. What do the following parts of a telephone do?

(1) Transmitter Picks up spoken sounds and changes them into electrical impulses.

(2) Receiver Changes electrical impulses back into sound waves.

E3. What kind of waves are radio waves?

Electromagnetic

E4, 5. What is the purpose of these radio parts?

(1) Microphone Vibrations are set up in the microphone which produce electrical impulses.

(2) Speaker Changes electromagnetic waves into sound waves.

Unit 3, Chapter 5
REVIEW

Matching Exercise Match each term in **Column A** with its description in **Column B**. Write the letter of the description in the space beside the correct term.

	Column A	*Column B*
a	**1.** noise	**a.** irregular vibrations at regular times
h	**2.** air	**b.** a reflected sound
b	**3.** echo	**c.** made by vibrating matter
e	**4.** loudness	**d.** produced when an object vibrates
f	**5.** pitch	**e.** depends on the energy of a sound wave
g	**6.** expansion	**f.** highness or lowness of sound
j	**7.** compression	**g.** air molecules are spread out
i	**8.** frequency	**h.** most common medium for sound waves
d	**9.** sound waves	**i.** number of vibrations per second
c	**10.** sound	**j.** air molecules are pushed together

True/False In the space provided, write **T** if a statement is true. If it is false, change the underlined word or words to make the statement true.

T	**1.** The <u>loudness</u> of a sound depends upon the energy of the sound wave.
higher	**2.** Sound travels faster when the temperature of the air is <u>lower</u>.
slower	**3.** The speed of sound is much <u>faster</u> than the speed of light.
T	**4.** The walls of a soundproof room are built of materials that <u>absorb</u> sound.
A speaker	**5.** <u>An antenna</u> in a radio changes electromagnetic waves into sound waves.
T	**6.** The speed of sound depends upon the <u>medium</u> through which it travels.
An echo	**7.** <u>Pitch</u> is a sound wave that is bounced off a reflecting surface.
apart	**8.** When molecules of air are pushed <u>together</u>, they are expanded.
Pitch	**9.** <u>Noise</u> refers to how high or low a sound is.
vacuum	**10.** Sound will not travel through a <u>solid</u>.
pitch	**11.** The length of a vibrating air column determines the <u>tone</u> of a sound.
T	**12.** The <u>tones</u> of sounds allow you to tell the difference between sounds produced by different instruments.
solid	**13.** Sound waves travel best in a <u>vacuum</u>.
frequency	**14.** The number of vibrations made by a vibrating object in one second is called its <u>tone</u>.
pitch	**15.** The <u>loudness</u> of a sound depends on how fast or slow an object is vibrating.

Multiple Choice Exercise Circle the letter of the word or phrase that best completes each statement.

1. Sound waves are changed to electromagnetic waves by **(a)** a receiver **(b)** a transmitter **(c)** a vacuum **(d)** an antenna.
2. Electromagnetic waves are picked up and changed into sound waves by **(a)** a receiver **(b)** a transmitter **(c)** a vacuum **(d)** an antenna.
3. Sound travels best in a **(a)** vacuum **(b)** gas **(c)** liquid **(d)** solid.
4. Sound waves cannot travel in a **(a)** vacuum **(b)** gas **(c)** liquid **(d)** solid.
5. The energy of a sound wave determines the sound's **(a)** pitch **(b)** tone **(c)** loudness **(d)** echo.
6. For sound waves to travel **(a)** molecules of matter must be present **(b)** no molecules of matter can be present **(c)** a vacuum must be present **(d)** a person must be present.
7. Sound is a form of energy caused by **(a)** matter **(b)** nuclear energy **(c)** vibrating objects **(d)** radiant energy.
8. Radio waves are **(a)** forms of radiant energy **(b)** forms of nuclear energy **(c)** forms of heat energy **(d)** electromagnetic.
9. A microphone changes **(a)** sound waves to electrical impulses **(b)** electromagnetic waves into sound waves **(c)** sound waves into electromagnetic waves **(d)** electrical impulses into sound waves.
10. The part of a telephone that picks up spoken sounds and changes them into electrical impulses is the **(a)** microphone **(b)** receiver **(c)** transmitter **(d)** speaker.
11. When irregular vibrations are produced at irregular times, **(a)** an echo is produced **(b)** noise is produced **(c)** music is produced **(d)** a vacuum is produced.

Short Answer Answer each of the following in complete sentences.

1. What is an echo? An echo is a sound wave that is bounced off a reflecting object.

2. What does a transmitter do? A transmitter picks up sound waves and changes them into electrical impulses.

You are looking at a sun tracker. It is one of the many new tools being used to try to meet the world's energy needs.

New science words: fossil, recycle, technology

Skills developed: observing, comparing

Chapter 6

THE GROWING NEED FOR ENERGY

The discovery of new knowledge is the chief function of *science. Technology* is the application of knowledge to practical purposes.

The world's energy needs are growing. The world population is growing very fast. Life is becoming more and more technological (machine run). To produce enough energy to keep on with this machine-run life is becoming a big problem. For example, between now and the beginning of the 21st century—less than 20 years away—the United States will use more energy than it has used so far in all its history.

New sources of fuels must be found. And new ways to carry fuel without polluting the air, land and waters also must be found. Right now there is need for hundreds of new electric power plants in the United States alone.

At present, it is hard to mine coal without damaging the surface of the earth, to ship oil without some spilling, to find the right places for power plants, and to store the wastes of fuel-burning machines and nuclear reactors.

Energy needs for the 21st century must be planned now. Planning means looking at our daily needs in terms of energy, and at how machines can help us. It also means looking at what our energy needs will do to the environment (our surroundings).

A. Machines and Energy

After doing this lesson you will be able:

A1 • to list the two things that energy supply depends on

A2 • to name two inventions that helped bring on the Industrial Revolution

In many parts of the world, energy and machines have helped to make a better way of life. Energy has helped to produce more and better food and to build better buildings. The use of energy depends on two things:
A1 **energy supply** and its use in machines called **technological skill.**

Early peoples used muscle energy to do their work. Later, they used animals to do work. Then came the windmill and water wheel. For years, these machines used the energy of wind and flowing water to do such work as grinding wheat and pumping water.

A2 The invention of the **steam engine** in the 18th century and the **gasoline engine** in the 19th century helped bring on the Industrial Revolution. This is the name given to the beginning of modern industry that uses machines to make things.

Since 1900, a large number of small and large power-driven machines have been invented. These machines are easy to handle and move about, and you use many of the smaller ones. With the invention of machines which used power, American industry was able to produce more goods. Mills and shops became mechanized (machine-run). Homes were filled with appliances. All this increased the need for energy. New machines lead to newer machines, and to greater demands for energy.

Check Up A

A1. What two things does the use of energy depend on? Energy resources and technological skill.

A2. What two inventions helped to bring on the Industrial Revolution? The steam engine and the gasoline engine.

B. Fossil Fuels

After doing this lesson you will be able:

- B1 • to tell what fossil fuels are and to name three
- B2 • to list two sources of oil besides oil wells
- B3 • to give the biggest pollution problem fossil fuels cause
- B4 • to explain what thermal pollution does to bodies of water

In the past, energy supplies have always been on hand. In 1850, wood supplied more than 90% of all energy used in the United States. By the beginning of the 20th century, coal had become the most used energy fuel. It supplied more than 70% of the total energy.

By 1950, coal was used to supply only 36% of the energy produced by fuels. The use of oil and natural gas for energy use had risen to 55%. In 1970, the use of coal amounted to about 20% of all energy consumed, oil and gas about 76%, waterpower almost 4% and nuclear energy about 0.3%.

Coal, petroleum and natural gas are **fossil**
fuels. They come from the remains of plant B1
and animal life, which lived millions of years ago, deep in the earth's crust. Fossil fuels cannot be replaced once they are used up.

Until nuclear power is completely safe to use, fossil-fuel burning plants must be built to meet the need for electrical power. Fortunately, recent findings tell us that supplies of fossil fuels—in the form of low-grade coal, oil shale, and tar sands—are enough to meet the country's energy needs for some time to come.

Low-grade coals have a lot of moisture and, therefore, are not good for heating. **High-grade coals** have been squeezed and heated by natural processes that go on within the earth's crust, and have lost much of their water. Thus, high-grade coals are good for heating.

 Words to review: mechanized; appliance

The fossil fuel coal is in large supply and cheap to use. But it creates an air pollution problem when burned.

Most **petroleum,** a natural oil that supplies oil and gas for fuels, comes from underground pools. The oil is brought to the earth's surface by wells. But, there are other places in which petroleum is found. One of these is a kind of rock called **shale.** Shale deposits are found in many parts of the world. **Oil-**
B2 **shale deposits** are found in several states of the United States. It costs more to get the oil from shale than from wells. The shale rock must be mined, crushed and refined to get the oil from it. A metric ton of shale makes only about 38 to 46 liters of crude oil. In the United States, a big shale-mining operation is going on in Colorado.

B2 There are large deposits of **tar sands** in the earth. (Tar sand is sandstone that has asphalt.) One tar-sand deposit now being mined is in Alberta, Canada. This area is thousands of kilometers wide, and the tar sands go down for a depth of 600 meters. To get oil from it, the sand is put through about ten different steps. As with shale, getting a little oil from a lot of sand costs a great deal.

No matter what kind of fuel is used to produce energy, there is a pollution problem. In the past, the biggest problem with fossil fuels has been that they have sulfur. So when B3
the fuel is burned, pollutants in the form of **sulfur dioxide** gas and **particles of sulfur** are given off into the atmosphere.

Different ways have been found to solve this problem. One way is to use a low sulfur-bearing coal from the western part of the United States. Another way is to change the coal to gas or gasoline and take out the sulfur gas before the fuel is burned. However, changing coal to gas or gasoline costs a lot and gives only a small amount of fuel.

Another problem of all power generating is **thermal pollution.** In a power-generating plant using fossil fuels, only about 40% of the heat energy generated produces electricity. The other 60% of the heat is waste. In a plant using nuclear fuels, the amount of waste heat is even greater. This waste heat, absorbed by cold water, is let out into rivers, seas and lakes. Thermal pollution often kills plant and B4
animal life when put into bodies of water. Scientists are trying to find better ways to generate power, ways which would do less damage to all forms of life.

One better way would be to cool the water after it is used in the power-generating

Inefficiency in boilers and other power equipment accounts for this waste.

Nuclear-powered plants have to operate at lower temperatures than fossil fuel plants. The fuel-to-electricity ratio is low. A nuclear plant produces almost 50% more thermal pollution than does a fossil fuel plant with the same electrical power output.

plant and **recycle** the water. Special water-cooling towers have been used for this in parts of Europe for the last 50 years. The tower was first used in the United States in 1963. Now, about fifty such towers are in use.

These giant cooling towers are 90 meters in diameter at the bottom and 150 meters or more high. The sides are made of concrete. In places they are as thin as 18 centimeters.

Inside the tower the hot water coming from the generator splashes down through wooden boards or plastic screens against an upward flow of air that comes from the base of the tower. The air absorbs heat from the water. As you know, hot air is lighter than cool air. So the warm, light air rises and cool, heavier air moves down to take the warm air's place. As the cool air meets the warm air, the warm air is pushed upward by the cool air. This draft keeps pulling cool air into the bottom of the tower and pushing hot air out of the top.

The cooled water from the tower is returned to the generating plant for reuse. A little "new" water has to be added, because in the cooling some water evaporates. So there is a great saving on water. And heated water is not let out into rivers and lakes to do harm to plants and animals.

Are there any problems with the towers? In some places the towers cause fogging of the atmosphere. In other places, frosting of the towers is a problem. Scientists are now working on these problems.

Check Up B

B1. What are fossil fuels? Name three of them. Fossil fuels come from the remains of plant and animal life that existed millions of years ago. Three fossil fuels are coal, petroleum, natural gas.

B2. What are two sources of oil besides oil wells? Oil-shale deposits and tar sand.

B3. What is the biggest pollution problem fossil fuels cause? When fossil fuels are burned, sulfur dioxide and particles of sulfur are given off into the atmosphere.

B4. What does thermal pollution do to bodies of water? Heated water often kills plant and animal life.

C. Energy Sources of the Future

After doing this lesson you will be able:

- C1 • to name the fuel used in nuclear reactors and the process that releases nuclear energy
- C2 • to list two problems nuclear reactors cause
- C3 • to explain what solar energy is and list three uses for solar energy
- C4 • to describe what a photovoltaic cell does
- C5 • to explain what geothermal energy is
- C6 • to explain what "hot spots" are

The world population probably will double by the year 2000. Machines are being used in almost all parts of the world. The energy needs of the mechanized (machine-run) world of the 21st century may be five times greater than they are now. The question is *not* whether energy production for the world should be increased, but *how* to increase it in the least harmful way.

Nuclear power is one of the most important new sources of energy. The "fuel" used in nuclear reactors is uranium. Nuclear en-

C1 ergy is released by **nuclear fission,** which, as
you learned, is the splitting of an atom of
uranium. Small amounts of uranium produce
a great amount of energy.

There are many people who think that
nuclear power should not be the energy
C2 source of the future. The hot water produced
as a waste in a nuclear power plant causes
damage if dumped into lakes and streams.
C2 Getting rid of the radioactive solid wastes
from nuclear power plants is another very
serious problem.

C3 **Solar energy** is energy from the sun. Sci-
entists have been able to change solar energy
directly into useful power. Solar energy is be-
ing used to heat buildings, to dry foods and
to generate steam for engines. **Photovoltaic**
C4 **cells** generate electricity directly from the
sun's energy. The widest use of photovoltaic
cells is in spacecraft.

France has built a huge solar furnace eight stories high. Many others may be built around the world. In the United States engineers are designing and testing plants that use electricity from solar energy. Many "test" homes and buildings are being heated and cooled by solar energy. At the Timonium elementary school (near Baltimore, Maryland) part of the school is heated and cooled by solar energy. The 180 solar panels on the roof collect the sun's rays which heat water. The heated water is used to heat classrooms. Then, the cool water is recirculated to the solar panels where it is reheated and used again. American scientists and engineers say that by the year 2020 solar energy may meet 25% of the country's energy needs.

The sun's energy is everywhere, is free of pollution, and has no supply or delivery costs. However, the weather with its seasonal, hourly, and daily changes, could affect its use. To get around weather problems, it may be possible to build large solar stations. These stations could store solar energy for future use.

Successful missions of unpiloted spacecraft in Earth's orbit and to other planets have shown that the direct conversion of solar energy in space to produce power in the spacecraft and telemeter signals back to earth is possible.

This one-story ranch home was built by students of a vocational school. The solar panels on the roof collect the sun's rays which heat water. The heated water is used to heat the house.

Scientists are working on ways in which satellites could collect solar energy and send it back to earth. But it has to be done in a way that would get rid of the problems created by our atmosphere and by the earth's rotation.

Some of these ideas are being successfully used. In Australia, Israel and Japan, solar energy is being used to heat several million homes. The sun's energy is collected in solar cells placed in the roofs of the houses. The energy is then picked up by air or water, stored in insulated tanks and used to heat the building.

With the increasing need for energy and the problem of pollution with both fossil fuels and nuclear energy, solar energy seems to be the answer to the world's energy needs.

C5 **Geothermal energy** is heat energy that can be collected from hot springs inside the earth. It collects in underground "hot spots." These
C6 are parts of the earth's crust where volcanoes are active and mountains are forming. In these places, either new material from the mantle is being added to the crust or crustal material is going down into the mantle. In both cases, hot molten (liquid) rock is formed and pushed up into the crust. You will learn more about the earth's crust and mantle in the next unit.

Much of Reykjavik, the capital of Iceland, is heated by geothermal waters. These waters are between 2 and 6 kilometers deep within the earth and are heated by the hot rock. This natural hot water is piped into office buildings, schools and homes.

Check Up C

C1. What is used as fuel in the nuclear reactor? What is the process of releasing nuclear energy called? Uranium is used as fuel in a nuclear reactor. Nuclear energy is released by nuclear fission.

C2. What two problems do nuclear reactors cause? (1) Release of hot water into lake and streams (2) getting rid of radioactive wastes.

C3. Explain what solar energy is and list three ways it has been used. Solar energy is energy obtained from the sun and used to heat buildings, dry foods, generate steam for engines.

C4. What does a photovoltaic cell do? Generates electricity from the sun's energy.

C5. What is geothermal energy? Geothermal energy is heat energy that can be obtained from hot springs inside the earth.

C6. What are "hot spots"? Parts of the earth's crust where volcanoes are active and mountains are forming.

Unit 3, Chapter 6
REVIEW

Matching Exercise Match each term in **Column A** with its description in **Column B**. Write the letter of the description in the space beside the correct term.

	Column A	*Column B*
d	**1.** uranium	**a.** volcanic activity and mountain formation
i	**2.** fossil fuels	**b.** energy from the sun
f	**3.** nuclear energy	**c.** problem caused by nuclear energy
a	**4.** hot spots	**d.** fuel used in most nuclear reactors
e	**5.** shale	**e.** source of oil
b	**6.** solar energy	**f.** energy created by fission reactions
j	**7.** geothermal energy	**g.** generates electricity from solar energy
g	**8.** photovoltaic cell	**h.** prevented by cooling towers
c	**9.** radioactive waste	**i.** coal, petroleum and natural gas
h	**10.** thermal pollution	**j.** energy from hot spots inside the earth

True/False In the space provided, write **T** if a statement is true. If it is false, change the underlined word or words to make the statement true.

Uranium **1.** Coal is the most common fuel used in nuclear reactors.

solar **2.** Photovoltaic cells generate electricity from geothermal energy.

solar **3.** The sun provides us with geothermal energy.

T **4.** Geothermal energy is the energy provided by hot springs within the earth.

nuclear **5.** Thermal pollution and radioactive wastes are two problems created by solar energy.

oil **6.** Coal, uranium, and natural gas are fossil fuels.

T **7.** Fossil fuels were made by the remains of plants and animals that died long ago and are deep in the earth.

fission **8.** Nuclear energy is produced by fusion reactions.

mountain **9.** Hot springs inside the earth are found near areas of volcanic activity and lake formation.

thermal **10.** Cooling towers are used to prevent air pollution.

Multiple Choice Exercise Circle the letter of the word or phrase that best completes each statement.

1. Heat energy that can be collected from hot springs inside the earth is **(a)** nuclear energy **(b)** fossil fuel **(c)** geothermal energy **(d)** solar energy.
2. Electricity is generated from the sun's energy by **(a)** fossil fuels **(b)** photovoltaic cells **(c)** uranium **(d)** natural gas.
3. The fuel most likely to be used in a nuclear reactor is **(a)** natural gas **(b)** coal **(c)** uranium **(d)** petroleum.
4. Energy from the sun is **(a)** solar energy **(b)** nuclear energy **(c)** geothermal energy **(d)** electrical energy.
5. One problem of all power-generating plants is **(a)** energy **(b)** thermal pollution **(c)** noise **(d)** nuclear reactions.
6. The splitting of an atom to produce two smaller atoms and energy is **(a)** fusion **(b)** fission **(c)** geothermal energy **(d)** solar energy.
7. Low-grade coals are not good for heating because they contain **(a)** carbon **(b)** moisture **(c)** oil **(d)** natural gas.
8. When fossil fuels are burned they give off **(a)** harmful water **(b)** sulfur dioxide and sulfur particles **(c)** harmful energy **(d)** solar energy.
9. Two problems caused by nuclear reactors are **(a)** thermal pollution and unused energy **(b)** thermal pollution and moisture **(c)** thermal pollution and radioactive wastes **(d)** radioactive wastes and unwanted energy.
10. All of the following are fossil fuels except **(a)** coal **(b)** petroleum **(c)** natural gas **(d)** uranium.

Short Answer Answer each of the following in complete sentences.

1. What is the biggest pollution problem created by the burning of fossil fuels? When fossil fuels are burned, they release sulfur dioxide and sulfur particles into the air. These chemicals can be harmful to humans.
2. What pollution problems are created by the use of nuclear energy? One problem with nuclear energy is the production of radioactive wastes. Another problem with nuclear energy is the production of great amounts of heat that can be released into lakes and streams as hot water. This thermal pollution is very hazardous to plant and fish life.
3. What is geothermal energy and where is it found? Geothermal energy is heat energy that can be obtained from hot springs, usually found near areas of volcanic activity or mountain formations, inside the earth.

Get Ready To Read UNIT 4

IDENTIFYING CAUSE AND EFFECT RELATIONSHIPS

In Unit 4 you will be learning new things about the earth we live on and the new frontiers in space and space exploration. You also will find out why things happen on earth, what makes, or causes, them to happen and what the results, or effects, will be.

Sometimes it is easy to recognize a cause *and* effect relationship. In written material word clues tell the reader that such a relationship exists.

Read the following group of sentences. Each one contains a cause *and* effect relationship. Circle the part of the sentence that contains the cause and underline the part of the sentence that tells the effect. Look for word clues that signal the cause *and* effect relationship.

CAN YOU SPOT THE WORD CLUES?

1. A total blackout of electrical power hit Haynes County residents because of the heavy rains from Tropical Storm George.
2. As a result of eating too much at the school fair, Jamie went home with a stomach ache.
3. Burning garbage caused the unpleasant smell to penetrate our neighborhood.
4. Arthur missed his plane due to the unusually heavy traffic on route to the airport.

Now make a list of the word clues you found in the above sentences. Often cause *and* effect relationships must be detected in your textbooks without the aid of word clues. In order to understand the relationship, you must read carefully the sentence containing the cause-effect relationship. And then you must think about what it says. Read this sentence.

Valuable crops were destroyed from the month long drought in the Southwest.

Which part tells the cause? Which part tells the effect?
Write an E on the line before the phrase if it tells the effect.
Write a C on the line if the phrase contains the cause.

__E__ Valuable crops were ruined.

__C__ There was a drought in the Southwest.

Sometimes one cause will bring about or lead to several effects. This next sentence is an example of that type of cause *and* effect relationship. Label the phrases C or E. Make sure you identify the cause.

> Last night's strong gale winds uprooted trees, knocked down telephone poles, and blocked several main roads in Bay City.

E Telephone poles were knocked down.

E Main roads were blocked.

C Strong gale winds occurred

E Trees were uprooted.

If you were to diagram this cause *and* effect relationship, it would look something like this.

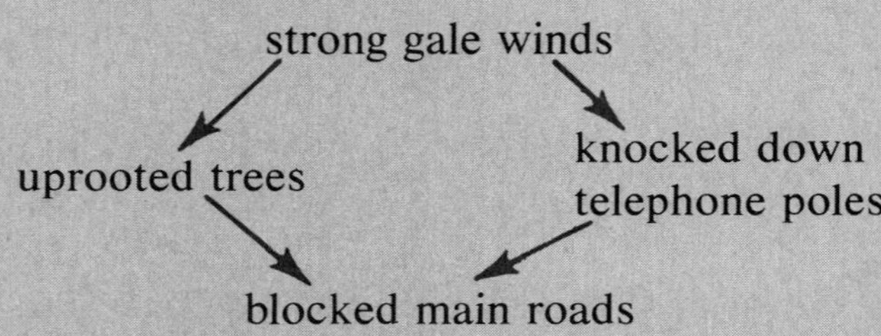

The following sentence is just the reverse of the last one. It contains several causes that lead to one effect.

> This year's heavy voter turnout is thought to be related to the candidate's enthusiastic campaigning coupled with voter interest in campaign issues.

Complete the diagram showing the causes that contributed to the heavy voter turnout.

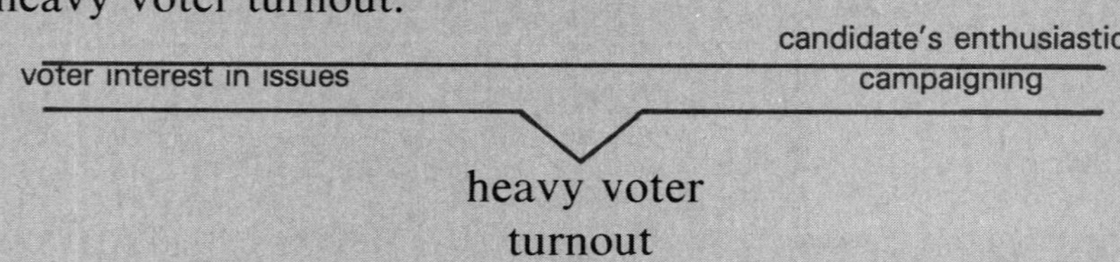

DIAGRAM EACH CAUSE AND EFFECT RELATIONSHIP.

Now read the next paragraph. Then label the sentences that follow as either a cause or an effect.

> Brad thinks his baby sister is very smart because she knows how to get everything she wants without talking. When Leslie is hungry, she bangs on the table and lets out a healthy cry. And if Leslie wants to go to sleep, she waves her hand and say, "Ma, ma, bye-bye." Brad doesn't think it's funny when

Leslie crawls into his room and stares at him while she pulls apart his things. But Brad figured out that that is Leslie's way of asking him to play with her. When Leslie gets tired of crawling and wants to be picked up, she extends both arms and reaches out. Then whoever is with her picks her up. Leslie can say a mouthful with very few words.

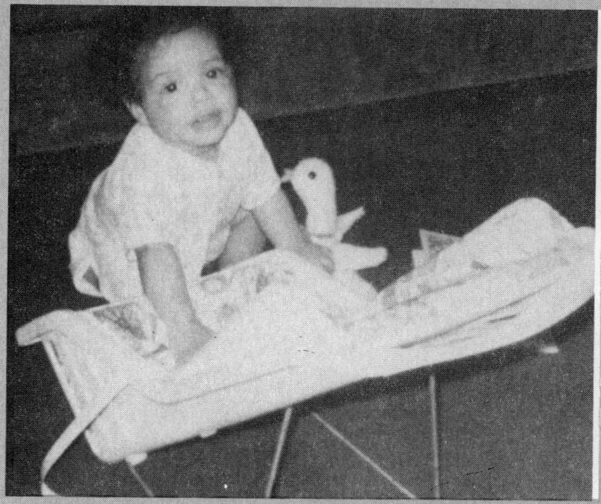

__E__ Leslie pulls apart Brad's things.

__C__ Leslie needs some sleep.

__E__ Leslie bangs on the table and cries.

__C__ Leslie thinks Brad should play with her.

__E__ Leslie waves bye-bye.

__C__ Leslie wants something to eat.

As you read further in Unit 4, you will discover the causes and effects on earth and in space that influence our lives.

RELATING NEW INFORMATION TO WHAT YOU ALREADY KNOW

In Unit 2 you learned that it is important to relate the new information you learn to what you already know. Read the next paragraph. Then answer the question based on what you know about today's energy shortage.

> An electric automobile has been designed that is almost completely pollution-free. It is powered by a battery that can be recharged by merely plugging it into any circuit. Imagine never having to worry about running out of fuel.

USE WHAT YOU KNOW.

1. Why is the invention of such importance? It is pollution free.
2. In what ways would it help reduce energy we are using now? ← It is powered by rechargable battery.
3. Do you think it would be as expensive to operate as the cars we are driving now? No
4. Besides saving energy, in what other ways can the electric automobile benefit the public? Drivers will not run out of fuel; air pollution will be reduced.
5. Do you think our environment would change with the use of the electric car? Why or why not? Answers will vary according to students' knowledge of energy problem.

UNIT 4 Introduction to Earth and Space Science

This is a view of Geiranger Fjord in Norway. Fjords probably are the result of glaciers passing through a valley.

New science words: geologist, hypothesis, theory

Skills developed: constructing a model, observing, inferring

Chapter 1

WHAT IS THE EARTH MADE OF?

How and when did the earth begin to exist? Has the appearance of the earth changed since it was first formed? Is it still changing?

Human beings have lived on earth for a long time. But they are just beginning to realize how little they know about the planet on which they live.

A. How the Solar System Was Formed

After doing this lesson you will be able:

- A1 • to find out what hypotheses are
- A2 • to describe three theories of how the solar system was formed

Have you ever wondered how the earth was formed? Scientists have had many different ideas about how the earth began.
A1 These possible ideas or explanations are called **hypotheses.** They are *not* facts.

In the late 1700s, the French scientist, Pierre Simon Laplace, thought that the earth began as a great cloud of dustlike particles of matter and gas, called a **nebula,** which was A2
spinning rapidly in space. The nebula formed into a "ball," and, as this ball cooled, it grew smaller and spun faster and faster. Small rings of gas and matter were pulled away from the spinning ball. Laplace thought that this ball became the sun and each of the rings of gas and matter became a planet.

The **collision theory** is another idea about A2
the beginning of the earth. This theory states that a near-collision (hitting together) with a passing star caused two long streams of matter to be pulled out from the sun. This matter broke up into small chunks some of which, in time, joined together, to form the planets.

Most of the recent theories about the beginning of our solar system are based on Laplace's idea that in the beginning there was a large spinning cloud of gas and dust. We now know that nebulae—tiny particles

Have students research the relationship between glaciers and fjords. What other formations are produced by glaciers?

Encourage students to discuss each hypothesis and measure each on its own merits.

of matter and gas, mostly hydrogen and helium—exist in space among the stars and
A2 planets.

One of the latest notions of how the solar system was formed is that the large, spinning cloud, in time, flattened into a disk and matter drifted toward the center to form a ball. While the ball and disk of nebulae spun in space, the ball was compressed and became very hot—a million degrees Celsius. Then nuclear fusion took place—the hydrogen was changed into helium releasing huge amounts of energy. And the ball began to shine. It was the "sun."

How did the planets form? Scientists do not agree on this question! But many now think that, with the rapid movement of the nebulae, gravitational and magnetic fields formed and caused the cooler outer parts of the nebulae to form small chunks of rocklike material called **planetesimals.** Over a period of 100 million years, these planetesimals joined together to form the planets.

Check Up A

A1. What do scientists call possible explanations? Hypotheses

A2. Describe briefly three theories on how the solar system was formed.

(1) Laplace's theory: Nebula formed a rotating ball. Small rings of gas and matter were pulled away from the ball. The ball became the sun and the rings became the planets.

(2) Collision theory: A near-collision with a passing star caused long streams of matter to be pulled out from the sun. This matter broke up into small chunks to form the planets.

(3) The latest theory: A large spinning cloud flattened into a disk and matter drifted toward the center to form a ball. Nuclear fusion took place to release energy. The sun was formed.

B. Planet Earth

After doing this lesson you will be able:

- B1 • to name the solid material that covered the earth when it was first formed
- B2 • to explain how the atmosphere was formed
- B3 • to find out what a geologist studies
- B4 • to identify the earth's crust and tell what it is made of
- B5 • to describe where the mantle is and what it is made up of
- B6 • to tell how many layers make up the earth's core and describe each layer
- B7 • to find out what happens to temperature as you go deeper below the earth's surface

Since scientists do not know for sure how the earth began, they do not know exactly what materials formed the earth. But they think that at one period in the earth's history the materials were hot and as they slowly cooled, the size of the earth became smaller.

Some scientists think that friction between the gases and dust particles caused the earth to be hot when it first formed. As the gases and dust cooled, the earth probably changed to a molten (liquid) state.

So far, geologists have not succeeded in drilling into the earth's mantle. Most of what they know about the mantle and the core comes from analyzing seismic waves.

As the surface of the earth slowly cooled, solid material was formed. This solid material
B1 is called **granite.** (You will learn more about granite later in this unit.) Some scientists think that at one time the entire surface of the earth was covered by a thin layer of granite.

The atmosphere around the earth was
B2 probably formed as gases escaped from the molten rock below the granite. After a long time, rain began to fall from the clouds in this atmosphere. As the surface cooled, some of the rain remained on the earth as water. Water began to fill in the low spots in the earth's granite covering.

B3 Scientists who study the earth and what it is made of are called **geologists.** In order to study the inside of the earth they have drilled holes in the earth's surface. They also study the ocean floors to find out what kind of plants and animals lived on the earth a long time ago.

Scientists think the earth is made up of three main layers. These layers are the *crust, mantle* and the *core*.

The outer surface of the solid earth is
B4 called the **crust.** The crust is between 4 and 16 kilometers thick under the oceans, and about 36 to 64 kilometers thick under the continents. The entire surface of the earth is made up of rock. In most places, however, the rock is covered with a layer of loose rock, sand or soil. The thickness of this layer of loose material is not always the same. In some places the loose covering may be hundreds of meters thick while in other places there is no cover material at all.

Because it is easy to study, geologists know more about the crust than any other layer of the earth. They are able to study the rocks and soil that make up the crust and the breaks found in it. From their findings, geologists think that the rocks in the crust are much lighter in weight than the rocks inside the earth.

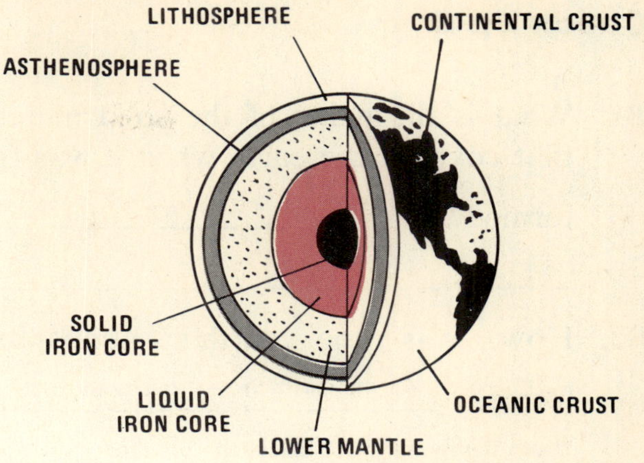

A model of the earth. The lithosphere is a plate of solid rock that "rides" on the partly molten asthenosphere below it.

Below the crust of the earth is a much B5
thicker layer called the **mantle.** It is thought that this layer, about 2,900 kilometers thick, is made up of solid rock. The mantle makes up four-fifths of the inside of the earth. This information comes from studying natural vibrations and artificially-produced vibrations that travel within the earth.

You've often seen the core of an apple. Well, the crust and the mantle of the earth are like layers around a core. The earth's **core** is believed to be about 3,400 kilometers across. Geologists also think that this core is made up of two layers. It is thought that B6
the outer layer makes up about 2,100 kilometers of the core while the inner layer is about 1,300 kilometers across.

The outer core is thought to be mainly B6
iron and nickel in a kind of liquid state like molten rock. The inner core appears to be iron and nickel in a solid state.

Temperatures deep in the earth are thought to be very high. Three kilometers down in a mine the temperature is 51°C. At a depth of 32 kilometers, the temperature reaches about 870°C. However, geologists B7
think that the temperature of the mantle is not greater than 2,190°C., and the temperature of the core is not greater than 5,524°C.

The thin region between the earth's crust and the mantle is known as the *Mohorovicic Discontinuity*. It was named after the Yugoslavian seismologist who discovered it in 1909. It is more commonly call the *Moho*.

Check Up B

B1. What is the name of the solid material that covered the earth when it was first formed? Granite

B2. How was the earth's atmosphere formed? As the earth cooled, gases escaped from the molten rock below the granite.

B3. What does a geologist study? The earth and what it's made of.

B4. Which part of the earth is the crust? What is it made of? The other surface of the earth is crust. It is made of rock.

B5. Where is the earth's mantle located? What is it made of? The mantle is located below the crust. It is made of solid rock.

B6. How many layers are there in the earth's core? What is each layer made of? The earth's core is made of two layers. The outer core is thought to be chiefly iron and nickel in a liquid state. The inner core appears to be iron and nickel in a solid state.

B7. How does temperature change as you go deeper below the earth's surface? The temperature increases as you go deeper.

As many as 500 earthquakes and tremors occur each day.

C. Earthquakes

After doing this lesson you will be able:

C1 • to tell what an earthquake is and what causes earthquakes

C2 • to describe what the seismograph tells about earthquake waves

The crust of the earth is always changing. These changes may be caused by **earthquakes.** An earthquake is a shaking or C1
trembling of the earth's solid crust. Earthquakes are caused by either the sudden movement of large blocks of solid rock along *cracks* in the crust or by changes beneath the earth's surface. The vibrations of an earthquake set up waves that transmit (carry) energy. These waves travel through the earth.

Scientists use an instrument (tool) called the **seismograph** to record earthquake waves. By studying records made by seismographs, C2
geologists are able to find out where the earthquake began, to study the direction the waves traveled, the speed at which the waves moved, and how long the earthquake lasted. These studies help scientists to answer some questions about the earth's interior.

Check Up C

C1. What is an earthquake? What causes earthquakes? An earthquake is a shaking or trembling of the earth's crust. It is caused by the sudden movement of large blocks of solid rock along cracks in the crust.

C2. What do seismograph records tell scientists about earthquakes? (1) The origin of the earthquake; (2) The direction of the waves travel; (3) The speed at which the waves move; (4) The time the earthquake lasted.

Unit 4, Chapter 1
REVIEW

Matching Exercise Match each term in **Column A** with its description in **Column B**. Write the letter of the description in the space beside the correct term.

	Column A	*Column B*
e	**1.** crust	**a.** scientist that studies the earth and what it is made of
c	**2.** seismograph	**b.** cloud of dustlike particles of matter and gas
j	**3.** planetesimal	**c.** tool that records earthquake waves
b	**4.** nebula	**d.** shaking or trembling of the earth's crust
a	**5.** geologist	**e.** outer layer of the earth
d	**6.** earthquake	**f.** iron and nickel are in a solid state here
i	**7.** outer core	**g.** solar system resulted from a near collision of the sun with another star
f	**8.** inner core	**h.** possible explanation
g	**9.** collision theory	**i.** iron and nickel are in a liquid state
h	**10.** hypothesis	**j.** chunk of rocklike material that came from nebulae

True/False In the space provided, write **T** if a statement is true. If it is false, change the underlined word or words to make the statement true.

crust	**1.** The core is the outer layer of the Earth.
atmosphere	**2.** The mantle was formed by gases escaping from molten rock.
hypothesis	**3.** A nebula is a possible explanation to a problem.
T	**4.** According to Laplace's theory, the solar system began as a great cloud of dust and gas.
T	**5.** According to the collision theory, the solar system resulted from a near collision of the sun with another star.
outer core	**6.** In the inner core, iron and nickel are in a liquid state.
crust	**7.** The mantle of the earth is located between the inner core and the outer core.
An earthquake	**8.** A seismograph is the shaking or trembling of the earth's crust.
seismograph	**9.** A nebula is a tool that records earthquake waves.
T	**10.** As you go deeper below the earth's surface, the temperature increases.
solid rock	**11.** The mantle is made up of liquid iron.
T	**12.** A planetisimal is a chunk of rocklike material in space.
granite	**13.** Scientists think that at one time the entire surface of the earth was covered with planetisimals.
lithosphere	**14.** The atmosphere is a plate of solid rock that rides on the asthenosphere below it.
crust	**15.** The mantle of the earth is between 4 and 16 kilometers thick under the oceans.

Multiple Choice Exercise Circle the letter of the word or phrase that best completes each statement.

1. Possible answers to scientific questions are called **(a)** facts **(b)** proofs **(c)** experiments **(d)** hypotheses.
2. The collision theory of the beginning of the earth states that a near-collision with a passing star caused two long streams of matter to be pulled from **(a)** the moon **(b)** a nebula **(c)** Jupiter **(d)** the sun.
3. The outer surface of the earth is the **(a)** mantle **(b)** core **(c)** asthenosphere **(d)** crust.
4. In Laplace's theory, as the spinning ball cooled it **(a)** grew smaller and spun faster **(b)** grew larger and lost speed **(c)** grew smaller and spun faster **(d)** grew smaller and lost speed.
5. A nebula is a **(a)** large planet **(b)** small star **(c)** cloud of dustlike particles and gas **(d)** cloud of dust.
6. According to the latest theory of the beginning of the solar system, the sun was formed by **(a)** nuclear fusion **(b)** a near-collision **(c)** nuclear fission **(d)** extremely low temperatures.
7. Some scientists think that the surface of the earth was originally covered by **(a)** solid iron **(b)** granite **(c)** molten iron **(d)** a mantle.
8. The outer core of the earth is thought to be made of **(a)** solid iron and nickel **(b)** molten iron and nickel **(c)** granite **(d)** more than half of the core.
9. The mantle of the earth is thought to be made of **(a)** solid rock **(b)** solid iron and nickel **(c)** rocks and soil **(d)** molten iron and nickel.
10. A seismograph records **(a)** hot spots **(b)** the temperatures of the core of the earth **(c)** earthquake waves **(d)** the changes in the make-up of the earth's surface.

Application Exercise Label the layers of the earth indicated in the diagram below. Write your answers in the provided spaces.

1. lithosphere
2. asthenosphere
3. inner core
4. outer core
5. lower mantle

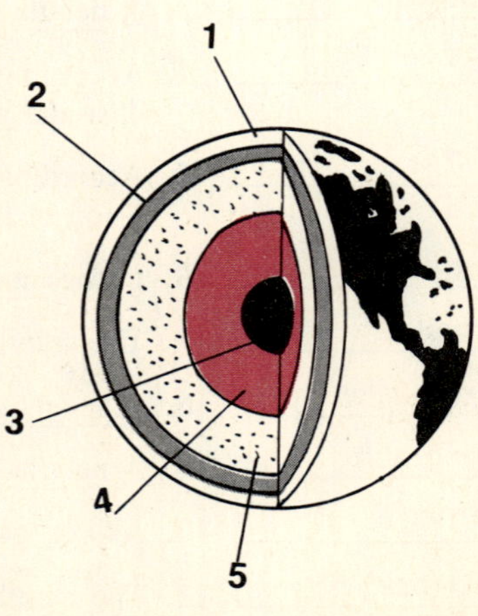

Part of the earth's crust is made of rocks. Where did the rocks in this photo come from?

They probably come from the mountains, in the background, as the products of weathering and erosion.

New science words: mineral, minerologist, radioactive

Skills developed: observing, describing, comparing, classifying

Chapter 2

ROCKS AND MINERALS

Have you ever taken a walk and picked up rocks, stones and pebbles? Have you wondered what rocks are made of? Geologists and rock collectors are always looking for and identifying different kinds of rocks. They know that the earth is a mixture of many **minerals** and **rocks.**

Display a mineral collection for students to examine.

A. Minerals

After doing this lesson you will be able:

A1 • to tell what a mineral is and how many have been found in the earth's crust

A2 • to find out what a mineralogist does

A3 • to list six physical properties used to identify minerals

A4 • to give the name of the scale used to measure the hardness of minerals

A5 • to name the two most abundant minerals in the earth's crust

The earth is made up of many different **elements.** The elements iron, aluminum, oxygen and silicon make up about nine-tenths of the mass of the earth's crust. You are familar with iron and aluminum because you use them in your daily lives. You also know about oxygen which you breathe in with air. You probably have never seen silicon because it is usually found chemically combined with oxygen. When silicon combines with oxygen, the *compound* (combination) is called **silica.** Sand is one example of silica. Few elements are found in nature in a pure form. They are almost always combined with other elements. For example, gold is an element found in a pure form. Iron is an element always combined with one or more other elements.

What is a **mineral?** A mineral is a single element or a combination of elements occurring naturally in the earth. A mineral is *inorganic;* that is, it does not contain carbon. Carbon comes from plants and animals that A1

There are 92 naturally occurring elements in the earth's crust. Eight common elements make up *98.5%* of the material found in the crust: oxygen, silicon, aluminum, iron, calcium, sodium, potassium and magnesium.

If you do not have a mineral collection for your students to examine, use a rock and mineral identification book that has color photographs.

lived millions of years ago. A mineral's chemical makeup is definite and never changes.

A1 About 2,000 minerals have been found in
A2 the earth's crust by **mineralogists,** scientists who study and identify minerals. Of these 2,000 known minerals, only ten make up most of the earth's rocks.

Because of the differences in their chemical makeup and the arrangement of their atoms, minerals have definite physical and chemical properties which identify them.
A3 Some of the physical properties are: (1) color, (2) luster, (3) streak, (4) hardness, (5) crystal shape and (6) cleavage.

1. Color. A mineral that is made up of only one element may have a **color** that never changes. Minerals of this kind are copper, gold and sulfur. Color alone is not a good way of identifying minerals because many minerals have the same color. Also, an impurity, such as a trace (little bit) of another element, can hide the true color of a mineral.

For example, pure quartz (silicon dioxide) is a colorless mineral. A trace of another element may make it green, pink or blue. Quartz with a little manganese dissolved in it is a purple-colored amethyst. Corundum (aluminum oxide) is a mineral which may be blue, pink, gray, white, green, red or brown. Blue corundum is a sapphire; red corundum is a ruby.

2. The **luster** of a mineral means its shine or lack of shine. Luster depends on how a mineral reflects light. If a mineral shines like a polished metal, its luster is called **metallic.** If it does not shine like a polished metal, it has a **nonmetallic luster.** Some kinds of nonmetallic luster are pearly, silky, greasy or dull.

3. Streak is the color of a finely-powdered mineral. The color of the powder may be different from the color of the solid mineral. Iron pyrite, a golden-yellow mineral, has a greenish-black streak. Although the color of a mineral may change greatly, its streak is usually the same. For example, the mineral hematite (iron oxide), is red or black in color. Its streak, however, is always reddish-brown. One way to identify minerals is the **streak test.** Rub an unknown mineral across a rough piece of porcelain tile. The color of the powder left behind is the mineral's streak.

4. The **hardness** of a mineral is measured by how it scratches or is scratched by a harder object. Hardness can be measured by scratching one mineral with another mineral, a steel knife blade, your fingernail or a shiny copper penny. The *diamond* is the hardest mineral because it can scratch all other minerals. *Talc* is the softest mineral because it can be easily scratched by all other minerals as well as by a fingernail.

A scale of hardness was devised (invented) by the mineralogist, Friedrich Mohs. A4
Mohs' Scale of Hardness is used to measure the hardness of minerals. Ten minerals are listed in the scale in order of hardness. Number 1 is the softest mineral and number 10 is the hardest:

1. talc	**6.** feldspar
2. gypsum	**7.** quartz
3. calcite	**8.** topaz
4. fluorite	**9.** corundum
5. apatite	**10.** diamond

If you have these minerals, you can identify the hardness of many other minerals. If a mineral is harder than **3** but not harder than **5**, it must have the hardness of **4.** Even if you don't have these minerals, you can get some idea of a mineral's hardness knowing that your fingernail has a hardness of about **2½,** a shiny copper penny about **3,** and a steel knife blade about **5.**

5. The **crystal shape** of a mineral often helps to identify it. A crystal is formed by the ar-

Minerals are divided into families according to chemical composition. Quartz and feldspar belong to the *silicates*. The other rock forming mineral groups are the *oxides* (magnetite, corundum), the *sulfides* (pyrite, galena), the *carbonates* (calcite), the *sulfates* (gypsum) and the *mineraloids* (bauxite, opal).

Use mineral specimens to demonstrate crystal shape.

Halite crystal

rangement (order) of the atoms or molecules of a mineral in a regular geometric shape (design). Each mineral always has the same crystal form.

Crystals of minerals may have cubed (four-sided), hexagonal (six-sided) or other shapes. In a mineral that has sulfur combined with another element the crystals are shaped like cubes. Pyrite has iron combined with sulfur. Pyrite is called "fool's gold" because it has the color of gold. However, gold is not found naturally in the shape of cubes as iron pyrite is.

6. Cleavage is the ability of a mineral to split or break apart along flat, smooth surfaces. Calcite splits in three directions, while mica splits in one direction. Cleavage does not happen in all minerals. In some minerals there is **fracture.** A fracture surface can be curved, splintery (like slivers) or irregular.

An important group of rock-forming minerals is the **silicates.** The silicates are made up of silicon and oxygen combined with one or more metals such as potassium, sodium and
A5 aluminum. **Feldspar,** an aluminum silicate mineral, is the most abundant (exists in great quantities) mineral in the earth's crust. It is used in making dishes and scouring, or scrubbing powders.

Use some of the minerals mentioned above to demonstrate cleavage.

Another abundant mineral is **quartz.** A A5
very large amount of quartz is found in the earth's crust. Sand is mostly made up of quartz crystals. Pure quartz is without color. Quartz is used in making glass, sandpaper and concrete. Colored quartz crystals are used as semi-precious stones. A special quartz is used in lenses and prisms.

Discover for yourself
how you can test some of the properties of minerals.

PURPOSE: To identify the physical properties of some common minerals • To develop skills of observation, comparison and description

You need: samples of quartz, mica, talc, pyrite, calcite, feldspar and other rock; shiny copper penny, knife, streak plate or unglazed tile

a. Scratch each sample with your fingernail (hardness of about 2½), the penny (hardness of about 3), and the knife (hardness of about 5). Write your findings on the chart on page 186.

1. Which minerals can be scratched by your fingernail? Talc, mica

2. Which minerals cannot be scratched by the knife? Quartz, feldspar, pyrite

b. Look at each sample. Write the color of each mineral on the chart.

c. Rub each sample over the streak plate or tile. Record the color of each streak on the chart.

How does the streak color compare with the color of the mineral? They are not always the same.

d. Close your eyes and feel each sample. Does the mineral feel cold, hard, greasy, etc.? Record your findings on the chart.

Check Up A

A1. What is a mineral? About how many have been found in the earth's crust?
A mineral is a single element or a combination of elements occurring naturally in the earth's crust. About 2000 have been found.

A2. What does a mineralogist do? Studies and identifies minerals.

A3. List six physical properties used to identify minerals.

(1) color

(2) luster

(3) streak

(4) hardness

(5) crystal shape

(6) cleavage

A4. Give the name of the scale used to measure the hardness of minerals. Moh's Scale of Hardness

Minerals	Hardness	Streak	Color	Feel
quartz	7	colorless	Answers will vary.	Answers will vary.
mica	2-3	colorless		
talc	1	white		
pyrite	6	greenish-black		
calcite	3	colorless		
feldspar	6	colorless		

A5. Name the two most abundant minerals in the earth's crust. ____________

Feldspar and quartz.

____

B. Rocks

After doing this lesson you will be able:

- B1 • to describe what rocks are and how rocks are classified
- B2 • to name the three classes of rocks and tell how each class is formed
- B3 • to name three kinds of igneous rocks and tell which is most abundant
- B4 • to explain why sedimentary rocks can be easily identified and to name five kinds of sedimentary rocks
- B5 • to name at least five kinds of metamorphic rocks
- B6 • to describe two kinds of artificial rocks

What is a rock? Now that you know what minerals are, it will be easier for you to understand what a rock is. You have learned that there are more than 2,000 different minerals in the earth's crust. This great number of minerals can be combined in many different ways. Because of this, there are many different kinds of rocks. This is why rocks have such a large number of different characteristics. Rocks differ in color, hardness and many other ways. Most rocks, however, have one characteristic in common: they are
B1 **mixtures of minerals** in that they have no fixed (definite) chemical makeup.

If you look closely at the crust of the earth, you will see that there is solid rock beneath the loose covering of sand and soil. This solid rock is called **bedrock.** The same kind of bedrock is usually found all over a particular area. Bedrock can be any kind of rock such as granite, slate, marble and sandstone.

How were rocks formed? Although there are many kinds of rocks in the earth's crust,

Words to review: classify, originate

geologists (scientists who study rocks) have been able to tell us how rocks were first formed. All rocks have been classified into three large groups based on where and how B2
they were formed. The three groups are: (1) igneous rocks, (2) sedimentary rocks and B1
(3) metamorphic rocks.

1. Igneous rocks. The earth's crust was formed when the hot, molten liquid of which it was made up cooled and hardened. The crust does not have the same thickness all over. The crust is also not as strong all over the earth. In some of the weak places, the hot, molten rock material called **magma** broke through the earth's crust and flowed onto the surface. The magma cooled on the earth's surface and became solid rock. Rocks that are formed from the cooling and hardening of molten magma are called **igneous rocks.** In the beginning of the earth's history, the only rocks on the earth were igneous rocks.

2. Sedimentary rocks. From the time the first rocks were formed, large rocks have been broken down into smaller rocks or particles. Sometimes these rock particles, called **sediments,** are cemented (glued) together in water to form a new rock. Rocks formed in this way B2
are called **sedimentary rocks.** Sedimentary rocks are also formed from materials dissolved and deposited (laid down) in the water or from the remains of plants and animals.

3. Metamorphic rocks are formed from igneous and sedimentary rocks. Because of heat B2
or pressure the original rocks changed into a new kind of rock.

Igneous rocks are made up of many different minerals. However, most igneous rocks are made up of the minerals quartz and feldspar. The minerals in igneous rock can be identified by either their crystal formation, their cleavage or chemical makeup. Granite, obsidian and basalt are examples of igneous B3
rocks.

Quartz and feldspar, two igneous rocks, are composed mainly of *oxygen* and *silicon*, the two most common elements in the earth's crust.

Steamboat rock is a red sedimentary rock formation in Oak Creek Canyon, Arizona. What do you see that tells you this is sedimentary rock?

Most sedimentary rocks are stratified or layered. Have students study the photograph. Ask: How do you think this rock was originally formed? (Layers of sediment were deposited by water, compacted by their own weight, and cemented by chemicals.) How did it get its present shape? (Probably by weathering and wind erosion.)

B3 1. **Granite** is the most abundant igneous rock on the earth. Granite was formed from magma which cooled very *slowly* beneath the earth's surface. Quartz and feldspar are always found in granite. Other minerals such as mica are almost always found in granite. Granite is light-colored and has large mineral crystals. Granite was formed below the earth's surface. So granite found on the surface has been exposed by weathering and the faulting and folding of mountains. You will learn more about these processes later. Granite is a good building stone because it cannot be damaged by temperature changes.

Granite

 Obtain samples of the rocks listed on these pages for students to examine.

A set of rocks native to your state may be available from your State Geology Survey Office.

2. Obsidian, another type of igneous rock, was formed when lava cooled very quickly. It is dark in color and looks glassy. Although it is usually made up of the same minerals as granite, obsidian has no crystals because the lava from which it was formed cooled very fast. Obsidian is very hard but can be split easily. Long ago people used obsidian for making sharp tools such as knives and arrowheads.

3. Basalt was also formed from lava. It does not contain quartz. It has a dark color and has fine grains unlike granite or obsidian. It can be dark green or black, depending on the minerals in it. Large amounts of basalt are found where volcanoes erupt such as in the Hawaiian Islands.

As you have seen, **sedimentary rocks** are made up of different materials. The rocks can contain pieces of other rocks, pebbles, sand and clay. They may be formed from minerals that were dissolved in water, or they may be made up of the remains of plants and animals. The particles or sediments that make up the rocks are cemented together by natural cements found in the water. When dissolved in water, minerals, such as quartz and lime, act as **natural cements.** Most minerals found in sedimentary rocks have no crystals.

As a river flows into an ocean or a lake, it carries a great amount of sediment. The sediment is usually made up of silt, clay, sand, gravel, pebbles and rocks. The first sediments dropped by the river are the heaviest materials, such as the rocks and pebbles. The next sediments dropped are gravel, silt and clay. Some of these sediments become cemented together to form rocks. Sedimen-
B4 tary rocks can be easily identified because they are usually formed in layers. These layers form because of the way different kinds of sediments dropped out of the running water. Fossil remains are found in sedimentary rocks, especially in shale, limestone and sandstone.

1. Conglomerate is made up of pebbles, B4
pieces of rock and gravel cemented together. Quartz pebbles are most often found in conglomerate. Many different kinds of conglomerate are used as building stone. They all look like concrete because of the pebbles and sand.

2. Sandstone is found in large amounts on B4
the earth. Can you tell why? Sandstone is made up of grains of sand (quartz) cemented together. Sandstone may be white, red or brown. The color of the sandstone depends on the mineral that cements the grains of sand together. Sandstone is used in glassmaking and in making sandpaper.

3. Shale is made up of fine particles of silt B4
and clay. When wet, it looks and smells like mud. Shale is smooth, soft and easily broken. It weathers (breaks down) fast when exposed to air. Shale is found in many colors, depending on the color of the sediments that formed the rock. It is used in making cement and bricks.

4. Limestone. Some types of limestone are B4
formed from the remains of the shells of sea animals. When these animals die, their shells become cemented and pressed together to

Limestone

Words to review: silt; fossil

form the rock. Since sea animals use the calcite dissolved in the water to make their shells, the limestone made from their shells is mostly calcite. Limestone is used in making cement. Farmers and gardeners use it in powdered form to cut down the amount of acid in the soil.

B4 **5. Coal.** When plants die in great numbers in shallow swamp waters, they turn into a brownish mass called **peat.** Further decay and pressure change the peat into **bituminous** (soft) **coal.** Changing of plant life into soft coal usually happens in layers of shale and sandstone.

Heat, pressure and chemical action are the main causes of changing igneous and sedimentary rocks into **metamorphic rocks.** In some ways they look like the rocks from which they were formed, but they also have some different characteristics. Most metamorphic rocks have larger crystals and usually new minerals in them. A few examples
B5 of metamorphic rocks are (1) marble, (2) slate, (3) schist, (4) quartzite, (5) anthracite coal and (6) gneiss.

1. Marble looks like the limestone from which it was formed, but it is harder and has more crystals in it. Pure marble is white, but other minerals mixed in may cause it to have many beautiful colors. In the United States marble is found in Alabama, Vermont, Georgia and Tennessee.

2. Slate looks like the shale from which it was formed, but it does not split like shale. It is much harder than shale. It splits into thin sheets, and is used for roofs and sidewalks.

3. Schist also looks like shale, but it has parallel strips of minerals which easily undergo cleavage.

4. Quartzite is harder than the sandstone from which it was formed. The pores (tiny holes) of original sandstone were filled with quartz crystals. It is used as building material.

5. Anthracite coal is black like bituminous coal, but harder and shinier. It is used for fuel.

6. Gneiss may be formed from many different rocks such as granite, sandstone, conglomerate and others. It has alternating strips of light and dark minerals. It is used for building and monuments.

Artificial rocks are manufactured or made by people. They are used mainly in the build-
ing industry. **Cement,** an artificial rock, is B6

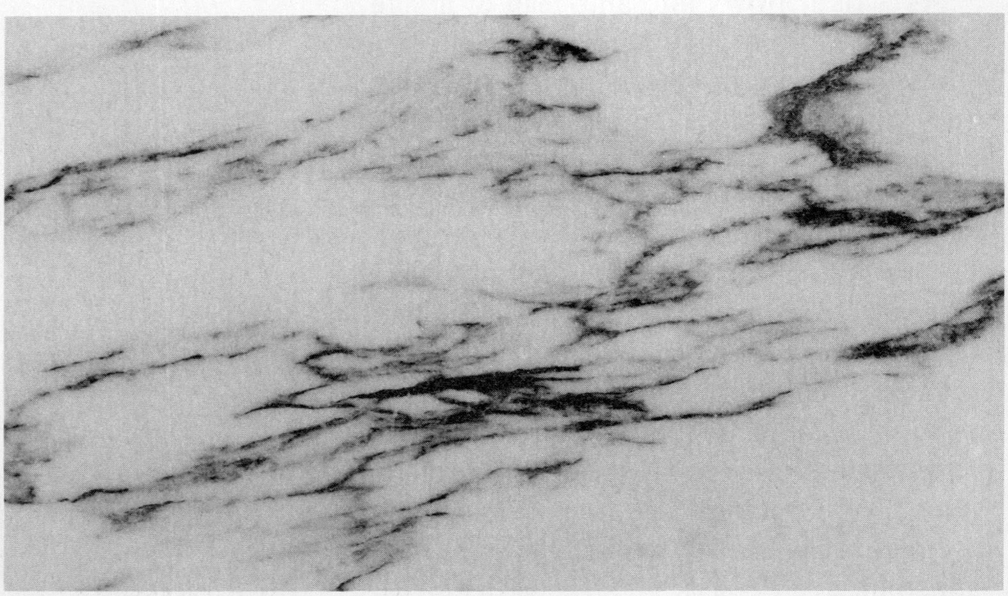

Marble is a hard metamorphic rock. Do you think that this is pure marble?

Encourage students to think of other examples of artificial rock: brick, blacktop, glass, porcelain.

made from limestone and shale crushed into small pieces which are then heated to a very
B6 high temperature. **Concrete,** another artificial rock, is made by mixing cement, sand and crushed rocks. Cement and concrete can be poured into any shape.

Discover for yourself

how you can identify the rocks in your neighborhood.

PURPOSE: To classify some common rocks • To develop skills of observation, comparison and classification

You need: rocks, labels, guide book to rocks in your area

a. Pick up some interesting rocks in your neighborhood. Label each rock with a specimen number and the place where you found it.

b. Identify the rocks by comparing them to the pictures in a guide book of rocks of your area. Label each rock you identify with a name. Group the rocks into three kinds: igneous, sedimentary and metamorphic.

1. Which rocks are igneous? ______
Answers will vary.

2. Which rocks are sedimentary? ______
Answers will vary.

3. Which rocks are metamorphic? ______
Answers will vary.

4. To which group do most of the rocks belong? Usually sedimentary.

Check Up B

B1. What are rocks? How are rocks classified? Rocks are mixtures of minerals. They are classified according to where and how they were formed.

B2. Name the three classes of rock and describe how each is formed. Igneous — formed from the cooling and solidifying of molten magma. Sedimentary — formed from sediments cemented together in water. Metamorphic — formed from heat and pressure on igneous and sedimentary rocks.

B3. Name the three kinds of igneous rock. Which is most abundant? Granite, obsidian, basalt. Granite is the most abundant.

B4. Why can sedimentary rocks be easily identified? Name five kinds of sedimentary rocks. Sedimentary rocks are usually formed in layers. Conglomerate, sandstone, shale, limestone, coal.

B5. Name five kinds of metamorphic rock. Marble, slate, schist, quartzite, anthracite coal, gneiss.

Describe two kinds of artificial rock. ___

Cement — made from crushed limestone and shale. Concrete — made by mixing cement, sand and crushed rocks.

C. How Old Is Planet Earth?

After doing this lesson you will be able:

C1 • to name three methods scientists use to figure out the age of the earth and tell which one is most used

C2 • to explain what the term half-life means

C3 • to indicate which isotope is most often used in radioactive dating and give its half-life

Geologists who study the history of the earth can describe very correctly what has happened on the earth in the past five billion years. They know about the plants, animals, weather, climate and the changes that took place on the earth's surface. And they can tell how old the earth is.

Before more exact modern methods were discovered, geologists used a number of ways to figure out the age of the earth. One way
C1 was based on how fast sediments dropped out of running water to form sedimentary rocks. Scientists estimate that it took between 12,000 and 30,000 years for enough sediment to collect to form a one-meter layer of sedimentary rock. By multiplying the number of meters of sedimentary rock on the earth by the time it took to form 1 meter, scientists figured out the age of the earth.

Another method was based on the amount
C1 of salt in the oceans. Scientists collected ocean water from all parts of the earth and figured out from these samples how much salt the oceans contained. Each year they collected samples and figured out how much salt had been added to the oceans by the rivers. Dividing the total amount of salt by the yearly increase of salt, they came up with an estimate of the age of the earth.

Today, the age of the earth is determined
mainly by using **radioactive elements.** These C1
are elements which give off atomic particles and decay (break down) into other elements.
The element most often used is the radio- C3
active **isotope** of uranium, **uranium 238.** (Isotopes are different forms of the same element.)

Radioactive elements, like uranium, decay and give off particles and rays at a steady rate from the nucleus of the atom. The decay of a radioactive element causes the formation of lighter elements. There are many radioactive elements in rocks. Because scientists have learned how to use these elements to find out the age of rocks, fossils and geologic time, they call them radioactive "clocks."

Each radioactive element breaks down or decays at its own steady rate. In many cases this rate is very slow. When uranium breaks
down, it decays so slowly that it takes $4\frac{1}{2}$ C3
billion years for *half* of the atoms to change into lead. The time it takes for half of the
atoms of a radioactive element to decay is its C2
half-life.

The decay of radioactive uranium ends in the formation of lead, helium and a great amount of energy. The reaction can be written as a word equation:

$$\text{uranium} \rightarrow \text{lead} + \text{helium} + \text{energy}$$

The helium, a gas, and the energy are given off into the atmosphere. Only the lead remains.

How do scientists use this information to figure out the age of rocks? First they find some igneous rocks which contain the element uranium. Then they figure out the amount of uranium in the rock and the *amount of the lead* formed from the *decay of the uranium.* The more lead in the rock, the older the rock is. Geologists have worked out a formula to help them determine almost ex-

The oldest rocks discovered in the U.S. are approximately 3 billion years old.

actly the age of the rocks. By using the uranium-to-lead method, geologists have discovered rocks that were formed almost 4 billion years ago. The oldest rocks were discovered in Africa.

The age of some meteors (rock-like objects that fall to earth from space) has also been figured out by the use of this **radiation dating process.** This is possible because some meteors contain uranium and lead. Scientists have estimated these meteors to be about 4½ billion years old. If these meteors were formed at the same time as the solar system, then the earth must also be about 4½ billion years old.

The uranium isotope which can be used for geologic dating is found only in igneous rocks. So scientists had to find other radioactive isotopes which could be used to discover the age of other rocks and objects.

A new method of radiation dating makes use of a radioactive isotope of potassium. When this **potassium isotope** decays it produces the gas argon which escapes into the atmosphere. The amount of potassium isotope left in the rock is then used to figure out the age of the rock. Potassium is a good tool to use for dating because it is found in many rocks other than those of the igneous group.

Check Up C

C1. Name three methods scientists use to figure out the age of the earth. Which method is used most? (1) Rate of sediment deposit; (2) Rate of salt deposit in the oceans; (3) Use of radioactive elements.

C2. What does the term *half-life* mean? The time it takes for half the atoms of a radioactive element to decay.

C3. Which isotope is most often used in radioactive dating? What is its half-life? Uranium 238. Its half life is 4½ billion years.

If these meteors are estimated to be 4¼ billion years old, this means that ½ of the uranium has decayed and ½ remains. Ask: How old would the meteors be if ¼ of the uranium remained? (9 billion years old)

D. Dating Living Things of the Past

After doing this lesson you will be able:

D • to name the isotope used to find out the age of things that lived long ago

Is everything on earth billions of years old? The answer, of course, is no. Certainly there are no living plants or animals that are that old. How do scientists find out the age of things that lived long, long ago?

Scientists have discovered a radiation dating method that uses radioactive **carbon-14.** D Radioactive carbon is always being formed in the upper atmosphere of the earth because of nuclear reactions. These reactions cause the nitrogen atoms in the air to be changed to carbon-14. The carbon-14 then chemically combines with oxygen to form radioactive carbon dioxide.

Green plants take in the carbon dioxide for use in the process of making food. They make carbohydrates which they use as food or which may be eaten by animals. Thus, the protoplasm in the cells of all living plants and animals has a small amount of carbon-14. The amount of carbon-14 compared to the amount of ordinary carbon stays the same as long as the plant or animal is alive. When the plant or animal dies, the carbon-14 begins to decay at a steady rate.

Carbon dating is especially useful in dating fossils (remains) of things that were once alive. All living matter contains carbon-14 and carbon-12. As carbon-14 decays, it changes to nitrogen-14. But, carbon-12 does not *decay. So by figuring out the ratio (proportion) of carbon-14 to carbon-12 in a fossil, a scientist is able to figure out how long ago it stopped living—or its age.*

The radioactive carbon-14 method can be used to date fossils from a few thousand to about 35,000 years old—such as this fossil skull. The facial parts of this skull were found by two famous anthropologists, Louis and Mary Leakey, at Olduvai Gorge, Tanzania. (Anthropologists are scientists who study the origin and development of the human race, Homo sapiens.) This skull, named Zinjanthropus's *skull, is thought not to be an ancestor of Homo sapiens. How is this skull different from yours?*

The forehead is slanted back, implying a very small brain. Cheek bones protrude, as does the lower jaw.

The half-life of carbon-14 is 5,700 years.

When carbon-14 decays, it gives off rays at a steady rate just like uranium. By measuring the amount of radioactive carbon left in a dead plant or animal, it is possible to tell how long ago the plant or animal died. Radioactive carbon dating is right from a few thousand to 35,000 years. After 35,000 years, the amount of radioactive carbon left is too little to be measured.

Carbon-14 dating has been used to find out the time of the last Ice Age, the age of parts of the Atlantic Ocean and the age of the wooden tools used by ancient peoples.

Check Up D

Name the isotope used to find out the age of things that lived long ago. Carbon-14

Unit 4, Chapter 2
REVIEW

Matching Exercise Match each term in **Column A** with its description in **Column B**. Write the letter of the description in the space beside the correct term.

	Column A	*Column B*
i	**1.** streak	**a.** most abundant mineral in the earth's crust
j	**2.** talc	**b.** formed from the cooling and hardening of magmas
f	**3.** diamond	**c.** most abundant igneous rock on earth
a	**4.** feldspar	**d.** scientist who studies and identifies minerals
h	**5.** cleavage	**e.** time it takes for one-half of the atoms of a radioactive element to decay
b	**6.** igneous rocks	**f.** hardest mineral
g	**7.** sedimentary rocks	**g.** formed from the cementing together of rock particles
c	**8.** granite	**h.** ability of a mineral to break apart along flat, smooth surfaces
e	**9.** half-life	**i.** color of a finely powdered mineral
d	**10.** minerologist	**j.** softest mineral

True/False In the space provided, write **T** if a statement is true. If it is false, change the underlined word or words to make the statement true.

Cleavage	**1.** Fracture is the ability of a mineral to split or break along flat, smooth surfaces.
inorganic	**2.** Minerals are organic.
luster	**3.** The shine, or lack of shine, of a mineral is its color.
a sedimentary	**4.** Conglomerate is an example of an igneous rock.
T	**5.** A mineral's color may be different than its streak.
T	**6.** Metamorphic rocks are formed from igneous or sedimentary rocks that have changed into a new kind of rock.
hardest	**7.** Diamond is the softest mineral.
Granite	**8.** Feldspar is the most abundant igneous rock on Earth.
geologists	**9.** Scientists who study rocks are called minerologists.
Talc	**10.** Gypsum is the softest mineral.
T	**11.** Feldspar is a silicate.
a sedimentary	**12.** Shale is an example of a metamorphic rock.

T 13. Slate is an example of a metamorphic rock.

below six 14. A mineral with a hardness of six can scratch all minerals with a hardness above six.

2,000 15. The earth's crust is made up of about 200 different minerals.

Multiple Choice Exercise Circle the letter of the word or phrase that best completes each statement.

1. The color of a mineral when it is powdered is the mineral's **(a)** luster **(b)** hardness **(c)** fracture **(d)** streak.
2. Gypsum, with a hardness of 2, can scratch **(a)** feldspar **(b)** diamond **(c)** quartz **(d)** talc.
3. How a mineral scratches, or is scratched by, other minerals determines the **(a)** fracture of the mineral **(b)** streak of the mineral **(c)** hardness of the mineral **(d)** cleavage of the mineral.
4. The most abundant mineral in the earth's crust is **(a)** granite **(b)** feldspar **(c)** aluminum **(d)** talc.
5. The most abundant rock on Earth is **(a)** granite **(b)** feldspar **(c)** coal **(d)** quartz.
6. The arrangement of the atoms or molecules in a mineral determines its **(a)** streak **(b)** color **(c)** crystal shape **(d)** luster.
7. Rocks formed from larger rocks that have broken down into smaller rocks or particles are **(a)** igneous rocks **(b)** sedimentary rocks **(c)** metamorphic rocks **(d)** organic rocks.
8. Granite, obsidian and basalt are examples of **(a)** igneous rocks **(b)** sedimentary rocks **(c)** metamorphic rocks **(d)** organic rocks.
9. Fossil remains are most likely to be found in **(a)** igneous rocks **(b)** metamorphic rocks **(c)** sedimentary rocks **(d)** marble.
10. An example of an artificial rock is **(a)** slate **(b)** basalt **(c)** conglomerate **(d)** cement.

Application Exercise Label each of the following rocks as **igneous, sedimentary** or **metamorphic**. Write your answers in the provided spaces.

sedimentary 1. shale

metamorphic 2. marble

igneous 3. obsidian

metamorphic 4. slate

sedimentary 5. conglomerate

igneous 6. granite

igneous 7. basalt

sedimentary 8. sandstone

metamorphic 9. quartzite

sedimentary 10. limestone

This is Mount St. Helens sending steam and ash into the air. This volcano began to erupt in March, 1980.

New science words: erosion, groundwater, weathering

Skills developed: observing, inferring

Chapter 3

THE EARTH'S CHANGING SURFACE

A. The Forces of Change

After doing this lesson you will be able:

A1 • to name at least four things that cause earth's surface to change

A2 • to name two kinds of forces that are always changing the earth's surface and explain how they differ

Scientists think that the earth is between *4 and 5 billion years old.* If this is so, does it mean that all the parts of the earth such as mountains, waterfalls, valleys and canyons are as old as the earth? Does it mean that these features, whatever their age, are all the same age? In this chapter you will learn that the answer to both questions is no!

The earth's surface keeps on changing both in size and in how it looks. The changing of a river's course by *flooding* may happen in a very short period of time. Great cracks in the earth's crust may appear in a few minutes because of an **earthquake.** However, most of the big changes in the earth's surface happen slowly over thousands, or even millions, of years. Many things change the earth's surface: *extreme cold, extreme heat,* the *pushing up of rocks* that were formed deep in the earth or in water, the *rising and the sinking of land.* A1

If the earth is about 5 billion years old, why haven't these things that are changing its surface worn it down completely? Geologists know that opposite forces are always at work changing the earth's surface. One kind of force breaks up and wears away rocks. The forces that wear away are called **destructive forces.** The opposite forces that are always raising and rebuilding the surface are called the **constructive forces.** Destructive forces and constructive forces must be in some way balanced since the earth's crust has not been worn down to the level of the sea. A2

This chapter offers a good opportunity to strengthen students' *observation* skills. Select an area (or areas) in the neighborhood that tudents may observe for change. Have them make a list of the changes they observe over a period of a week or longer.

Check Up A

A1. Name four or more things that change the earth's surface.

(1) Extreme cold

(2) Extreme heat

(3) Pushing up of rocks

(4) Rising and sinking of land

A2. Name two kinds of forces that are always changing the earth's surface. How do they differ? Destructive forces wear away the earth's surface. Constructive forces rebuild the earth's surface.

B. Destructive Forces

After doing this lesson you will be able:

B1 • to name the two main destructive forces that change the earth's crust

B2 • to name four agents of destructive forces

B3 • to explain what mechanical weathering is and list three ways it is carried on

B4 • to explain what chemical weathering is and list three ways it is carried on

B5 • to tell how erosion is different from weathering

B6 • to list three agents of erosion and tell which is the main one

B1 **Weathering** and **erosion** are the most important *destructive forces* that change the rocks of the earth's crust. Wind, weather, running water and the moving ice of glaciers are the *agents* of these destructive forces. An agent is something that causes things to happen.

Weathering is the *breaking of rocks* into smaller pieces without the removal of the broken pieces. Weathering is caused mainly by conditions in the atmosphere. The agents of weathering may break up rocks either by *physical* (mechanical) or *chemical* means.

Mechanical weathering is the physical B3
breaking up of rocks. Mechanical weathering takes place when rocks are broken into small pieces but keep the *same* chemical makeup as the original rock. Some examples of mechanical weathering are: the breaking of shale into smaller pieces of shale, or the breaking of a large granite boulder into small granite pebbles. Mechanical weathering is carried on in many ways, some of which are:

1. Temperature changes. Rocks are broken B3
because of changes in temperature. The sun *heats up* the rock to a high temperature during the day. At night the rock *cools off.* The heating of the rock causes it to *expand* and the cooling causes it to *contract.* These changes cause the minerals to loosen. This loosening crumbles the rock and makes it fall apart.

2. Frost action. Water seeps into the cracks B3
and pores of rocks. When the temperature falls below freezing, the water freezes and expands. The melting of the ice and refreezing of the water in the cracks cause the rocks to split into smaller pieces. The results of frost action can be seen whenever the temperature goes below freezing.

3. Plant action. Many kinds of small plants B3
grow on rocks. They help crumble rocks by forcing their roots into small openings in the rocks. Trees also cause rocks to split apart by sending their roots through cracks in them.

Chemical weathering is the breaking up, B4
or decomposition, of rocks into materials that have a chemical makeup *different* from the original rocks. As a result of chemical weathering the minerals of a rock are changed into new substances. Chemical weathering of rocks is carried on by chemical materials in the atmosphere. Oxygen, water and carbon dioxide are the main agents of chemical weathering.

Heat a small rock over an alcohol burner for several minutes. Immediately immerse the rock in cold water. Often, it will crack or split apart.

Carbonic acid gives carbonated soft drinks a "tart" taste, and dissolved carbon dioxide causes the fizz.

B4 1. **Oxygen** in the air easily combines (joins together) with many minerals in rocks. The chemical reaction in which oxygen combines with another substance is called **oxidation.** The minerals that easily combine with oxygen are compounds of iron. Many rocks have different compounds of iron. Oxidation of these compounds forms iron rust. Rocks in which rust has formed fall apart easily because rust is very soft.

B4 2. **Water.** Many minerals combine with water to form new compounds. These new compounds expand slowly and soon crumble causing the rocks to fall apart. Water also speeds up the chemical weathering process of oxidation.

B4 3. **Carbon dioxide,** found in the atmosphere, dissolves easily in water to form a weak acid called **carbonic acid.** This acid removes some of the minerals in rocks and forms new compounds called **carbonates** which easily crumble. When water with this acid in it flows over limestone, the acid removes the calcite which makes up the rock and it is carried away by the water. This is how underground limestone caves are formed.

Weathering is a slow process; it takes many years. Mechanical and chemical weathering of rocks take place all the time.

Erosion is the name given to all processes by which soil and rock are loosened and then carried to another place. *Wind, running water* and *moving ice* are agents of erosion that can move small pieces of rock from one place to
B5 another. Erosion is really weathering plus removal. Weathering breaks up the rocks to make them ready for transport (removal) by the agents of erosion.

B6 1. **Wind** is able to pick up and carry small pieces of rock, sand and soil. As the wind carries these particles with great speed, they (mostly the sand) cut into other rocks and surfaces. Cutting away the softer minerals causes the rock to crumble into small pieces. Wind-blown particles also file (rub away) and polish rock surfaces. Sometimes rocks are formed into strange shapes before they fall apart completely.

2. **Running water** is the chief agent of erosion B6
on the earth. Running water erodes (wears down) the earth's surface (a) mechanically, and (b) chemically.

a. Mechanical erosion. Running water carries large amounts of sand, pebbles, pieces of rock and boulders. The sharp materials cut into, wear down and file away the rocks and land over which the water flows. It erodes the land, and the materials carried by the river along its bottom are rounded and smoothed. This erosive action of running water has cut through high mountains to form beautiful gorges and canyons such as the Grand Canyon formed by the Colorado River.

b. Chemical erosion. Plain water removes very few minerals from rocks and soil. However, when the water contains carbonic acid, it removes calcite from many rocks.

3. **Moving ice.** Large sheets of ice and snow B6
called **glaciers** are formed high in the mountains and move to lower levels. As the glaciers move down, they pick up and carry large amounts of loose rock material of all sizes. The glaciers use these materials as tools to erode the surface over which they pass. In carrying on the process of erosion, glaciers polish and smooth the bedrock and cut deep, parallel scratches in rocks. Some glaciers round off mountains, deepen and smooth out valleys, and even dig holes into the earth's surface. These holes sometimes become filled with water to form lakes. When the glacier melts, it drops the load of rocks, ground-up particles and clay it has been carrying. Glaciers can cause great changes in the earth's surface.

To show how a glacier can erode, freeze some ice cubes with sand on the bottom of the tray. Rub the ice cubes over an old piece of window glass. Note the scratches on the glass caused by the ice-carried sand.

This glacier is a river of ice. How can it widen the valley and smooth rocks?

A glacier moves very slowly. As it moves, it scrapes the sides of the valley, breaking off rocks and grinding them into smaller pieces. It also drags rocks across the valley floor, smoothing the bedrock.

Check Up B

B1. What are the two main destructive forces that change the earth's crust? Weathering and erosion.

B2. Name four agents of destruction. (1) Wind; (2) weather; (3) running water; (4) glaciers.

B3. What is mechanical weathering? List three ways it is carried on. Physical breaking up of rocks. (1) Temperature changes; (2) Frost action; (3) Plant action.

B4. What is chemical weathering? List three ways it is carried on. Chemical decomposition is the breaking up or the decomposition of rocks that have a chemical composition different from the original rocks. (1) Oxygen; (2) water; (3) carbon dioxide.

B5. How is erosion different from weathering? Erosion is weathering plus removal.

B6. What three agents cause erosion? Which is the main agent? (1) Wind; (2) running water; (3) moving ice. Running water is the chief agent.

C. Constructive Forces

After doing this lesson you will be able:

- **C1** • to name two ways constructive forces build up the earth's surface
- **C2** • to explain what plates are and what plate tectonics is
- **C3** • to tell what faulting is and name the kind of mountains formed by this process
- **C4** • to tell what folding is and name the kind of mountains formed by this process
- **C5** • to tell what a volcano is
- **C6** • to explain the difference between magma and lava

Constructive forces help build up the earth's surface. These forces which can lift parts of continents and which build mountains come from inside the earth. Two ways constructive forces make themselves felt are C1
as **earthquakes** and **volcanoes.**

Some scientists think that at one time the continents were all one big land mass. A gigantic movement of the earth's crust may have caused this land mass to split into large C2
pieces that scientists call **plates.** When this split happened, each continent was a part of one of these plates. The plates began to move

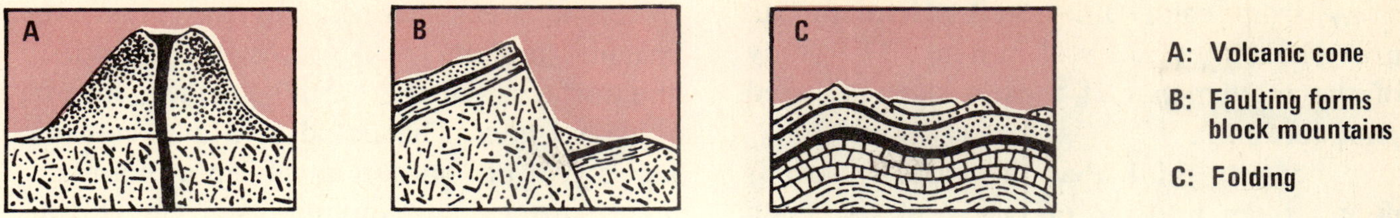

Three ways mountains are formed.

View of the active crater of Mount St. Helens.

Northwest view of the Spanish Peaks in Colorado. Note the faulting movement in the earth's crust.

Appalachian Mountains below, Jackson County, North Carolina. They are examples of folded mountains.

The highest mountain in the world, Mt. Everest, stands 8,711 meters above sea level. It is part of the Himalayas, which is a folded mountain range.

away from each other and scientists can prove that they are still moving. The study
C2 of the movements of these plates is called **plate tectonics.**

Scientists think that the earth's crust is now broken into **ten** or **twelve plates.** Each solid plate slides as a *separate unit* over the partly molten material below it. The plates change in size and shape. As they move they collide (hit against) each other. Magma from within the earth rises up in certain places to build the plates. The movement of these plates can explain the earth's constructive forces. When a plate floats to a deep trench in the ocean floor, it falls deeper into the earth. As it falls, it starts earthquakes and pushes up mountains.

A **mountain** is a raised part of the earth's crust. Mountains are formed in many ways. Opposing forces (forces that work against each other) can cause the rocks of the earth's crust to break. One side of the cracked rock slides past the other along the break. This
C3 kind of breaking followed by the sliding movement of rock is called **faulting.** Sometimes one side of the rock moves down while the other side moves up. Mountains formed
C3 by faulting are called *fault* or *block* mountains. The Sierra Nevadas and the Block Mountains of Oregon were formed in this way.

Mountains are also formed when layers of
C4 sedimentary rock are bent out of shape. Because of pressure from the side on the crust of the earth, sedimentary rock layers are forced upward. Some of the layers will bend up in some places and down in others. This
C4 process is called **folding** and the mountains that are formed are **folded mountains.** The Appalachian Mountains and part of the Rockies in the United States were formed from the folding of rocks.

Volcanoes also form mountains. A volcano
C5 is an opening in the earth's crust through which pass hot, molten materials and gases. When a break happens in the crust, the **magma** (molten rock) forces its way out to
the surface. On the earth's surface, magma C6
is called **lava.** The liquid lava and volcanic ashes pile up, harden and form a mountain. There are volcanic mountains on the Hawaiian Islands, on Iceland, and on the west coast of the United States.

Check Up C

C1. Name two ways the earth's surface is built up. Earthquakes and volcanoes.

C2. What do scientists say plates are? What is plate tectonics? Large sections of land formed by a gigantic movement of the earth's crust. Plate tectonics is the study of the movement of these plates.

C3. Explain what faulting is. What kind of mountains are formed by this process? The breaking and sliding movement of rock. Fault or block mountains are formed by this process.

C4. Explain what folding is. What kind of mountains are formed by this process? The bending of layers of sedimentary rock. Folded mountains are formed by this process.

C5. What is a volcano? An opening in the earth's crust through which hot molten materials and gases pass.

C6. What is the difference between magma and lava? Magma is molten rock below the surface of the earth. Once on the earth's surface, magma is called lava.

D. The Formation of Soil

After doing this lesson you will be able:

- **D1 • to tell what soil is and name the two layers of soil**
- **D2 • to explain what groundwater is**
- **D3 • to describe what the water table is and where it is located**

You have learned that wind, running water and glaciers are destructive forces. All these agents usually drop the rocks, pebbles and sand they have been carrying. Thus, while they are wearing down one part of the earth, they are building up another part of the earth.

Running water is the most effective destructive force.

D1 What is soil? The loose material found on top of the crust of the earth is **soil.** It can be made up of sand or clay. Soil is made by the weathering and erosion which break down rocks into sand and clay. When quartz is broken into tiny particles, sand is formed. Clay is a fine mud which results from the breaking up of other minerals besides quartz. Soils are grouped according to their arrangement in layers as (1) topsoil, and (2) subsoil.

1. Topsoil is the upper, loose, dark material that makes up the top of the earth's surface. Topsoil is made up of *sand, clay* and *humus.* **Humus** is the organic material that results from the decay of dead plants and animals. All of these materials — sand, clay and humus — are needed to make good soil in which plants can grow properly.

Sand makes the soil loose so air and water can enter it easily. **Clay** contains some of the minerals plants need, and holds the sand grains together so that water is held in the soil. As the humus keeps on decaying, it adds minerals to the soil. The best kind of soil for plant growth is one with the right mixture of sand, clay and humus.

2. Subsoil is the layer of soil just below the topsoil. This layer is usually reddish-brown in color because it has a great deal of clay. Subsoil also contains sand. However, since subsoil does not contain humus, plants do not grow well in it. Both the topsoil and the subsoil rest on solid rock called **bedrock.** This is the material that was broken down to make sand and clay.

Plants need many kinds of minerals to grow. These minerals are always being removed from the soil. For plants to grow well they must be given back to the soil. Farmers and gardeners add **fertilizer** to the soil. Fertilizers contain the phosphorous, potassium and nitrogen that plants need. Fertilizers can be made from chemicals or from the decaying remains of plants and animals.

Farmers also allow some of their fields to remain **fallow,** or unplanted. By not planting

Layers of soil resting on bedrock.

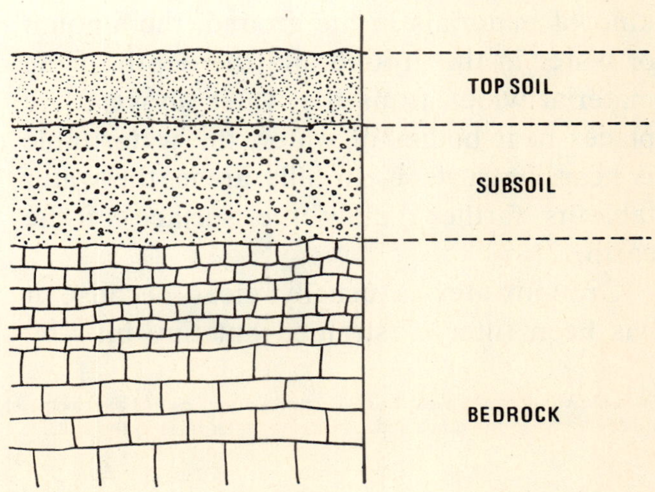

Examples of crops that restore minerals to the soil are clover, peas and other members of the legume family. These plants have nodules on their roots containing colonies of nitrogen-fixing bacteria which restore nitrates to the soil.

these fields for a season, minerals remain and some are stored up again by natural processes. Farmers also **rotate** their crops. *Crop rotation* means that farmers change the kind of crops they plant in a field. One crop is a "money" crop. The other helps to build up soil in the field. Some crops take minerals from the soil, while other crops give the minerals back to the soil.

Plants also need water to grow. **Irrigation** brings water to the land. The federal government has built many dams to irrigate the dry lands of our country. Everything possible is being done to save our topsoil and make it fertile (able to make things grow). In order to live, animals must have the food made by plants. With cities spreading out all over, you can see why it is difficult to save topsoil and farmlands.

What happens to the water after a rainstorm? When it rains, the soil absorbs some
D2 of the water. **Groundwater** is the water absorbed by the soil. The water leaks into the ground until it reaches a material through which it cannot pass. It then fills all the empty spaces and rises higher and higher as the rain comes down. This layer holds all the water it can. The upper part of this water-
D3 holding layer of soil is called the **water table.** The water table may be a few meters or many hundreds of meters below the surface. It depends on the thickness of the ground, the kind of materials in the ground, the amount of water in the ground and the depth of the material which traps the water above it. In
D3 places near bodies of water, the water table is near the surface. In hilly areas, the water table is farther below the surface of the earth.

Groundwater is usually clear because it has been filtered (strained out tiny bits) by the soil. Wells and springs bring this clear groundwater to the surface. It is used as drinking water by people and other animals. Plants also depend on groundwater.

All kinds of controls have been placed on the use of water in order to save it. Better conservation of national forests and other natural plant life will help keep groundwater levels high. Forests hold rain water much better than farmed or cultivated lands. The government is damming up small streams so that rain will seep (leak through) into the ground instead of filling up the streams. You can help in this effort by using water wisely and carefully.

Check Up D

D1. What is soil? Name the two layers that make up soil. Soil is the loose material found on top of the earth's crust. Topsoil and subsoil.

D2. What is groundwater? Water absorbed by the soil.

D3. What is the water table? At what depth is it found when near bodies of water? When in hilly areas? The upper part of the water-holding layer of soil is called the water table. Near bodies of water, it is near the surface. In hilly areas, it is farther below the earth's surface.

Fill a jar with clean gravel. You will find that you can fill approximately 1/3 of the jar with *water* after it is "full" of gravel. Water fills the spaces between the gravel just as ground water does in the soil.

Unit 4, Chapter 3
REVIEW

Matching Exercise Match each term in **Column A** with its description in **Column B**. Write the letter of the description in the space beside the correct term.

	Column A		Column B
b	**1.** lava	**a.**	bending of sedimentary rock layers
e	**2.** magma	**b.**	molten rock on the earth's surface
f	**3.** soil	**c.**	forces that build up the earth's surface
i	**4.** groundwater	**d.**	process by which soil and rock are loosened and carried to another place
h	**5.** plates	**e.**	molten rock inside the earth
a	**6.** folding	**f.**	loose material on the earth's crust
j	**7.** faulting	**g.**	forces that wear away the earth's surface
d	**8.** erosion	**h.**	solid part of the earth's crust on which the continents drift
c	**9.** constructive forces	**i.**	water found in the soil
g	**10.** destructive forces	**j.**	breaking and sliding of rock

True/False In the space provided, write **T** if a statement is true. If it is false, change the underlined word or words to make the statement true.

Lava	**1.** <u>Magma</u> is molten rock on the earth's surface.
constructive	**2.** Earthquakes and volcanoes are <u>destructive</u> forces.
Folding	**3.** <u>Faulting</u> is the bending of sedimentary rocks.
T	**4.** <u>Plates</u> are the solid part of the earth's crust on which the continents drift.
Faulting	**5.** <u>Folding</u> is the breaking and sliding of rock.
T	**6.** <u>Topsoil</u> is made up of sand, clay and humus.
water table	**7.** The upper part of water-holding soil is called the <u>groundwater</u>.
bedrock	**8.** The layer of solid rock below the subsoil and the topsoil is <u>fallow</u>.
destructive	**9.** Weathering and erosion are <u>constructive</u> forces.
T	**10.** A <u>volcano</u> is an opening in the earth's surface through which molten rocks and gases pass.
soil	**11.** The loose material found on top of the crust of the earth is <u>bedrock</u>.
T	**12.** <u>Mountains</u> can be created by both folding and faulting.
weathering	**13.** The breaking down of rocks into smaller pieces without the removal of the pieces is <u>erosion</u>.
weathering	**14.** Frost action is an agent of <u>erosion</u>.
T	**15.** Large sheets of moving ice and snow are called <u>glaciers</u>.

Multiple Choice Exercise Circle the letter of the word or phrase that best completes each statement.

1. Scientists think that the earth is between **(a)** 3 and 4 million years old **(b)** 30 to 40 million years old **(c)** 3 to 4 billion years old **(d)** 3 to 4 trillion years old.
2. The breaking down of rocks into smaller pieces without the removal of the pieces is **(a)** mechanical erosion **(b)** weathering **(c)** chemical erosion **(d)** faulting.
3. An opening in the earth's crust through which hot molten materials and gases pass is **(a)** a volcano **(b)** a folded mountain **(c)** erosion **(d)** a mountain.
4. Topsoil is made of sand, clay and **(a)** rock **(b)** lava **(c)** groundwater **(d)** humus.
5. The molten rock inside the earth is **(a)** lava **(b)** bedrock **(c)** groundwater **(d)** magma.
6. Forces that build up the earth's surface are **(a)** constructive forces **(b)** destructive forces **(c)** irrigation **(d)** erosion.
7. The breaking and sliding of rock is **(a)** folding **(b)** faulting **(c)** irrigation **(d)** magmas.
8. The upper part of water-holding soil is **(a)** groundwater **(b)** the water table **(c)** bedrock **(d)** irrigation.
9. Wind, running water and glaciers are **(a)** constructive forces **(b)** forms of weathering **(c)** destructive forces **(d)** forms of irrigation.
10. The process in which soil and rock are loosened and carried to another place is **(a)** chemical weathering **(b)** mechanical weathering **(c)** erosion **(d)** irrigation.

Short Answer Answer each of the following in complete sentences.

1. What is the difference between faulting and folding? Faulting is the breaking and sliding movement of rock. Folding is the bending of layers of sedimentary rocks. Faulting may result in the formation of fault or block mountains. Folding may result in the formation of folded mountains.

2. What is the difference between a constructive force and a destructive force? A constructive force is one that builds up the surface of the earth. A destructive force is one that wears away or breaks down the earth's surface.

This photograph was taken 96 kilometers above the earth's surface. The white ripples are not waves; they are clouds.

Do you know what kind of ocean you are looking at in this photograph?

New science words: atmosphere, humidity, precipitation

Skills developed: observing, measuring, comparing

Chapter 4

THE EARTH'S ATMOSPHERE

Have you ever wondered what makes the wind blow? Why it rains? What causes a thunderstorm? The answers to all these questions are the same. These things happen because of changes in the atmosphere. The atmosphere is always changing. Directly or indirectly, the atmosphere's source of energy for all these changes is the sun.

A. What Is the Atmosphere?

After doing this lesson you will be able:

A • to name the five layers of the atmosphere and tell what each layer is like

The earth is surrounded by an ocean of air without which life could not be. This invisible ocean is called the **atmosphere.** It is made up
A of five layers: (1) the troposphere, (2) the stratosphere, (3) the mesosphere, (4) the ionosphere and (5) the exosphere.

1. The troposphere. This is the layer of the atmosphere *closest* to the earth's surface. It is the layer of air that we live in. It stretches from the earth's surface to a height of about 8 kilometers over the earth's poles and to about 16 kilometers over the equator. Weather and changes in temperature take place in this layer. For about every 350 meters of altitude (height), the temperature drops 6°C.

2. The stratosphere. This is the layer of air above the troposphere. It rises from 16 to 60 kilometers above the earth's surface. The air in the stratosphere has little oxygen. There are no clouds, and it is very cold. High winds called **jet streams** are found in this layer.

3. The mesosphere. This layer of the atmosphere rises to an altitude of about 80 kilometers. Part of this layer overlaps the upper part of the stratosphere. The temperature in the mesosphere is not the same. From about 40 to 50 kilometers, temperature increases. Above 50 kilometers temperature drops as altitude goes on increasing.

4. The ionosphere. This layer rises from about 80 kilometers to more than 480 kilometers above the surface of the earth. The air in the ionosphere is very thin and has almost no oxygen. In this layer are electrically charged particles called **ions.** The ions act like mirrors and reflect radio waves back to earth. This is one of the ways the National Weather Service is able to collect information about the weather.

The ionosphere is influenced by electric storms in the earth's high atmosphere caused by sunspot activity, solar flares and chromospheric eruptions of the sun. All of these affect our weather.

5. The exosphere. Above the ionosphere is the exosphere, the outer region of the earth's atmosphere. This is the layer in which space satellites orbit (circle) the earth. The outermost boundary of this region is not known. There is little matter in this region. It is nearly a perfect vacuum.

Check Up A

List the five layers of the atmosphere and tell what each layer is like.

(1) Troposphere — layer closest to the earth's surface. Weather takes place here.

(2) Stratosphere — contains little oxygen, no clouds, and is very cold.

(3) Mesosphere — temperature is changeable.

(4) Ionosphere — layer of electrically charged ions.

(5) Exosphere — nearly a perfect vacuum.

B. The Weather

After doing this lesson you will be able:

- B1 • to explain the difference between weather and climate
- B2 • to list six things that have an effect on weather
- B3 • to explain the difference between a wind and a convection current
- B4 • to explain the difference between a sea breeze and a land breeze
- B5 • to explain the difference between absolute and relative humidity

Do you know what the word *weather* means? **Weather** tells us what is going on in B1 the atmosphere, or air, at a particular place and time. If we are told that it is cloudy and raining in New York City today, we know what is happening in the atmosphere over New York City today.

Climate, on the other hand, tells us about B1 the weather in a particular place over a long period of time. For example, Cleveland, Ohio, has an *average* rainfall of 80 centimeters per year. The *average* summer temperature in Rome, Italy, is 25°C.

Weather may change quickly, depending on conditions (what is going on) in the air. Therefore, to understand the conditions, you need to know where the changes take place and why. The weather is controlled by many different things, some of which are (1) air B2 temperature, (2) air pressure, (3) air currents and winds, (4) humidity, (5) air masses and fronts and (6) precipitation.

1. Air temperature. The part of the atmosphere nearest the earth picks up a great deal of heat from the earth which picked up some of the heat energy given off by the sun. The *top layer* of the stratosphere is *warmer* than other layers of the troposphere. Why is this so? As air close to the earth is heated the gas molecules move farther and farther apart and the air is lighter. The lighter, warmer air rises. The heavier, cooler air above moves down toward the surface of the earth.

The time of day and the time of year also control temperature. The sun is the earth's main source of heat. In the Northern Hemisphere, it is warmer in the summer than in the winter. This is because the hours of sunshine are longer and the sun's rays hit the

Your local weather station can give you information on the average rainfall for your area.

Observe that during winter, high pressure regions are normally clear and cold; low pressure regions, however, are cloudy and warmer. *Pressure-altitude* relationships depending on temperature also exist: pressure decreases more rapidly with altitude when temperature is low than when temperature is high.

earth more directly. So, more light energy is received and changed into heat.

2. Air pressure. Because air has weight, it presses on all objects it touches. Air pressure is controlled by the temperature of the air, the amount of water vapor in the air, and the altitude.

Warm air is lighter than cold air and, therefore, has less pressure than cold air. An increase in pressure can be caused by a decrease in temperature.

3. Air currents and winds. Movements of air in the atmosphere are called **air currents.**
B3 Horizontal (parallel to ground) movements of air close to the earth's surface are called **winds.** A wind is always named *for the direction from which it blows.* For instance, a wind *coming from the west* and blowing toward the east is called a *west wind.*

B3 When the movements of air are *vertical,* either as updrafts or downdrafts, they are called **convection currents.** Warm, light air rises upward as cooler air, which is denser and heavier, moves down to take its place. As warm air rises it cools off, becomes heavier and sinks back to earth. Thus, air rises and falls in a *circular* flow.

a. Land and sea breezes. Land heats up faster than water. On land, radiation from the sun is absorbed by a thin surface layer of soil or rock. The heat spreads out through a layer that is only a few centimeters or a meter thick. Temperatures in this thin layer can get very high. Much of this heat escapes into the air above. At night, the sun's rays no longer heat the land. Then the land begins to cool as it transfers (loses) heat to the air above. Thus, there is usually a large difference in the temperatures of land, and the air above it, from day to night.

When the sun's rays hit water, the water begins to heat up. Water takes in heat more slowly than does land. Water also loses heat more slowly than does land. Therefore, *the change in temperature of water,* and air above it, from day to night is small.

Direction of land and sea breezes.

Because of this difference in heating, the temperature of the air over the land is higher during the day than the temperature of the air over the water. The air over the land, being warmer and lighter, rises upward and the cooler air flows down to take its place. Such air movements are called **local winds.**

When cool air blows *from the water* to- B4
ward the land it is called a **sea breeze.** This cool air pushes up and takes the place of the warm air over the land. Then, toward sundown, the opposite takes place. Because all the heat is near the surface, the air over the land cools faster than the air over the deep water. This cool air begins to move *toward* B4
the *water* and is called a **land breeze.**

b. Planetary winds. There is a great wind system that blows over large parts of the earth. They move in cycles like the local winds. Differences in temperatures of air over land and over water cause some of these winds.

As you know, it is hotter at the equator than it is at the poles. The warm air around the equator rises and is replaced by polar (from earth's poles) air. The warm air flows toward the poles. You would think that it would travel northward and southward. But, because the earth rotates, moving air is shifted to the right in the northern hemisphere and to the left in the southern hemisphere.

The amount of heat needed to raise the temperature of 1 gram of a substance 1 degree C is known as its *specific heat.* Water has the highest specific heat of any known substance. Consequently, it heats up and cools down more slowly than land.

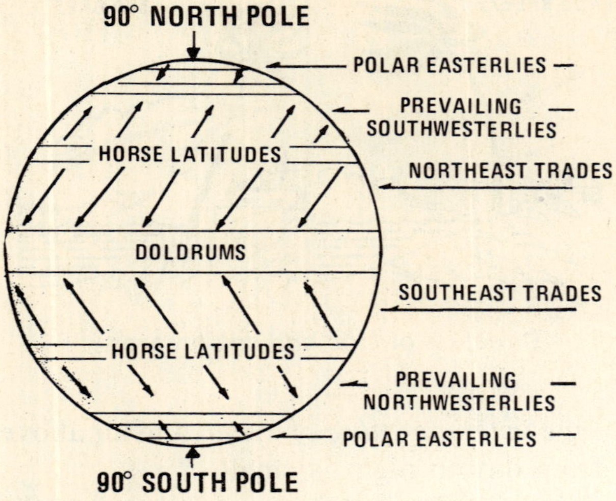

Planetary wind belts.

These winds marked the major *trade* routes for the old time sailing ships.

After traveling from the equator at about 30° latitude most of the air cools and falls back to the earth on both sides of the equator. Of the air that falls back, some flows back to the equator, forming the **trade winds.** Some of the air rises and again flows toward the poles, forming winds known as the **prevailing westerlies.** The prevailing westerlies of both hemispheres flow from west to east.

What happens to the air that rises at the equator and does reach the poles? At the poles it falls and flows back toward the equator in the form of winds called the **polar easterlies.**

4. Humidity. Water vapor in the air is called **humidity.** Water vapor is an invisible gas that is lighter than air.

How does water vapor get into the air? Water is always evaporating from oceans, lakes and rivers into the atmosphere. In fact, about one million metric tons of water are changed into water vapor every day. The amount of water that evaporates depends on the temperature, the wind and the dryness of the air.

Warm air can hold more water vapor than cold air. As the temperature rises, water evaporates faster and enters the air as water vapor. The actual amount of water vapor in the air at a certain temperature is called the B5
absolute humidity. The greatest amount of water vapor the air can hold at a certain temperature is called its **capacity.** When air is holding all the water vapor it can possibly hold at a certain temperature, the air is said to be **saturated.**

The ratio between the amount of water vapor in the air (absolute humidity) and the greatest amount the air can hold at a given temperature without the vapor becoming visi- B5
ble (capacity) is **relative humidity.** This ratio, or relative amount, tells how close the air is to being saturated.

5. Air masses and fronts. An **air mass** is a very large amount of air that has the same temperature and humidity. The temperature and humidity of an air mass are controlled by the part of the earth over which the mass forms.

An air mass can cover thousands of kilometers of the earth's surface, and it can be many kilometers high. Once formed, the air mass begins to move through the atmosphere. It keeps the same temperature and humidity for long distances as it moves. The mass may stay the same for several days. However, it may be changed by the conditions of the region over which it passes. The speed at which an air mass moves changes. It may move as slowly as 8 kilometers an hour, or as fast as 40 kilometers an hour.

An air mass formed over the cold, polar regions of the earth is called a **polar air mass,** and is dry and cloudless. This cold air mass usually brings fair weather with gentle to strong winds.

An air mass that forms over humid, hot parts of the earth is called a **tropical air mass.** This warm air mass holds a great deal of water vapor and has many clouds. It brings dew, fog and steady rains.

As air masses move, sometimes a cold air mass meets a warm air mass. The boundary or surface where the two air masses meet is

Relative humidity can be determined by this formula:

$$\frac{\text{amount of water vapor in the air}}{\text{amount that could be held by air at a particular temperature}} \times 100.$$

It is expressed as a percentage.

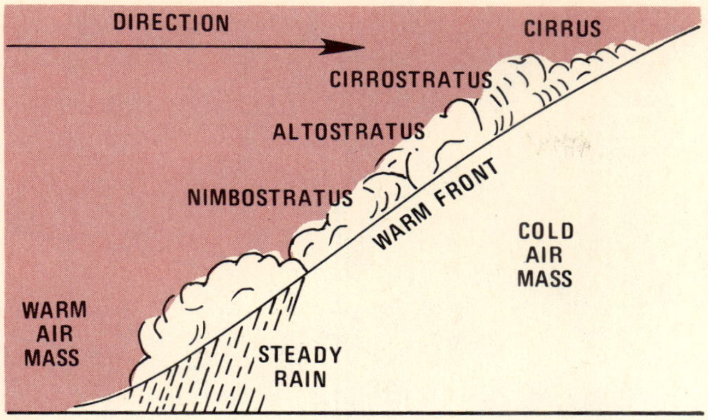

Warm front.

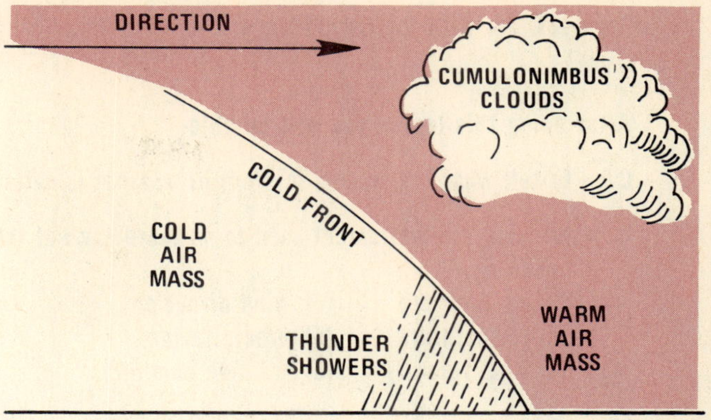

Cold front.

called a **front.** If a cold air mass flows under a warm air mass a **cold front** is formed. A cold front brings heavy rain and thunderstorms. If a warm air mass takes the place of a cold air mass, a **warm front** is formed.

6. Precipitation. Water vapor in the air condenses (changes from gas to liquid) to form tiny droplets of water. When these droplets become too heavy to hang in the air they fall to the earth as rain, snow, sleet or hail. This is known as **precipitation.**

Check Up B

B1. What is the difference between weather and climate? Weather tells us what is going on in the atmosphere at a particular place and time. Climate tells us about the weather in a particular place over a period of time.

B2. List six things that control weather.

(1) Air temperature
(2) Air pressure
(3) Air currents and winds
(4) Humidity
(5) Air masses and fronts
(6) Precipitation

B3. How is a wind different from a convection current? Winds are horizontal movements of air. Convection currents are vertical movements of air.

B4. How is a sea breeze different from a land breeze? A sea breeze brings cool air from the water toward the land. A land breeze brings cool air from the land toward the water.

B5. What is the difference between absolute humidity and relative humidity? Absolute humidity is the actual amount of water vapor in the air at a particular temperature. Relative humidity is the ratio between absolute humidity and the maximum amount the air can hold without becoming visible.

Meteorology originally meant a study of anything *high in the air*. (This is how meteors were named.)

C. What Is Meteorology?

After doing this lesson you will be able:

C1 • to tell what the scientific study of weather is called

C2 • to name the instrument used to measure each of the following

(1) air pressure	(4) wind direction
(2) temperature	(5) precipitation
(3) wind velocity	(6) relative humidity

C3 • to name two types of barometers and tell which is the one weather forecasters usually use

C4 • to be able to tell what a feathered arrow symbol means

C5 • to explain what cloud seeding is

C6 • to explain what Tiros I was

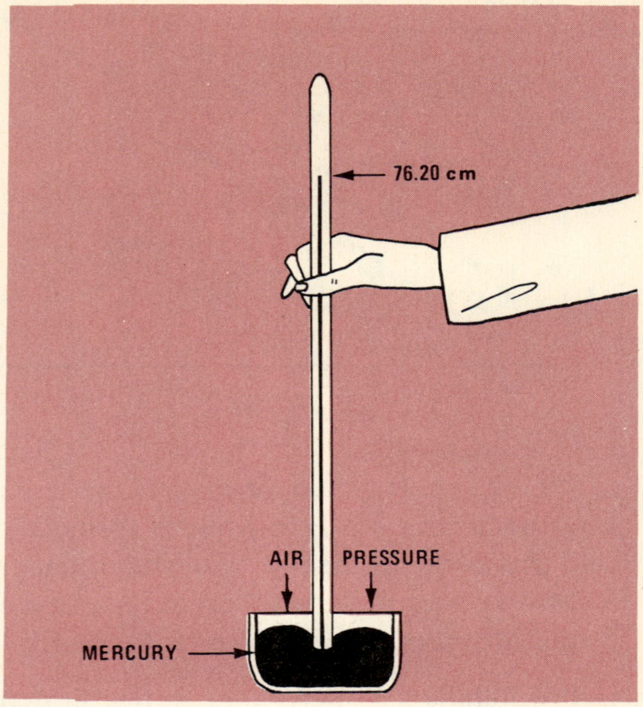

A mercury barometer.

C1 The scientific study of weather is called **meteorology.** Weather forecasters are **meteorologists.** They use many instruments to learn what is happening in the atmosphere so that they can better understand and predict (forecast) the weather. With these instruments they can measure (1) air pressure, (2) temperature, (3) speed and direction of the wind, (4) rainfall and (5) relative humidity.

C2 (1) **1. Measuring air pressure.** Changes in the pressure of the air can be measured by a **barometer.** The two barometers most often
C3 used are: (a) the mercury barometer and (b) the aneroid barometer.

C3 **a. The mercury barometer.** The first **mercury barometer** was made by an Italian scientist, Evangelista Torricelli, in 1643. Torricelli believed that air had weight and, therefore, causes pressure.

He filled a glass tube, about one meter long and closed at one end, with mercury. He held his finger over the open end and turned the tube upside down into a dish of mercury. When he took his finger away from the open end, some of the mercury ran out into the dish. However, a column of mercury about 76 centimeters high stayed in the tube. What kept the mercury in the tube? Why didn't it all run out?

Torricelli saw that as atmospheric pressures changed, the height of the mercury column in the tube changed, too. When the air pressure fell, the mercury in the tube ran out into the dish. As the air pressure increased, the mercury went higher in the tube.

At sea level there was enough air pressure to keep 76 centimeters of mercury in the tube. Or, to put it another way, there was enough air pressure to keep the mercury from running out of the tube. A mercury barometer gives a very accurate (exact) reading of air pressure.

b. The aneroid barometer. Weather forecasters usually use an **aneroid barometer** to C3
measure air pressure. It is easy to handle and does not contain liquid.

The aneroid barometer has a small, round metal can from which most of the air has been taken out. The top of the can moves outward

If you construct a mercury barometer, work in a well ventilated area and cover the top of the mercury in the dish with a thin layer of water. This will reduce the danger of poisonous *mercury vapor.*

Aneroid means *without air*. Note the vacuum box in the drawing below. Small aneroid barometers are available at low cost in department stores.

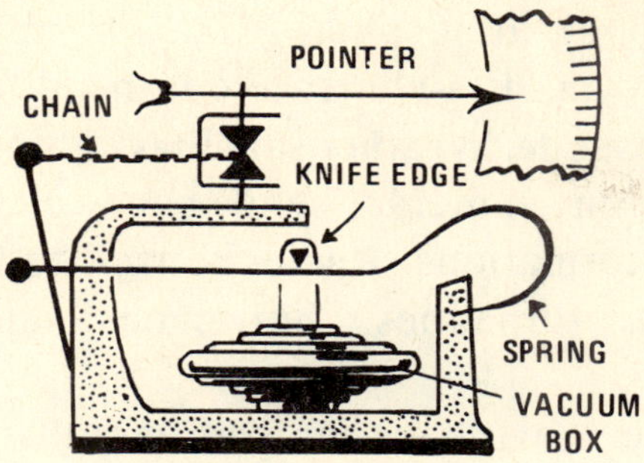

An aneroid barometer.

or inward depending on changes in atmospheric pressure. This movement causes a pointer to move across a scale and tell the air pressure.

A high-pressure reading or rising barometer means fair weather. A low-pressure reading or falling barometer usually means poor weather, with a chance of clouds and rain.

2. Measuring temperature. One of the instruments most often used in meteorology is the
C2 (2) **thermometer.** Thermometers measure temperature changes in units called **degrees.** On the Celsius (centigrade) scale the freezing point of water is 0°C, and the boiling point is 100°C.

3. Measuring wind velocity and direction. Winds always blow from a region of high pressure to a region of low pressure. The difference in air pressure between these two regions controls the speed of the wind.

Anemometer (left); weather vane (right).

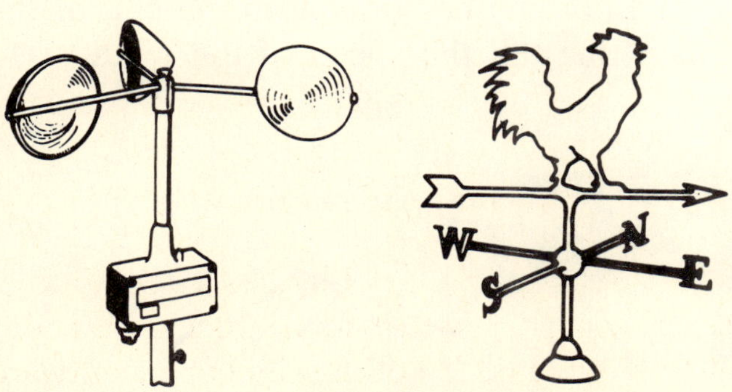

The speed of a wind is measured by an C2 (3)
anemometer in kilometers per hour or knots. The top part of the anemometer is made out of cups attached to arms. They are attached to a pole and turn freely. As the wind blows, it hits the cups causing the top part of the anemometer to spin. The pole is connected to a speedometer which tells the speed of the wind.

To find the direction of a wind, a **weather** C2 (4)
vane or **wind vane** is used. The weather vane changes direction with the wind. It points toward the direction in which the wind is moving. A wind, as you know, is named for the direction from which it blows.

4. Measuring precipitation. A **rain gauge** is C2 (5)
used to measure precipitation. The rain gauge is placed outdoors to catch the rainfall. It is a container holding a narrow tube which is connected to a funnel. The top of the funnel is 10 times larger than its narrowest part. When one centimeter of rain falls into the funnel, the narrow tube is filled to a depth of 10 centimeters. Suppose the tube has only five centimeters of rain. How much rain actually fell?

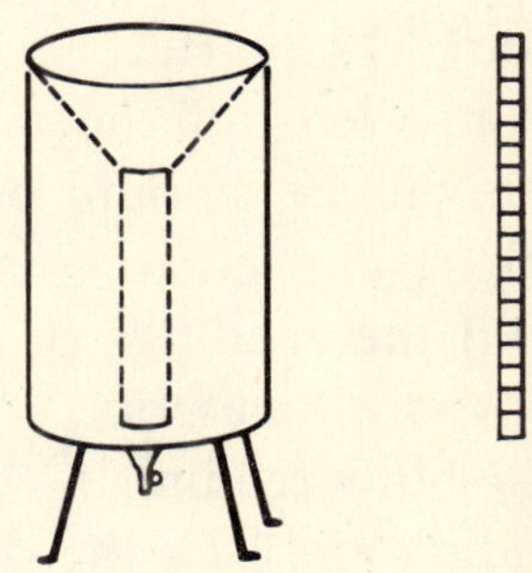

Rain gauge.

To tell the amount of snowfall, snow is collected in the rain gauge. After the snow melts, it is measured in the same way as the rainwater. Snowfall is also measured by placing a ruler in a spot in which the snow has not drifted.

You can construct a simple rain gauge by using a large juice can and a small soup can. Pour 2 centimetres of water into the large can. Then pour this into the small can. Mark this "2 centimetres" and divide into 0.5 centimetre fractions. Collect rain in the large can and pour it into the small one for greater measuring accuracy.

Using blank maps of the United States, encourage students to make their own "weather maps." Practice using the symbols.

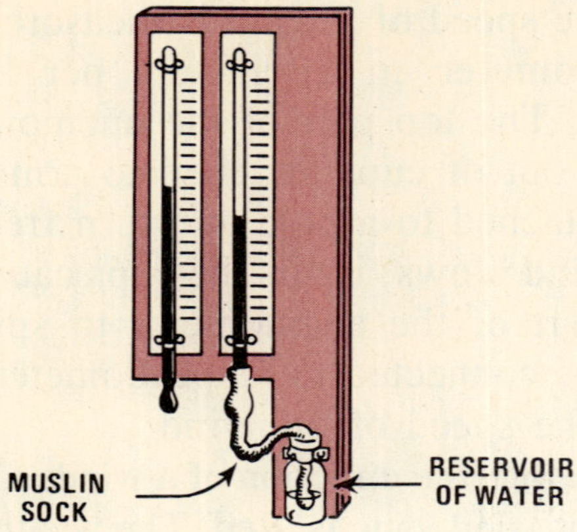

Wet and dry bulb thermometer.

5. Measuring relative humidity. An instru-
C2 (6) ment used to measure humidity is called a **hygrometer.** A common type is the **wet-bulb and dry-bulb thermometer** called a **psychrometer.** One of its two thermometers has wet cloth wrapped around its bottom. This is the wet-bulb thermometer. As the water evaporates from the wet cloth, the bulb is cooled and the thermometer measures a lower temperature than the dry-bulb thermometer. The less moisture there is in the air, the faster the water evaporates from the cloth and the lower the reading on the wet-bulb thermometer will be.

The dry-bulb thermometer measures the air temperature. The difference between the temperature readings of the two thermometers depends on the amount of moisture in the air. A chart is used to find the **relative humidity** from the readings of the wet-bulb and dry-bulb thermometers.

If the dry-bulb reading is 16°C and the wet-bulb reading is 13°C, the difference between the two temperatures is 3°C. Using a chart to locate the dry-bulb reading, you will find that the relative humidity is 84%.

All of the instruments studied so far are used on the ground to collect weather information. But there are other ways to get such information. Gas-filled balloons sent up into the atmosphere send back information about temperature, air pressure and the humidity of the upper atmosphere. The information is sent by an ultra-shortwave transmitter called a **radiosonde. Weather satellites** equipped with television cameras send back pictures of cloud formations. **Radar** is used to find rain clouds, tornadoes, hurricanes and other storms.

There are weather stations all over the world that gather information. This information helps meteorologists predict the weather. The National Weather Service charts all the information on a weather map. The official weather map shows the weather in the United States at 7:00 A.M. Eastern Standard Time.

Meteorologists put each of the barometer readings on a map of the United States and Canada. Each point is marked with its barometric pressure which is measured in **millibars.** Points having the same barometric pressure are joined by a solid black line called a **height contour line.** The words *high* and *low* on the weather map stand for high pressure or low pressure areas.

Weather maps help meteorologists figure out how the winds in a pressure area are moving. Winds move in a clockwise direction from a high pressure area. Winds move in the opposite, counterclockwise, direction toward the center of a low pressure area. In the United States the pressure systems move in a west-to-east direction because of the prevailing westerlies.

Weather maps have many symbols. Some symbols look like musical notes and are
called feathered arrows. The line points in C4
the direction from which the wind is blowing. Shorter lines drawn on the end of the main line tell the speed of the wind.

The feathered arrow.

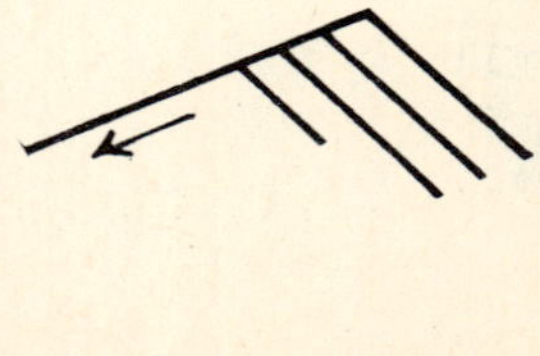

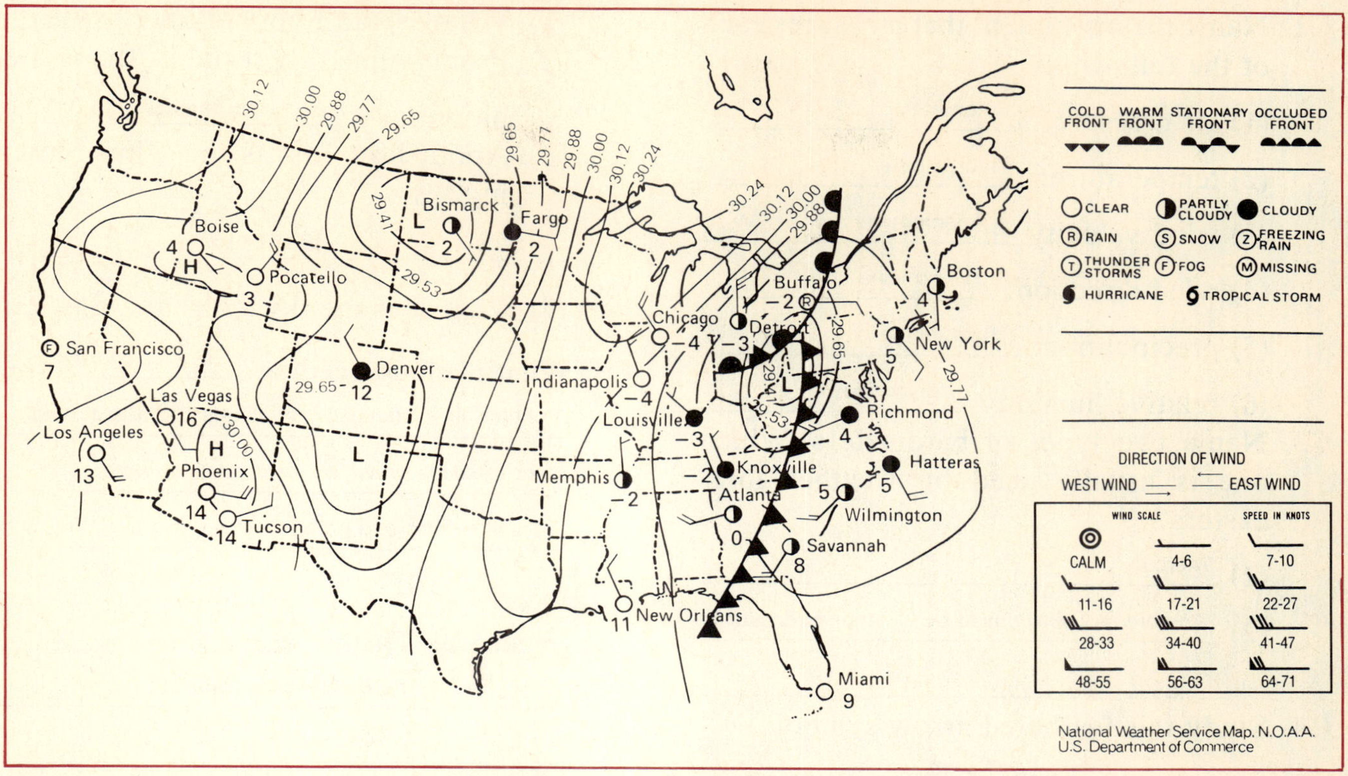

National Weather Map. What were the temperature and weather conditions for your area on February 27, 1981?

Study the key to the symbols used on the map. Locate the weather station closest to your city or town. The temperature is the number adjacent to the station name. Weather conditions are indicated by the feathered arrow symbol.

Some of the circles are filled in, and some are only partly filled in. Some are not filled in at all. The circles stand for the part of sky covered by clouds. The number next to each circle tells the air temperature in degrees *Fahrenheit* (F). Sometimes, below the temperature number, there is another number written in decimals. This tells the number of centimeters of precipitation that fell in the last six hours.

Can the weather be controlled? Scientists are carrying on all kinds of experiments so that they may one day control the weather. Scientists have made rain by **cloud seeding.**
C5 Dry ice or other chemicals are put into the clouds to cool the air. They act as "seeds" on which ice crystals or water droplets can form and grow.

Pilots *track* (follow) storms in planes and even fly ino the "eyes" of hurricanes so they can learn more about them. Scientists also study high-altitude winds and violent storms hoping to learn how to control them.

Daily weather observation by satellite began in 1960 when the national space C6
agency sent the 121 kilogram satellite, **Tiros I**, into orbit. This satellite carried two television cameras, and took thousands of pictures of the cloud formations over many parts of the earth. Since then larger and more complicated spacecraft have been launched to observe the weather.

Check Up C

C1. What is the scientific study of weather called? Meteorology

C2. Name the instrument that measures each of the following:

(1) air pressure: Barometer

(2) temperature: Thermometer

(3) wind velocity: Anemometer

(4) wind direction: Wind vane

(5) precipitation: Rain gauge

(6) relative humidity: Hygrometer

C3. Name two types of barometers. Which one is usually used by weather forecasters?

(1) Mercury

(2) Aneroid — usually used by weather forecasters

C4. Study this feathered arrow symbol:

West

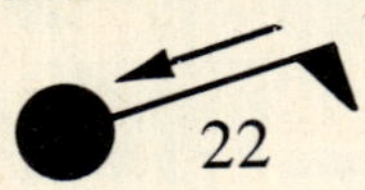

East

(1) What is the temperature? 22°C

(2) From what direction is the wind blowing? Northeast

(3) About how fast is the wind blowing? 48 to 55 knots

(4) What does the sky look like? Cloudy

C5. What is cloud seeding? Dry ice or other chemicals introduced into clouds cool the air and act as seeds on which ice crystals or water droplets form and grow.

C6. What was Tiros I? The first weather satellite.

THINK ABOUT THIS

Cleaning Up the Air

Exhaust from automobile engines is a large source of air pollution. Some manufacturers are looking for ways to decrease exhaust. The rotary engine is one of these ways. This engine runs in a circular motion instead of the back-and-forth piston motion of most automobile engines. It runs at a lower temperature and produces fewer oxides of nitrogen that pollute the air.

The electric automobile is another way to decrease exhaust. It is almost completely pollution free. It is battery-powered and can be recharged by simply plugging it into any circuit. But it cannot operate at the high speeds or long distances of highway driving. It can, however, be used in the city where air pollution is a serious problem.

Unit 4, Chapter 4
REVIEW

Matching Exercise Match each term in **Column A** with its description in **Column B**. Write the letter of the description in the space beside the correct term.

	Column A		Column B
d	**1.** atmosphere	**a.**	layer of the atmosphere closest to the earth's surface
f	**2.** exosphere	**b.**	instrument that measures air pressure
h	**3.** ionosphere	**c.**	layer of the atmosphere where jet streams are found
i	**4.** mesosphere	**d.**	ocean of air that surrounds the earth
c	**5.** stratosphere	**e.**	instrument that measures wind velocity
a	**6.** troposphere	**f.**	satellites orbit the earth in this layer of the atmosphere
b	**7.** barometer	**g.**	the scientific study of the weather
e	**8.** anemometer	**h.**	layer of the atmosphere that reflects radio waves
j	**9.** hygrometer	**i.**	temperature in this layer is very changeable
g	**10.** meteorology	**j.**	instrument that measures relative humidity

True/False In the space provided, write **T** if a statement is true. If it is false, change the underlined word or words to make the statement true.

Answer	Statement
direction	**1.** A weather vane is used to find wind speed.
air pressure	**2.** An anaroid barometer is used to measure relative humidity.
T	**3.** A high-pressure reading or rising barometer indicates fair weather.
meteorology	**4.** The scientific study of the weather is called astronomy.
an air mass	**5.** A large amount of air that has the same temperature and humidity is a cold front.
T	**6.** Climate is the weather in a particular place over a long period of time.
ionosphere	**7.** The air in the mesosphere consists of electrically charged particles.
troposphere	**8.** The layer of the atmosphere closest to the earth is the mesophere.
T	**9.** Air currents are movements of air in the atmosphere.
a west wind	**10.** A wind that comes from the west and blows toward the east is an east wind.
precipitation	**11.** Rain, snow, sleet and hail are forms of relative humidity.
absolute	**12.** The actual amount of water vapor in the air at a particular temperature is called the relative humidity.
rain gauge	**13.** Precipitation is measured with a barometer.
T	**14.** Relative humidity can be measured with a wet-bulb and dry-bulb thermometer.
T	**15.** Scientists have made rain by cloud seeding.

Multiple Choice Exercise Circle the letter of the word or phrase that best completes each statement.

1. The outer region of the earth's atmosphere is the **(a)** troposphere **(b)** mesosphere **(c)** ionosphere **(d)** exosphere.
2. Jet streams are found in the **(a)** troposphere **(b)** stratosphere **(c)** ionosphere **(d)** exosphere.
3. The layer of the atmosphere that is closest to the earth's surface is the **(a)** troposphere **(b)** stratosphere **(c)** ionosphere **(d)** mesosphere.
4. The ocean of air that surrounds the earth is the **(a)** stratosphere **(b)** mesosphere **(c)** ionosphere **(d)** atmosphere.
5. Water vapor in the air is **(a)** air mass **(b)** precipitation **(c)** humidity **(d)** a sea breeze.
6. An air mass that forms over humid, hot parts of the earth is called a **(a)** cold front **(b)** warm front **(c)** polar air mass **(d)** tropical air mass.
7. Air pressure is measured with a **(a)** thermometer **(b)** barometer **(c)** rain guage **(d)** anemometer.
8. An anemometer measures **(a)** wind speed **(b)** wind direction **(c)** humidity in the air **(d)** millibars.
9. Relative humidity is measured with a **(a)** hygrometer **(b)** anemometer **(c)** rain guage **(d)** weather vane.
10. Rain clouds, tornadoes, hurricanes and other storms can be detected with **(a)** a hygrometer **(b)** a rain guage **(c)** radar **(d)** an anemometer.

Application Exercise Tell what each of the weather symbols below stands for. Write your answers on the line below each symbol.

1. wind speed

2. cold front

3. cloudy skies

4. hurricane

5. warm front

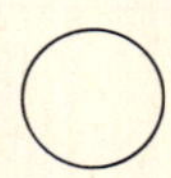

6. clear skies

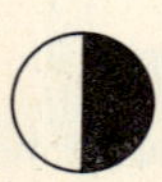

7. partly cloudy

8. stationary front

9. tropical storm

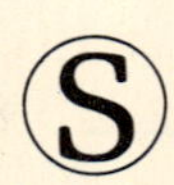

10. snow

This is how earth would have looked to you had you been on the moon with the Apollo 17 space crew.

New science words: eclipse, galaxy, orbit, revolution, rotation

Skills developed: comparing, classifying

Chapter 5

EARTH AND ITS NEIGHBORS IN SPACE

Have you looked into the sky on a clear night and wondered about all those shining objects you saw? For centuries people have watched the sky and have been puzzled about the moon, sun, stars and planets. For a long time, people believed that the earth stood still and all the other bodies moved around it. Today we know this is not true.

A. The Stars

After doing this lesson you will be able:

- A1 • to describe Copernicus' theory of the solar system
- A2 • to tell how Galileo helped prove this theory
- A3 • to list some objects in the universe
- A4 • to define the term galaxy and to tell what galaxy the earth is a part of
- A5 • to define the term light year and tell how fast light travels
- A6 • to explain what the color of stars means
- A7 • to describe what a constellation is
- A8 • to define nebulae

Astronomers, scientists who study the solar system and the universe, have given us the correct facts about the earth's place in the universe.

One of the earliest astronomers, Nicholas Copernicus (1473–1543), thought that the A1
sun was the center of the solar system and certain other objects in the sky revolved (circled) around the sun. When Galileo built the first telescope in about 1609 Copernicus' A2
theory was proved right. The invention of the telescope and other tools helped scientists to collect a lot of information about outer space.

The universe takes in the stars, moon, A3
earth, the other planets, the planetoids (as-

The Milky Way was so named because it appears as a white band of stars across the sky. There are about 100 billion stars in the Milky Way Galaxy.

teroids), meteors and comets—everything in space that was formed there naturally.

There are billions of stars in outer space. The sun is a star and the closest one to the
A4 earth. You live in a **galaxy,** or group of stars, called the **Milky Way.** The sun and the planets are part of this galaxy.

Stars come in many sizes. The sun, with a diameter (distance through center) of about 1,400,000 kilometers is only a medium-sized star. But the sun looks like it is very large. Do you know why? It looks brighter and larger because it is closer to us than any other star. It is about 150 million kilometers away.

The second closest star to the earth is about 40,000 billion kilometers away. Because the distance of stars and other distances in the universe are so great, astrono-
A5 mers use a special unit to measure distance. The unit is called a **light-year.** A light-year is the distance light travels in one year. Light travels at about 297,600 kilometers per second. It takes the light rays from the sun about eight minutes to reach the earth. This means that the sun is only 8 light minutes away from the earth.

Stars also differ in their **color.** Some look white, red, yellow or blue. The color of a star
A6 tells astronomers the temperature of the star. The hottest stars have a blue or white color. Yellow stars have medium temperatures, and the coolest stars are red. The colors of stars go from blue, to blue-white, white, white-yellow, yellow, orange or red. The sun is yellow in color and is, therefore, only a medium-temperature star.

Stars are made up mainly of the gases hydrogen and helium. Stars give off their own light and heat. Many astronomers think that the heat and light come from a type of nuclear reaction in which hydrogen atoms fuse (join together) to form helium atoms. When all the hydrogen has changed to helium, the star's core shrinks and becomes very hot. When this happens the outer part expands and gets redder and brighter. Much later, after more changes, the star burns itself out and gives off little or no light. Will our sun always give off light and heat? The sun is as old or older than the earth. Scientists think that the sun will keep on giving off heat and light for another four to five billion years.

Long ago, the early Greeks and Romans gave names to the stars that always seemed to be in the sky. They grouped stars into forms of animals, objects or famous people. A group of stars that appears to form pat- A7
terns or pictures in the sky is called a **constellation.** One of the easiest constellations to locate in the sky is *Ursa Major* or the Great Bear. Part of the tail of *Ursa Major* is the Big Dipper in whose bowl are the pointer stars. Sailors use the pointers of the Big Dipper to find the North Star, or *Polaris*. This star helps sailors to tell where north is.

Now that you know a little more about stars, look at the **Milky Way.** All the stars you see in the sky (plus our solar system) belong to the Milky Way galaxy. It was named the Milky Way because there are so many stars they give the entire galaxy a milky look. There are other galaxies or groups of stars outside the Milky Way. Most of these galaxies are very far away from the earth. With the help of powerful telescopes, astronomers have taken photographs of masses of A8
gas and dust between the stars. These masses are called **nebulae.**

Eighty-eight constellations have been named.

The constellations Great Bear and Orion.

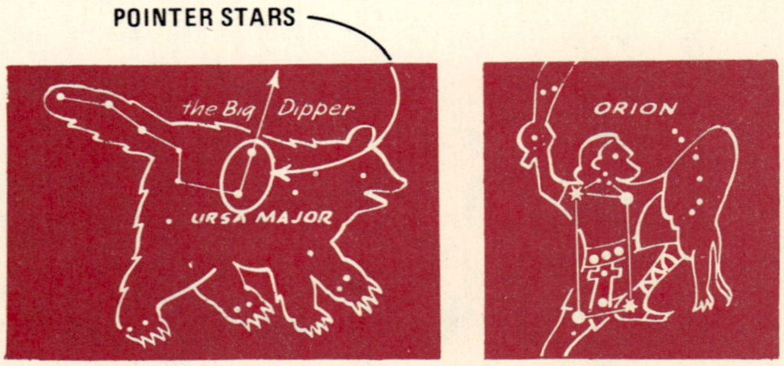

On a clear night in open country, 2,000 to 3,000 stars can be observed. The number visible is greatly reduced in the city because of bright lights and the dusty atmosphere.

The most distant objects visible in telescopes, *quasars*, have been observed as far as 15 billion light-years away. A quasar is similar in appearance to a star, but is brighter than an entire galaxy. For an interesting article on quasars, see *Smithsonian* magazine, April 1977, pps. 40-47.

Check Up A

A1. How did Copernicus describe the solar system? The sun was the center of the solar system and certain other objects revolved around it.

A2. How did Galileo help prove Copernicus' theory of the solar system? Built the first telescope that helped to prove Copernicus' theory.

A3. What are some objects in the universe? Stars, moon, earth and other planets, planetoids, meteors, comets.

A4. What is a galaxy? What is the name of the galaxy you live in? A galaxy is a group of stars. We live in the Milky Way Galaxy.

A5. What is a light-year? How fast does light travel? A light-year is the distance light travels in one year. Light travels at 297,600 km/sec.

A6. What does a star's color tell you about the star? The color tells the temperature of the star.

A7. What is a constellation? Name two often seen. A constellation is a group of stars that forms a pattern or picture in the sky. Answers will vary for two constellations most often seen.

A8. What are nebulae? Masses of gas and dust

B. The Solar System

After doing this lesson you will be able:

- B1 • to name the six kinds of bodies that make up the solar system
- B2 • to tell where the sun is located in the solar system and how planets travel in the solar system
- B3 • to explain the difference between a planet's revolution and its rotation
- B4 • to name the nine planets in the solar system and give some of their characteristics
- B5 • to tell what a satellite is

The solar system is part of the galaxy called the Milky Way. It is made up of the B1 sun, planets, moons, comets, asteroids and meteors. Solar means *of the sun*.

Astronomers cannot answer many questions about the planets. What questions would you ask? Could life exist on any of these planets? What kind of life would it be? Would it be like life on earth? There is no doubt that these questions will be answered someday.

The sun, as you already know, is the center B2 of the solar system. Nine planets travel or revolve around the sun. Each planet revolves around the sun in its own path called its **orbit.** There is a force in the universe that tries to pull every particle of matter toward every other particle of matter. This is called **gravitational force.** The sun's gravitational force tends to pull the planets toward the sun. But there is another force that keeps an object moving in a straight line. The balance between this force and the gravitational pull of the sun keeps the planets from spinning off into space, and causes them to revolve around the sun.

The planets revolve around the sun at different speeds. The closer a planet is to the sun, the faster it moves. A planet that is close to the sun will go completely around the sun once, or complete one *revolution,* in

The order of planets beginning with the *smallest* is: (1) Mercury, (2) Pluto, (3) Mars, (4) Venus, (5) Earth, (6) Neptune, (7) Uranus, (8) Saturn, (9) Jupiter.

The planet Venus as photographed by Pioneer Venus Orbiter, February 14, 1979.

less time than a planet that is farther from the sun.

The time it takes a planet to travel around the sun once is called its **period of revolution.** The period of revolution of a planet controls
B3 the length of the planet's **year.** A planet also spins or **rotates** on an imaginary axis that goes from the planet's north pole to its south pole. The time it takes a planet to rotate once on its axis controls the length of a **day** on the planet.

The satellites revolve around the planets; planets revolve around the sun; the sun revolves within the Milky Way.

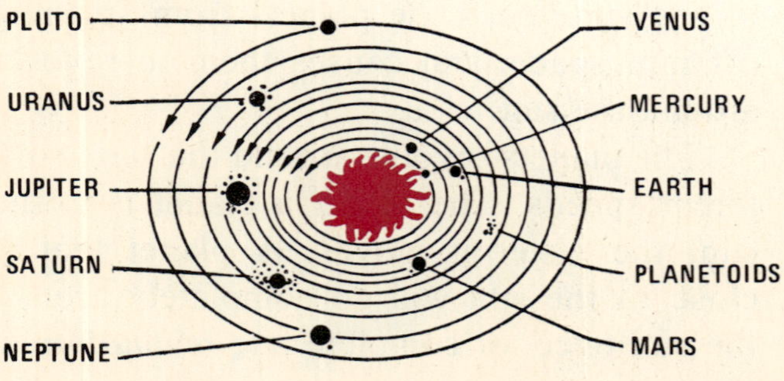

Most of you know about the planets in the solar system from information that has been collected with optical and radio telescopes. In the years to come, space exploration will supply much more information. Remember that planets have no light of their own. They reflect the light of the sun.

The order of planets beginning with the planet closest to the sun is: (1) Mercury, B4
(2) Venus, (3) Earth, (4) Mars, (5) Jupiter, (6) Saturn, (7) Uranus, (8) Neptune and (9) Pluto.

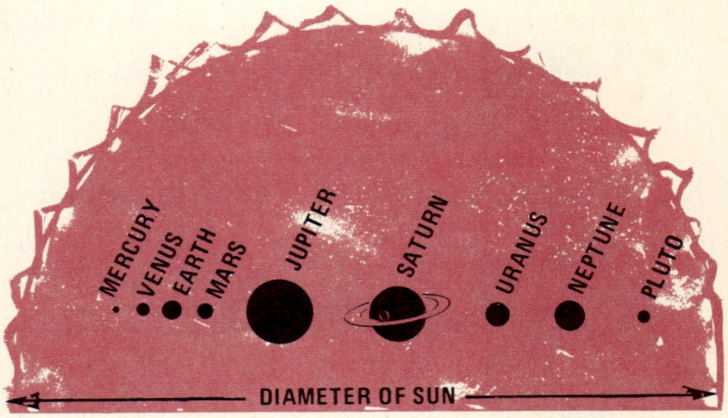

Size of the planets compared to the sun.

1. Mercury is the smallest planet and the closest to the sun. It takes Mercury 88 days to revolve around the sun. Mercury makes one rotation on its axis every 59 earth days. Because its "day" is so long, one side of Mercury becomes very hot while the other side becomes very cold. The atmosphere on Mercury is very thin.

2. Venus seems to be covered by thick clouds. But recent photographs show that the clouds are really a haze. The surface of Venus is very hot and dry. Scientists think that the atmosphere is made up mostly of carbon dioxide with very little oxygen. Venus is called earth's "twin planet" because the two are almost the same size. Venus rotates in the opposite direction from the earth at the rate of once in 243 days.

Photographs from the Mariner 9 spacecraft show that the surface of Mars is marked by high volcanoes, large faults, rift valleys and channels that may have been formed by water erosion. Many scientists think these channels indicate that Mars once had a watery environment. Scientists also know that under the polar ice caps of frozen carbon dioxide (dry ice) there is some water ice.

Voyager 1 spacecraft came within 78,000 miles of Saturn. The equipment on board is very sophisticated. Yet the probe operates on less than 400 watts which is about what a color television set operates on.

3. The **earth** is the next planet in distance from the sun. It has the highest forms of life. However, since this is the planet you live on, it will be studied more closely in other parts of this book.

4. Mars. This red-colored planet comes after the planet Earth in line to the sun. Until the Viking probes (exploring devices) landed on Mars, little was known about this planet. Important information has been collected by these probes. It was found that the planet's upper atmosphere is made up of two to three percent nitrogen and very little oxygen. These two gases are needed in the atmosphere to support life as we know it. From photographs taken, parts of the Martian surface looks like deserts on Earth. Most of the soil is a reddish fine-grained material which gives the planet its red color.

5. Jupiter and **Saturn** are the largest planets. Jupiter's surface is completely covered by an atmosphere of heavy, dense clouds which are thought to be made up of ice crystals. Jupiter has sixteen moons—the last two were discovered recently.

Voyager 1 made some startling discoveries about Saturn. What was thought to be three or four rings turned out to be hundreds of rings. Sixteen moons were found orbiting Saturn.

Information sent back to earth by Pioneer 10 shows that both Jupiter and Saturn are more like stars than planets. Like stars, they seem to give off their own energy. Also they seem to be made of more hydrogen and helium gas than solid material. However, Pioneer 10's findings also show that Jupiter has an inner layer of liquid metallic hydrogen with a rocky core.

6. Uranus and **Neptune** are very much like Saturn and Jupiter in their makeup. However, Uranus and Neptune are more solid than gaseous. Uranus has five moons and Neptune has two. The larger planets are colder because they are so far from the sun. There is very little chance that there is life as we know it on these planets.

This view of Saturn and its rings was taken by Pioneer Saturn spacecraft.

7. Pluto is the farthest planet from the earth. It is also solid like the earth. A large part of its orbit lies near or even within Neptune's orbit. Because of this, some astronomers think that Pluto may be an escaped satellite of Neptune.

Moons, or satellites, are small bodies that B5
revolve around larger bodies such as planets. The satellites of some planets are thought to be made of the same materials as the planets. Satellites revolve around the planets and travel through space with them. Mercury and Venus have no moons. The natural satellites that revolve around the planets should not be confused with artificial satellites which you will study later.

Recent discoveries have shown that Uranus has more than 5 moons Uranus has a belt of satellites containing up to as many as 18 moons. This belt of satellites travels in an orbit closer to the planet than are the 5 main moons. These satellites are smaller than the known moons.

Check Up B

B1. List six kinds of bodies found in the solar system.

(1) Sun

(2) planets

(3) moons

(4) comets

(5) asteroids

(6) meteors

B2. Where is the sun located in the solar system? How do planets travel in the solar system? The sun is the center of the solar system. The planets revolve around the sun in their orbits.

B3. Explain the difference between a planet's revolution and its rotation. Revolution refers to a planet traveling around the sun. Rotation refers to a planet turning on its axis.

B4. Name the nine planets in our solar system and give one fact about each.

(1) Mercury — smallest; closest to sun

(2) Venus — Earth's twin planet

(3) Earth — has highest forms of life

(4) Mars — red planet

(5) Jupiter — has 16 moons

(6) Saturn — has hundreds of rings

(7) Uranus — has 5 moons

(8) Neptune — has 2 moons

(9) Pluto — farthest from the sun

B5. What is a satellite? A small body that revolves around a larger body.

C. The Moon

After doing this lesson you will be able:

- C1 • to list and describe three features of the moon's surface
- C2 • to explain why only one side of the moon is seen from the earth
- C3 • to list the main phases of the moon and explain what causes them
- C4 • to describe the causes of tides

The **moon** is the earth's nearest neighbor in space. It is the earth's only natural satellite. A lot has been learned about the moon from astronauts who landed on the moon and brought back rock samples.

The moon is about 384,000 kilometers from earth. The moon is much smaller than earth with a diameter a little less than 3,500 kilometers. There is no water or atmosphere on the moon. Not having an atmosphere causes the temperature to be very hot during the day and very cold at night. Its surface is covered with rocks and thick dust. The kinds of rocks are glass type rocks, breccia (made of other rock fragments) and anorthosite. The age of lunar (moon) rock samples, brought back by the astronauts, has been found to be three to four billion years old.

The surface of the moon is made up of plains, mountains and craters. The plains, which are called lunar "seas," are broad, flat stretches of land that look like dark patches on the moon's surface. The mountain ranges are like those on earth. But scientists are not certain how the mountains were formed. The craters are holes probably caused either by collisions with meteors or volcanic explosions. C1

U.S. astronauts made the first lunar landing on the "Sea of Tranquility."

Have students compare the photographs on this page and on page 226 to the chart on top of page 227. Have them number these phases corresponding to the numbers on the chart.

8

2

(First Quarter) 3

4

Which photographs show the new moon and the first quarter?

5 (Full Moon)

6

7

1 (New Moon)

Which photographs show the full moon and the last quarter?

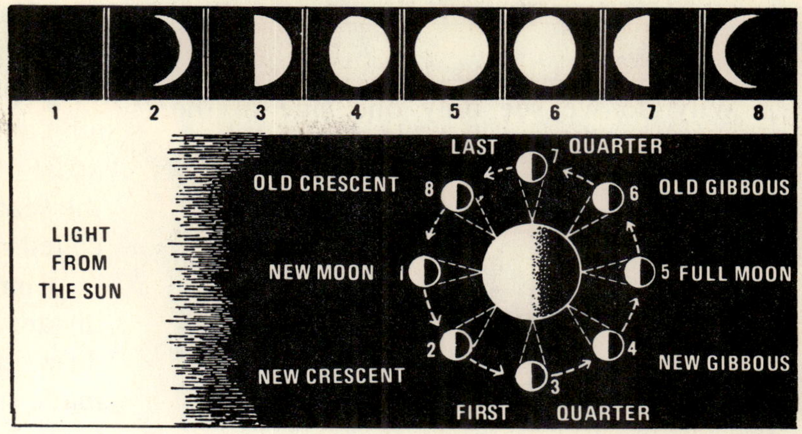

The moon's phases as seen from earth. The moon gets its light from the sun. During its revolution around the earth, only portions of its lighted half can be seen from the earth. The entire lighted half can be seen during the full moon phase.

The moon, like the earth, rotates on its axis from west to east. It takes the moon 27⅓ days to make a complete *rotation.* A complete *revolution* of the moon around the earth from west to east in its orbit also takes 27⅓ days. Because the periods of rotation and revolution are the same, we see only one side of the moon.

The moon, like the planets, has no light of its own. It reflects the light it receives from the sun. As the moon revolves around the
C3 earth, you see different amounts of its surface lighted by the sun. These different parts of the moon that you see from the earth are called the **phases** of the moon.

There are four main phases. It takes about one week from one main phase to the next main phase. The main phases are called **new moon, first quarter, full moon** and **last quarter.** During the new moon, the moon cannot be seen because the lighted side is turned away from the earth. **New crescent** occurs (happens) between the new moon and first quarter. From the earth you see only one-half of the moon. In the first quarter, only one-half of that is visible. **New gibbous** occurs between the first quarter and full moon. It looks like a full moon except for a crescent-shaped "bite" out of its side. **Old gibbous** occurs between full moon and the last quarter. Before the new moon, **old crescent** occurs.

Have you ever wondered what causes the
C4 **tides?** Tides are the lowering and rising of the

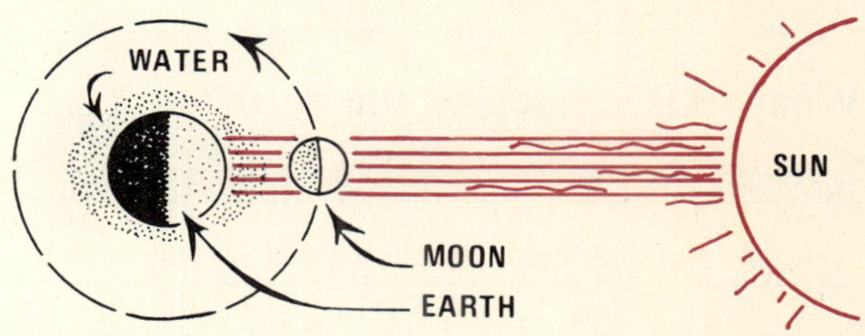

When the sun, moon and earth are in a straight line, high tides occur on both sides of the earth.

waters in the ocean. They are caused by the gravitational pull between the earth and the C3
moon. The gravitational pull between the sun and the earth also helps cause tides.

The sun and the moon both pull the waters of the earth toward them. The pull of the moon on the waters is greater than that of the sun. Do you know why? The moon is closer to the earth than the sun is. It takes a little over six hours from high tide to low tide. As the earth turns on its axis (once every 24 hours), there are two high tides and two low tides each day.

Check Up C

C1. List three features of the moon's surface and give a brief description of each.

(1) Plains — broad, flat stretches of land.

(2) Mountains — similar to those on earth.

Gibbous means *hunchbacked* or *humped*. In gibbous phases, the moon is more than half full, but not completely round in appearance.

(3) Craters — holes caused by meteors or volcanoes.

C2. Why do we see only one side of the moon? Because the periods of rotation and revolution are the same.

C3. List the main phases of the moon and explain what causes the phases. (1) New moon (2) First quarter (3) Full moon (4) Last quarter. As the moon revolves around the earth, different parts of its surface are illuminated by the sun .

C4. What causes tides on the earth? The gravitational pull between the earth and the moon, and also the gravitational pull between the sun and the earth.

D. Planetoids, Meteors, Comets

After doing this lesson you will be able:

D1 • to explain what planetoids (or asteroids) are and how they may have been formed

D2 • to describe the difference between a meteor and a meteorite

D3 • to describe what a comet is made of and how it behaves near the sun

Over 50,000 small masses of stone called **planetoids** or **asteroids** travel around the sun. Most of these small masses are found between Mars and Jupiter, and they act like
D1 planets. They revolve around the sun in their orbits and rotate on their axes. The planetoids come in many sizes from 1/100 millimeter to 750 kilometers in diameter. They
D1 are thought to be pieces of a planet that was pulled apart by Jupiter's strong gravitational force.

You are all familiar with "falling stars" or "shooting stars." These stars are not stars at
D2 all; they are **meteors.** Meteors are pieces of metal and rock that move through space at great speed. As they come into the earth's atmosphere, friction between them and the air causes them to burn and give off light. Most of them burn up before they ever reach
the earth. Those that do reach earth are D2
called **meteorites.**

Comets move toward the sun from great distances in the solar system and from all
directions. A comet is a chunk of material D3
made up of dust, ice and frozen gases. Comets have orbits that are flatter than the orbits of planets. As a comet gets close to the sun, the cloudlike haze of its head grows larger and brighter. A tail formed of fine dust and gas streams out from the side away from the sun.

A very bright comet called *Halley's Comet,* can be seen every 75 years. It is supposed to appear again in 1986.

Check Up D

D1. What are planetoids and how were they probably formed? Masses of stone found between Mars and Jupiter. They are thought to be pieces of a planet that was pulled apart by Jupiter's strong gravitational force.

D2. What is the difference between a meteor and a meteorite? Meteors are pieces of metal and rock that move through space at great speed. A meteor that reaches earth is a meteorite.

D3. What is a comet made of? What happens to it as it gets closer to the sun? A comet is made of dust, ice and frozen gases. As it nears the sun, its head grows larger and brighter; its tail streams out on the side away from the sun.

This difference is less than 40 kilometers from a true sphere. It is essentially a sphere, flattened at the poles, with a slight bulge in the southern hemisphere.

E. Planet Earth

After doing this lesson you will be able:

- E1 • to describe the shape of the earth's orbit
- E2 • to explain what rotation and revolution cause to happen on the earth
- E3 • to explain why the northern hemisphere is hotter in the summer although it is farther away from the sun
- E4 • To describe the difference between a solar eclipse and a lunar eclipse

The earth was once thought to be shaped like a ball, but newer measurements show that it is pear-shaped. The earth rotates on its axis and revolves around the sun like the other planets. The earth revolves around the sun
E1 in an **elliptical,** or oval, path or orbit. The earth is about 150 million kilometers away from the sun. It revolves from west to east around the sun. It takes the earth 365¼ days or one year to make a complete revolution. Because the orbit of the earth is elliptical, its distance from the sun is always changing.

While the earth is *revolving* around the sun, it also spins or *rotates* on its axis. It takes the earth 24 hours to make its rotation. The earth rotates from west to east.

The rotation of the earth causes day and
E2 night. As the earth rotates, part of its surface faces the sun. This part has daytime while the

The earth's orbit is oval in shape, so the earth is always at different distances from the sun.

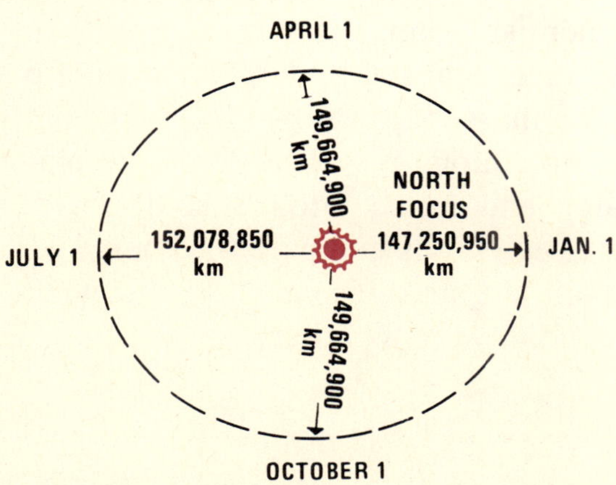

part of the earth turned away from the sun has night. The length of the days and nights on the earth are not the same because of the tilt of the earth's axis. Because of this tilt, one part of the earth is bent more toward the sun than another part. As the earth revolves around the sun, it passes through four periods E2
of time which have different temperatures. These periods are called **seasons.**

In summer, although the sun is farther away from the earth, the days in the northern hemisphere are longer and hotter. The earth is tilted toward the sun and, the straight rays E3
of the sun hit this hemisphere directly. Now, it is able to take up more of the sun's energy than it can in winter. As the earth revolves around the sun toward the winter season, the earth is tilted so that the northern hemisphere is leaning *away* from the sun. The rays of the sun can no longer shine on the earth directly in the northern hemisphere. The rays that hit this hemisphere in winter are *slanted.*

Do both northern and southern hemispheres have the same seasons at the same time? No. If the northern hemisphere is tilted toward the sun, the sun's rays strike the earth directly, and it is summer there. At the same time, the southern hemisphere is leaning away from the sun. It does not receive direct rays and it is winter there.

As the earth revolves, that part turned toward the sun changes. The earth reaches a point where the southern hemisphere is tilted toward the sun and the northern hemisphere is tilted away from the sun. Then it is summer in the southern hemisphere and winter in the northern hemisphere.

Long ago people used to think that the world was coming to an end when an **eclipse** happened. What is an eclipse? When the moon or the earth gets in the way of light from the sun and blocks off some of that light, there is an eclipse.

In a **solar eclipse,** the moon gets in the E4
way of the sunlight and throws a shadow on

Note: The earth is *farthest* from the sun on July 1 and *closest* on January 1. The tilt of the earth's axis, however, turns the Northern Hemisphere *toward* the sun in summer and *away* in winter.

Eclipses of the sun and moon—when part of the earth comes within the moon's umbra, or shadow, the sun is eclipsed. When the moon passes within the earth's umbra, the moon is eclipsed.

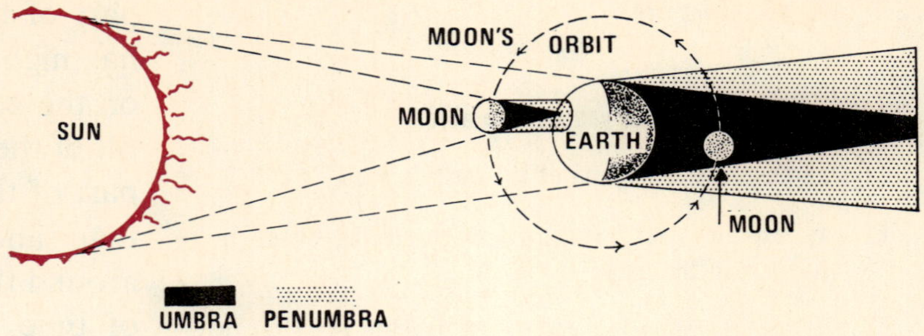

the earth. Solar eclipses take place during the new moon when the moon is between the earth and the sun. Solar eclipses happen about 2 to 5 times a year.

A **lunar eclipse** happens when the earth gets in the way of the sunlight and throws a shadow on the moon. This happens when the moon is on the opposite side of the earth from the sun. The moon is in this position when it is full.

Check Up E

E1. What shape is the earth's orbit? Elliptical

E2. What do revolution and rotation cause to happen on earth? Revolution of the earth around the sun causes the seasons. Rotation of the earth causes day and night.

E3. Why is it hotter in the northern hemisphere in summer than in the winter? The earth is tilted toward the sun and the straight rays of the sun hit the northern hemisphere.

E4. What is the difference between a solar eclipse and a lunar eclipse? In a solar eclipse, the moon casts a shadow on the earth. In a lunar eclipse, the earth casts a shadow on the moon.

THINK ABOUT THIS

Dust Storms on Mars

When the Mariner 9 spacecraft reached Mars, the planet was almost covered by a gigantic dust storm. Reports from Mariner 9 showed that the dust clouds towered as high as 70 kilometers above the surface. They swept across the planet at speeds greater than 500 kilometers per hour.

Scientists think that these violent storms on Mars are caused by large masses of polar air. The masses flow down from the Martian north pole along a valley, cross the planet's equator and move into its southern hemisphere. Scientists compare the energy of this flow of cold air to the energy of a waterfall and a tidal wave together!

Unit 4, Chapter 5
REVIEW

Matching Exercise Match each term in **Column A** with its description in **Column B**. Write the letter of the description in the space beside the correct term.

	Column A	*Column B*
e	**1.** astronomer	**a.** small body that revolves around a larger body
j	**2.** galaxy	**b.** distance that light travels in one year
a	**3.** satellite	**c.** pieces of metal and rock that travel through space and sometimes land on earth
f	**4.** planetoids	**d.** lowering and rising of the waters in the ocean
c	**5.** meteorites	**e.** scientist that studies the solar system
g	**6.** constellation	**f.** small masses of stone that orbit the sun
b	**7.** light year	**g.** group of stars that forms a pattern in the sky
h	**8.** nebulae	**h.** masses of gas and dust
d	**9.** tides	**i.** satellite of the earth
i	**10.** moon	**j.** groups of stars

True/False In the space provided, write **T** if a statement is true. If it is false, change the underlined word or words to make the statement true.

one year	**1.** A light-year is the distance that light travels in one second.
T	**2.** The color of a star is an indication of its temperature.
constellation	**3.** A group of stars that forms a pattern in the sky is a galaxy.
Galaxies	**4.** Nebulae are groups of stars.
Mercury	**5.** Earth is the smallest planet and the planet that is closest to the sun.
T	**6.** The path a planet takes around the sun is called its orbit.
year	**7.** The period of revolution of a planet is the length of a planet's day.
Pluto	**8.** Neptune is the farthest planet from Earth.
phases	**9.** The different parts of the moon that can be seen from Earth make up the constellations of the moon.
new crescent	**10.** The Old gibbous occurs between the new moon and the first quarter.
asteroids or planetoids	**11.** Small masses of stone that travel around the sun are meteors.
T	**12.** When the Earth gets in the way of sunlight and throws a shadow on the moon, a lunar eclipse occurs.
T	**13.** Ursa Major is a constellation.
sun	**14.** The center of our solar system is Earth.
elliptical	**15.** The earth has an irregular orbit.

Multiple Choice Exercise Circle the letter of the word or phrase that best completes each statement.

1. A small body that revolves around a larger body is a **(a)** satellite **(b)** meteor **(c)** constellation **(d)** galaxy.
2. The only satellite of Earth is **(a)** Ursa Major **(b)** the nebulae **(c)** the moon **(d)** Mars.
3. The planet farthest from the sun is **(a)** Jupiter **(b)** Mercury **(c)** Pluto **(d)** Neptune.
4. Small masses of stone that orbit the sun are called **(a)** satellites **(b)** planetoids **(c)** meteorites **(d)** constellations.
5. Ursa Major and Orion are **(a)** satellites **(b)** planetoids **(c)** meteorites **(d)** constellations.
6. The phase of the moon that occurs between the new moon and the first quarter is the **(a)** old gibbous **(b)** new gibbous **(c)** new crescent **(d)** old gibbous.
7. Jupiter has **(a)** one moon **(b)** hundreds of rings **(c)** 16 moons **(d)** 2 moons.
8. The masses of stone found between Jupiter and Mars are **(a)** meteorites **(b)** satellites **(c)** constellations **(d)** planetoids.
9. The surface of Venus is characterized as being **(a)** hot and wet **(b)** cold and wet **(c)** cold and dry **(d)** hot and dry.
10. The gravitational pull of the sun and the moon on the waters of Earth causes **(a)** eclipses **(b)** nebulae **(c)** tides **(d)** seasons.

Application Exercise Label each of the numbered bodies in the diagram below. Write your answers in the provided spaces.

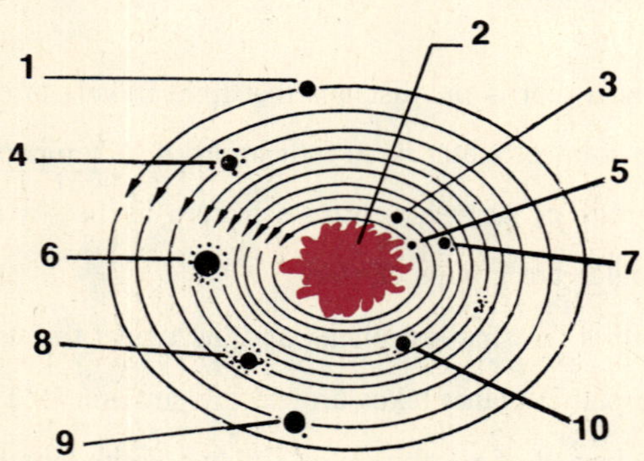

1.	Pluto	**6.**	Jupiter
2.	sun	**7.**	Earth
3.	Venus	**8.**	Saturn
4.	Uranus	**9.**	Neptune
5.	Mercury	**10.**	Mars

This is a photograph of the Andromeda Galaxy. Its shape may be the shape of our own galaxy, the Milky Way.

New science words: oxidizer, thrust, weightlessness

Skills developed: constructing a model, observing, describing, inferring

Chapter 6

SPACE EXPLORATION

People always have had a strong desire to fly. Before the airplane was invented, some persons made wings. They strapped them to their bodies and tried to fly as a bird flies. Their experiments always failed.

Then, on December 17, 1903 the Wright brothers made the first successful flight. Since then there have been great improvements not only in air travel but in space travel. Scientists learned a lot about rockets during World War II. Werner von Braun, a German scientist who came to work in the United States in 1945, and Dr. Robert Goddard, an American scientist, helped to build a rocket that could carry a spaceship around the earth. Spacecrafts that can do this are called **artificial satellites.** The moon is a **natural satellite.**

A. Getting Off Into Space

After doing this lesson you will be able:

- A1 • to explain how a rocket engine is different from a jet engine
- A2 • to name and describe the four main parts of a rocket
- A3 • to tell what escape velocity is
- A4 • to tell what "lox" is
- A5 • to explain what an oxidizer does
- A6 • to tell the difference between a manned and unmanned spacecraft

A rocket engine is different from a jet engine in that it does not depend on an outside source of oxygen to run; it carries its A1

Dr. Goddard built the first liquid-fueled rocket. Werner von Braun helped design the WWII German V-2 rocket, and the Saturn 5 moon rocket. Dr. von Braun was head of the Army Ballistics Missile Agency which later became the civilian NASA.

own oxygen supply. This is important because outer space has almost no oxygen.

A rocket is made up of four main parts: A2 (1) the **airframe,** a lightweight outer part that can stand great changes in pressure; (2) the **propellant,** a mixture containing the fuel and the oxidizer (the source of oxygen for burning the fuel); (3) the rocket **engine** or power plant; (4) the **payload** which holds the passengers, instruments and satellites, if any.

How do you get a rocket into space and out of the earth's pull of gravity? How do you get it to circle the earth in an **orbit,** or path. Do you remember that the earth's gravitational field pulls everything toward it?

If you blow up a balloon and then let it go, the balloon will go forward while the air comes out the other end. A rocket works in the same way. The fuel in the **combustion chamber** of the rocket engine is burned. The hot gases that come from the burning of the fuel are pushed out through an **exhaust nozzle** in the back of the rocket. They are pushed with such force that the rocket is shot forward. This principle is known as **action-reaction:** for each action, there is an equal and opposite reaction.

Before a rocket can leave the earth's atmosphere, it must escape the earth's strong gravitational field. A very strong force and high speed are needed for this to be done. A
A3 rocket must have a speed of about 40,000 kilometers an hour to escape the gravity of the earth. This high speed is called the **escape velocity** of the rocket.

How is a rocket made to reach escape velocity? The rocket is streamlined (smoothly shaped) to cut down friction with the atmosphere. The most important thing, however, is the rocket fuel. The kind of fuel and oxidizer control the rocket's **thrust.** Thrust is the *force* produced by the rocket's engine when it fires.

The fuel most often used in rockets is called LOX—*l*iquid *ox*ygen and alcohol.

"Escape" in this case refers to overcoming the tendency of the earth's gravitational field to attract objects toward itself.

The oxygen in the fuel is needed because there is no oxygen in space. Without oxygen, nothing can burn. A chemical which A5
is needed for the burning of the fuel is an **oxidizer.** Scientists have invented a solid fuel and are experimenting with nuclear fuels. Liquid fuel takes up more space than solid or nuclear fuels. Liquid fuel also explodes more easily.

A rocket must also be able to orbit the earth, sun or moon or other planets. Many-staged rockets are those in which each stage is a rocket by itself. They are used to get a satellite into space and into orbit around a body in space. As the fuel of each stage is used up, the stage drops off and burns up.

The part of the rocket that stays in space after the rocket engines fall off is the **satellite.** It carries the astronauts and all the instruments. If a spacecraft does not carry astronauts, it is an unmanned, or unpiloted, satel- A6
lite and is controlled by radio signals from the ground. Radio signals are also used in a piloted satellite for control in an emergency and for the astronauts to receive instructions.

Discover for yourself
how you can show how a rocket works.

PURPOSE: To show the principle of rocketry • To develop skills of constructing a model, observation, description and inference

You need: Pyrex bottle with a rubber stopper, water, thin wire, scissors, stick, Bunsen burner, matches.

Caution: Use only Pyrex glassware and two tablespoons water when conducting this experiment.

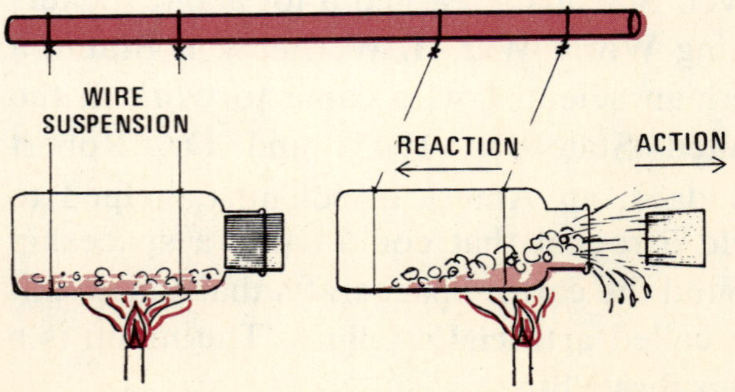

a. Put about two teaspoonfuls of water in a Pyrex bottle and put a rubber stopper on firmly but not too tightly.

b. Hang the bottle by 2 wires tied to a stick and put a lighted Bunsen burner under the bottle. See diagram.

1. After a few minutes of heating, what happened? The stopper popped out of the bottle.

2. What caused this action? Steam pressure from the heating of water.

3. What happened to the bottle? It swung backwards.

4. What principle is at work here? For each action there is an equal and opposite reaction.

Check Up A

A1. How is a rocket engine different from a jet engine? A rocket engine does not depend on an outside source of oxygen. It carries its own supply.

A2. Describe the four main parts of a rocket.

(1) Airframe — lightweight outer structure

(2) Propellant — mixture containing rocket fuel and oxidizer.

(3) Engine — power plant

(4) Payload — passengers, instruments, satellites, if any.

A3. What does the term *escape velocity* mean? The speed a rocket must have to escape the gravity of the earth.

A4. What is LOX? One of the fuels most frequently used in rockets made up of liquid oxygen and alcohol.

A5. What does an oxidizer do? Makes it possible for a fuel to burn.

A6. What is the difference between a piloted and an unpiloted spacecraft? Space vehicles which do not carry astronauts are called unpiloted.

B. Hazards to Life in Space

After doing this lesson you will be able:

B1 • to explain what causes weightlessness in space

B2 • to state two problems scientists had to think about in building spacecraft

B3 • to list three life processes that instruments in a spacecraft check on

Before sending astronauts into space there were many important things scientists had to think about. How could humans live in a small space with high temperatures, no gravity, no air, and no pressure?

A **space suit** was designed to protect the astronaut against high temperatures. This suit, as well as the spacecraft, is pressurized. If something goes wrong with the oxygen supply in the spacecraft, this suit can take over the job. When the suit is blown up, the air pressure is increased to a point the astronaut can stand. Contour seats (shaped to fit the body) were built to protect the astronauts and make them comfortable.

Scientists also had to deal with another problem — **weightlessness.** An astronaut is weightless in the spacecraft because *both* the astronaut and the spacecraft are "falling" *around* the earth. Everything in space is
B1 weightless once it overcomes the Earth's gravitational field. All objects, if not tied down, fly around. The astronauts' equipment and all objects in the capsule must be tied down. Special containers were invented to keep the food and liquid from floating around.

Teams of scientists worked together to solve problems astronauts would have to face. The astronauts were trained to deal with weightlessness and to work in a small
B2 space. There were also problems in building the spacecraft. Its outer covering had to be strong enough to take the bombardment (attack) of meteors which could puncture it. Its also had to stand up under the heat and cold of outer space.

During a flight, astronauts are carefully monitored (watched) by instruments at a "flight center." Electrical systems keep the astronauts in contact with the earth, and operate the instruments that tell how the spacecraft is working. In the space suit there are instruments that check the way the astro-
B3 naut's body is working — heartbeat, breathing, temperature.

Check Up B

B1. What causes weightlessness in space?

Once an object overcomes the earth's gravity, there is no force or weight on it.

B2. What two problems did scientists have to think about when building a spacecraft?

(1) Its covering had to withstand bombardment of meteors;

(2) Its covering had to withstand extreme heat and cold.

B3. Name three life processes that instruments in a space suit check.

(1) Heartbeat

(2) Breathing

(3) Temperature

C. Space Probes

After doing this lesson you will be able:

- C1 • to give the name of the first space satellite to be launched
- C2 • to give the name of the first American to orbit the earth
- C3 • to give the name of the first American to "walk in space"
- C4 • to give the names of the first human beings to walk on the moon
- C5 • to explain what Skylab is
- C6 • to name the spacecraft that first explored Mars
- C7 • to name the spacecraft that explored Jupiter and its moons
- C8 • to explain what the discovery of active volcanoes on, Jupiter's moon, Io means

The first satellite, Sputnik I, was launched C1
by the Soviet Union in 1957. A few months later, the United States launched a satellite, Explorer I. These were unpiloted spacecraft.

Since then all kinds of spacecraft have been successfully launched. Some have photographed the moon, Mars, Venus and Jupiter. One satellite had a digging arm which was photographed as it dug into the "sandy"

This picture of Jupiter was taken by the Voyager 1 spacecraft. On the extreme right, you can barely make out the Great Red Spot.

Voyager 1 weighs less than a ton. It carries cameras and instruments that collect and relay information to earth.

surface of the moon. In 1976, the United States bicentennial year, Viking 1 and 2 landed on Mars and carried out experiments. Other satellites are sending back to earth information about the earth's weather. Special satellites send television pictures and messages from one continent to another.

The first American astronaut to go up in a satellite was Alan Shepard. Astronaut John
C2 Glenn was the first American to orbit the earth. With James McDivitt at the controls
C3 of a Gemini capsule, Edward White became the first American to walk in space. He moved himself around in space with a compressed air gun.

Before the United States could land astronauts on the moon, scientists had to make a satellite **rendezvous** (meet) and **dock** (hitch) to a fuel-carrying rocket in space. In 1966,
C4 Neil Armstrong and David Scott were finally able to do this. On the Gemini 10 flight, Michael Collins and John Young hooked up with an Agena rocket, rode it to a new altitude, dropped away from it and docked with another rocket. The flight of Charles Conrad and Richard Gordon in Gemini 11 gave them smooth docking practice and set still another altitude record – 1360 km. Edwin Aldrin and James Lovell, in Gemini 12, had a trouble-free space walk and four successful dockings.

All of this helped prepare for the Apollo
11 landing on the moon in July, 1969. Neil C4
Armstrong and Edwin Aldrin made their trip to the moon in the lunar module *Eagle* while the command ship stayed in orbit around the moon. During the two and a half hour stay outside the module, the astronauts carried out tests on equipment and clothing. They collected samples of lunar materials and set up tests on the materials of the moon's interior.

On May 14, 1973, Skylab, an orbiting C5
laboratory, was launched unmanned into space. It was damaged during the launch and

For information and catalogs for films, filmstrips and other teaching materials available from NASA, write: National Aeronautics and Space Administration, Code AFEE-4, Washington, D.C. 20546 serious problems almost ended the project. Two weeks after the launching, astronauts Charles Conrad, Paul Weitz, and Joseph Kerwin were launched into space to correct Skylab's main problems. They spent 28 days in the space laboratory carrying out such experiments as studying the sun and the resources of the earth.

In July, 1979, Skylab fell to Earth over Australia. It had been in orbit for six years. Three different crews had spent a total of 171 days working in the space laboratory. Skylab should have been able to stay in space longer. But too many sunspots (magnetic storms on the surface of the sun) cut short its orbit. Sunspots heat the earth's atmosphere and cause it to expand. This made Skylab come down much sooner than it was meant to.

The search for life on Mars began with
C6 the Viking 1 landing on July 20, 1976. Sixty minutes after landing, the unpiloted spacecraft sent back photographs of rocks that made geologists very happy. Besides looking for some form of life, the Viking mission found out many things about the surface of Mars. The Soviet landings on Venus and the United States' landing on Mars were just the beginning of the exploration of the solar system.

C7 On September 5, 1977, Voyager 1 was launched from Earth. Voyager 2 was actually launched fifteen days earlier. However, Voyager 1 led the way and Voyager 2 checked on the findings of Voyager 1. The two spacecraft flew closer to Jupiter than any other spacecraft had ever done. Their cameras discovered a very narrow ring made up of large boulders around Jupiter's middle. On March 9, 1979, pictures of six volcanoes erupting on Io, Jupiter's second closest moon, were sent back to Earth. These pictures caused great excitement. For the first time active volcanoes were found in a place other than Earth. This means that there is some other part of our solar system C8
that is geologically alive. Remember, volcanoes are constructive (building up) forces!

Voyager 1 flew past Saturn in October of 1980. Voyager 2 flew past Saturn in August of 1981. If all goes as planned, Voyager 2 will reach Uranus in 1986 and Neptune in 1989. At some point, both probes will leave our solar system.

Check Up C

C1. Name the first spacecraft to be placed in orbit. Sputnik I

C2. Who was the first American astronaut to orbit the earth? John Glenn

C3. Who was the first American to "walk" in space? Edward White

C4. Who were the first astronauts to walk on the moon? Neil Armstrong and Edwin Aldrin

C5. What is Skylab? An orbiting space laboratory.

C6. What spacecraft was the first to land on and explore Mars? Viking 1

C7. Name the spacecraft that explored Jupiter and its moons. Voyager 1, Voyager 2

C8. Why is the discovery of active volcanoes on Io so important? It shows that some other part of our solar system is geologically alive.

Unit 4, Chapter 6
REVIEW

Matching Exercise Match each term in **Column A** with its description in **Column B**. Write the letter of the description in the space beside the correct term.

	Column A		Column B
g	**1.** Aldrin and Armstrong	**a.**	speed needed to leave the earth's gravitational field
d	**2.** John Glenn	**b.**	rocket fuel and oxidizer
i	**3.** Sputnik I	**c.**	moon of Jupiter
j	**4.** Skylab	**d.**	first American to orbit the earth
e	**5.** Viking 1	**e.**	first spacecraft to land on and explore Mars
c	**6.** Io	**f.**	outer part of a rocket
b	**7.** propellant	**g.**	first astronauts to walk on the moon
a	**8.** escape velocity	**h.**	liquid oxygen and alcohol
f	**9.** airframe	**i.**	first spacecraft to be placed into orbit
h	**10.** LOX	**j.**	an orbiting space laboratory

True/False In the space provided, write **T** if a statement is true. If it is false, change the underlined word or words to make the statement true.

alcohol	**1.** LOX is made up of liquid oxygen and water.
payload	**2.** The part of a rocket that carries the passengers and instruments is the airframe.
airframe	**3.** The lightweight outer part of a rocket is the propellant.
T	**4.** The space suit was designed to protect astronauts against high temperatures.
combustion chamber	**5.** The fuel of a rocket is burned in the exhaust nozzle.
exhaust nozzle	**6.** The hot gases that result from burning fuel in a rocket are pushed out through the airframe.
escape velocity	**7.** The speed that a rocket must reach to overcome the earth's gravity is the thrust.
orbit Earth	**8.** Sputnik was the first spacecraft to land on the moon.
T	**9.** Skylab was an orbiting space laboratory.
oxidizer	**10.** Propellant is a mixture containing rocket fuel and alcohol.
T	**11.** One difference between a rocket engine and a jet engine is that the rocket engine carries its own supply of oxygen.
An oxidizer	**12.** A propellant makes it possible for fuel to burn.
T	**13.** The kind of fuel and oxidizer a rocket uses controls its thrust.
a natural	**14.** The moon is an artificial satellite of Earth.
Viking 1	**15.** Voyager 1 was the first spacecraft to land on and explore Mars.

Multiple Choice Exercise Circle the letter of the word or phrase that best completes each statement.

1. The lightweight outer structure of a rocket is the **(a)** payload **(b)** engine **(c)** propellant **(d)** airframe.
2. The part of a rocket that carries the passengers, instruments and satellites is the **(a)** payload **(b)** engine **(c)** propellent **(d)** airframe.
3. A rocket engine is different from a jet engine because it **(a)** requires an outside source of oxygen **(b)** does not require an outside source of oxygen **(c)** runs on helium **(d)** runs on hydrogen.
4. The mixture containing the rocket fuel and oxidizer is called the **(a)** LOX **(b)** propellant **(c)** airframe **(d)** combustion chamber.
5. The speed that a rocket must reach to overcome the gravity of the earth is the **(a)** combustion chamber **(b)** thrust **(c)** escape velocity **(d)** oxidizer.
6. The first spacecraft to land on and explore Mars was **(a)** Sputnik **(b)** Voyager 1 **(c)** Voyager 2 **(d)** Viking 1.
7. The first satellite to orbit the earth was **(a)** Sputnik **(b)** Gemini **(c)** Skylab **(d)** Apollo 11.
8. Skylab was **(a)** a lunar landing module **(b)** a rocket **(c)** an orbiting laboratory **(d)** a space probe.
9. The Voyager spacecraft explored **(a)** the moon **(b)** Jupiter and its moons **(c)** Mars **(d)** Earth.
10. The fuel in a rocket is usually made of **(a)** oxygen and hydrogen **(b)** oxygen and water **(c)** water and alcohol **(d)** oxygen and alcohol.

Short Answer Answer each of the following in complete sentences.

1. What causes weightlessness in space? An object becomes weightless after it overcomes the gravity of Earth. Weightlessness is caused by the lack of a pull of gravity (force) on the object.

2. Why is a space suit important to an astronaut? A space suit protects an astronaut from high (and low) temperatures in space. Space suits are also designed to monitor the astronauts heartbeat, breathing and body temperature.

PRONUNCIATION GUIDE

When you read science material, you often come across words that are unfamiliar to you or difficult to pronounce. In this book, new or difficult science words appear in a special kind of type called **boldface** type. The Glossary/Index in the back of this book contains a list of these boldface words. Those words which are difficult to pronounce are spelled phonetically, that is, according to the way they are said. The key below will help you say these words correctly.

ă	pat	î	dear, deer, fierce,	p	pop	zh	garage, pleasure;
ā	aid, fey, pay		mere	r	roar		vision
â	air, care, wear	j	judge	s	miss, sauce, see	ə	about, silent,
ä	father	k	cat, kick, pique	sh	dish, ship		pencil, lemon,
b	bib	l	lid, needle	t	tight		circus
ch	church	m	am, man, mum	th	path, thin	ər	butter
d	deed	n	no, sudden	*th*	bathe, this		
ĕ	pet, pleasure	ng	thing	ŭ	cut, rough		
ē	be, bee, easy,	ŏ	horrible, pot	û	circle, firm, heard,		
	leisure	ō	go, hoarse, row,		term, turn, urge,		
f	fast, fife, off,		toe		word		STRESS
	phase, rough	ô	alter, caught, for,	v	cave, valve, vine		Primary stress′
g	gag		paw	w	with		**bi·ol′o·gy**
h	hat	oi	boy, noise, oil	y	yes		[bī ŏl′ə jē]
hw	which	ou	cow, out	yo͞o	abuse, use		Secondary stress′
ĭ	pit	o͝o	took	z	rose, size,		**bi′o·log′i·cal**
ī	by, guy, pie	o͞o	boot, fruit		xylophone, zebra		[bī′əlŏj′ĭkel]

GLOSSARY/INDEX

A

absorption (ăb **sôrp′**shən): the process that allows digested foods to be used by an organism, 9

air, pressure, 208-209; current, 209; mass, 210; frame, 234

algae, 14

alimentary canal (ăl′ə **mĕn′**tə rē kə **năl′**): the tube in which food is digested, 52

alpha rays or particles (**al′**fə rās; **al′**fə **pär′**tĭ kəls): a form of radioactive energy given off naturally by atoms of radium and uranium, 115

alternating current (A.C.) (**ôl′**tər nāt ĭng **kûr′**ənt): an electric current that reverses its direction many times per second, 131

alveoli (ăl **vē′**ə lī′): the tiny air sacs in the lungs, 70

amoeba (ə **mē′**bə): a very small one-celled protist without definite shape, found in water, 8-9

ampere (**ăm′**pîr′): the unit used to measure the rate of electron flow, 124

Ampere, André, 124, 130

amphibia, 17

amphibians (ăm **fib′**ē əns): animals that spend their young life in water breathing with gills, and their adult life on land breathing with lungs, 17, 35

anal spot (**ā′**nal spät): an opening in the cell membrane of the paramecium through which solid wastes are excreted, 9

anemia (ə **nē′**mē ə): a disease caused by a lack of hemoglobin in the blood, 100

anemometer (ăn′ə **mŏm′** î tər): an instrument used to measure the velocity of the wind, 213

aneroid barometer, 212-213

animal cell, 6, 7

animal kingdom, see *kingdom*

animals: classification of, 13, 14-18; multicellular, 13; kingdom, 14-18, 15, 32; hollow-bodied, 15; jointed-leg, 16; cold-blooded, 17; warm-blooded, 17, 35

ankle (**ăng′**kəl): the joint between the foot and the leg, 91

U

V

W

X

Name ______________________ **Class** __________ **Date** __________

Activity 1. Living Things Word Search

You will need: Pen or pencil

Do the following: Below are listed 12 clues. Find the word in the puzzle that matches each clue. Circle the word and write it on the line next to the clue.

1. All living things are made of it. ______________
2. Nonliving matter in a plant cell ______________
3. Living building blocks ______________
4. Control center in a cell ______________
5. Fluid-filled cell spaces ______________
6. Plant green matter ______________
7. It surrounds its food. ______________
8. It has an oral groove. ______________
9. Outside layer of a plant cell ______________
10. Paramecium's hands and feet ______________
11. Green plants ______________
12. Element in all organic matter ______________

N	V	A	C	C	E	L	S	T	U	A
O	P	R	O	T	O	P	L	A	S	M
B	V	C	E	L	L	U	L	O	S	E
R	A	C	U	O	E	S	E	M	P	B
A	C	E	S	U	E	L	C	U	N	A
C	H	L	O	R	O	P	H	Y	L	L
H	N	L	V	U	F	E	O	G	N	J
L	R	W	C	I	L	I	A	O	S	A
O	P	A	R	A	M	E	C	I	U	M
P	V	L	Y	D	R	S	S	E	L	A
N	A	L	E	N	A	M	O	E	B	A

Name ______________________ **Class** __________ **Date** __________

Activity 2. Classifying Living Things

You will need: Your textbook
Reference books about animals
Pen or pencil

Do the following:

- Select 3 of these 5 chordates.

  1. fish
  2. amphibians
  3. reptiles
  4. birds
  5. mammals

- List 5 ways in which all are alike.

____

____

____

____

____

____

____

____

- List 3 ways in which each is different.

____

____

____

____

____

____

____

____

Name ______________________ **Class** __________ **Date** __________

Activity 3. Flowering Plants

You will need: Your textbook
Pictures of flowering plants

Do the following:
- Draw a flowering plant.
- Label the following parts.

anther	filament	leaf	ovary
petal	sepal	stamen	stem
stigma	ovule	style	

Name ____________________ **Class** __________ **Date** __________

Activity 4. The Cave Dweller in Me

You will need: Your textbook
Pen or pencil
Your imagination

Do the following:

- Pretend you lived 400,000 years ago. Your home was somewhere in central Asia.
- Write a diary of a typical day in your life. Begin with waking up in the morning. You probably would not have brushed your teeth. Remember to include meal times. Decide whether you were a child or an adult. (Do not say "I wouldn't have known how to write!")

Name ______________________ **Class** __________ **Date** __________

Activity 5. Digestive System Word Search

You will need: Pen or pencil

Do the following: Find the words hidden in the puzzle below. When you find them, circle them. The words you should look for are these:

1. rennin
2. digestion
3. enzymes
4. esophagus
5. gastric juice
6. glands
7. incisors
8. intestine
9. liver
10. molars
11. mouth
12. pancreas
13. pepsin
14. ptyalin

N	O	I	T	S	E	G	I	D	E	R	E
O	S	E	M	Y	Z	N	E	O	N	T	S
G	A	S	T	R	I	C	J	U	I	C	E
S	E	O	I	E	N	A	L	B	T	T	E
R	R	P	W	N	I	D	I	P	S	U	S
O	C	H	O	N	L	Y	V	Y	E	H	D
S	N	A	N	I	A	M	E	F	T	N	N
I	A	G	T	N	Y	L	R	U	N	E	A
C	P	U	T	E	T	L	O	O	I	G	L
N	I	S	P	E	P	M	R	M	T	A	G
I	T	R	A	V	E	S	R	A	L	O	M

Name ______________________ **Class** __________ **Date** __________

Activity 6. The Circulatory System

You will need: Your textbook
Pencil or pen

Do the following:
- Pretend you are a blood cell in the left ventricle of the heart. You are about to take a trip around your body.
- Describe the path that you as a blood cell will take.
- Name at least 15 sights you will see or pass through along the way. Include the names of blood vessels, organs and heart-chambers. Be sure to put them in the order in which you will see them. Your last stop will be your starting point.

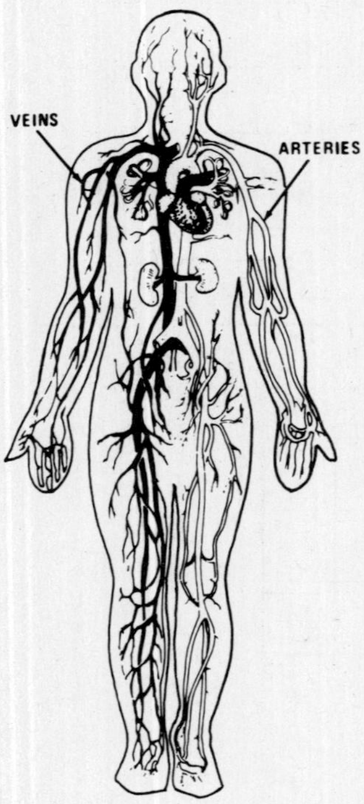

Name ______________________ **Class** __________ **Date** __________

Activity 7. Skeletal System Word Search

You will need: Pen or pencil

Do the following: Find the word in the puzzle that matches each clue. Circle the word in the puzzle and write it on the line next to the clue.

1. Supports the body ______________________
2. Kneecap ______________________
3. Bone below the knee ______________________
4. Material inside bones ______________________
5. Bone in upper leg ______________________
6. Collar bone ______________________
7. Shoulder blade ______________________
8. Brain protector ______________________
9. Back of the skull ______________________
10. Type of joint ______________________
11. Upper arm bone ______________________
12. Ties bones together ______________________
13. Ties muscles to bones ______________________
14. Voluntary muscle in arm ______________________

T	S	B	R	F	H	U	M	E	R	U	S
E	C	L	A	V	I	C	L	E	L	P	S
N	A	W	E	A	N	B	G	L	E	K	C
D	P	O	N	N	C	N	U	C	E	A	R
O	U	R	A	Y	I	K	I	L	J	I	A
N	L	R	K	H	S	B	E	O	A	C	N
S	A	A	L	L	E	T	A	P	A	I	I
F	E	M	U	R	O	R	E	N	N	T	U
I	A	O	T	N	E	M	A	G	I	L	M

Name ____________________ **Class** __________ **Date** __________

Activity 8. A Healthful Menu

You will need: Your textbook
Pen or pencil

Do the following:
- List all the foods you ate yesterday and today.

____

____

____

- What essential nutrients are missing from the foods in your list?

____

____

- Plan a menu for 3 days' meals. Make the menu foods that you like to eat. Try to include foods from all the basic groups. The menu should be healthful.

	Day #1	Day #2	Day #3
Breakfast			
Lunch			
Dinner			
Snacks (Here's your chance.)			

Can you stick to this diet? ____________________

Name ______________________ **Class** __________ **Date** __________

Activity 9. Observe a Chemical Change

You will need: Paper cup
Steel wool
Water
Pencil or pen

Do the following:
- Wet a small piece of steel wool.
- Place it in the cup.
- Allow it to stand for several hours.

What change has taken place in the steel wool?

____

Did you notice a color change? ______________________
(If you did not, wet the steel wool again and leave for several more hours.)

Did you notice a change in the texture? (Compare the feeling of the piece that you wet with a fresh piece of steel wool.)

____

How is the reddish brown substance different from the steel wool? ____

This new substance is iron oxide. It is a compound formed out of iron and oxygen. The presence of water speeds the chemical change. What is the *common* name for this new substance?

____

How can we tell that this new substance is a compound rather than a mixture? ____

____

Iron is a gray solid. Oxygen is a gas that has no color. Rust is a reddish brown solid that looks different from both iron and oxygen. A chemical change has taken place. What other chemical changes do you know of? ____

How do you know that they are chemical changes?

____

Name ______________________ **Class** ________ **Date** ________

Activity 10. Series and Parallel Circuits

You will need: Your textbook
Pencil or pen

Do the following:
- Draw wires to complete each circuit so that the bulbs will light.
- Wire each circuit to agree with the label.

Batteries in series
Bulbs in series

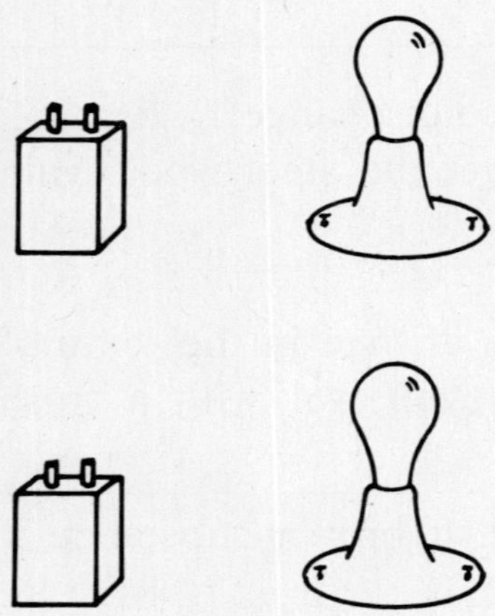

Batteries in parallel
Bulbs in parallel

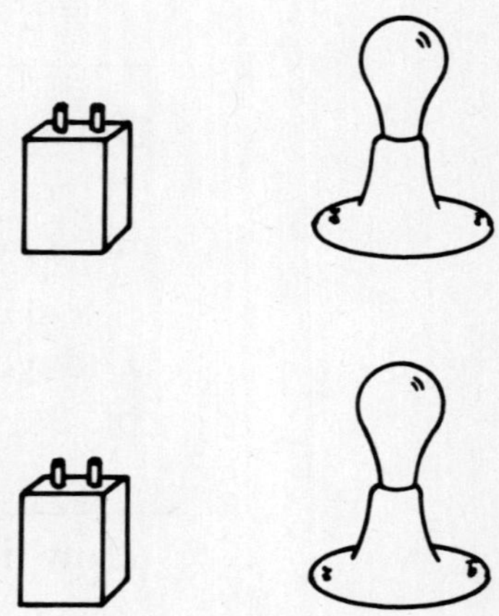

Batteries in series
Bulbs in parallel

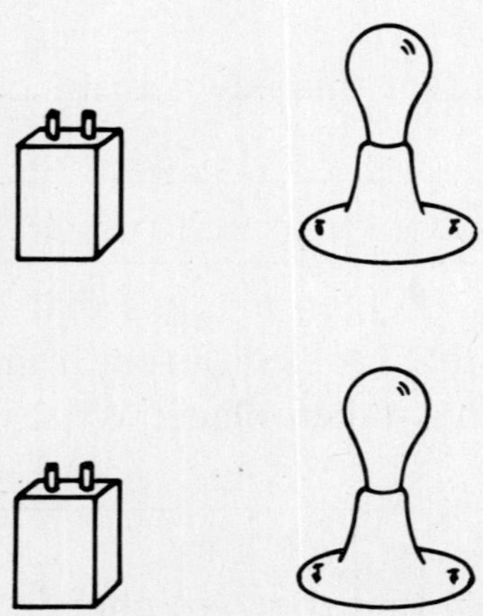

Batteries in parallel
Bulbs in series

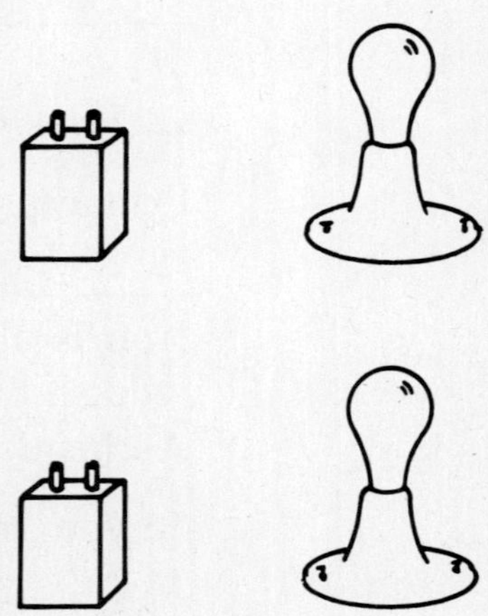

Name ____________________ Class __________ Date __________

Activity 11. Make a Psaltery

You will need: Rubber bands of different sizes
Piece of wood (about 16 cm x 30 cm x 2 cm)
6D finishing nails
Hammer
Meter ruler

Do the following:

- Drive 8 nails in a row.
- Leave at least half of each nail sticking out.
- Space the nails about 2 centimeters (cm) apart.

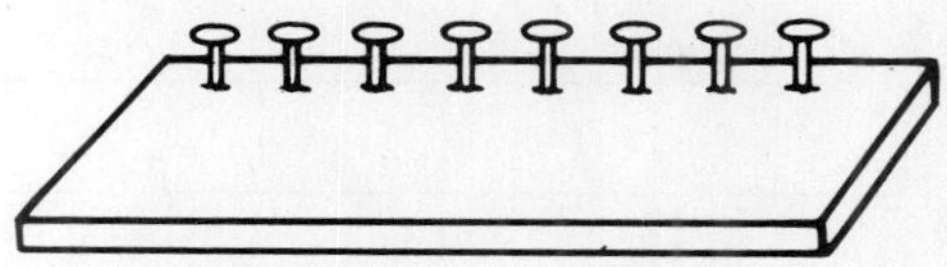

- Drive 8 nails the same way, but on the opposite edge of the board.

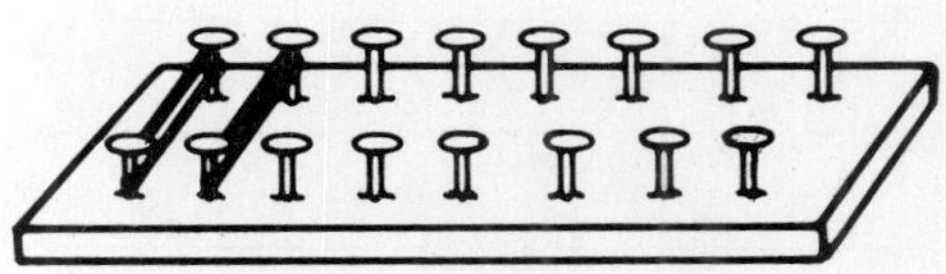

- Stretch rubber bands between pairs of nails to make a musical scale. Remember, a tightly stretched rubber band makes a higher pitched sound than a loose one. And, a thick rubber band makes a lower pitched sound than a thin one.
- When you have your instrument completed, number the strings. Make "1" the lowest note.
- Try playing this:
1-1-5-5-6-6-5; 4-4-3-3-2-2-1;
5-5-4-4-3-3-2; 5-5-4-4-3-3-2;
1-1-5-5-6-6-5; 4-4-3-3-2-2-1.

Name ______________________ Class __________ Date __________

Activity 12. Changing Energy to Electrical Energy

Generating plants do not actually create electricity. They change energy from one form to another. Some electrical generators get energy from moving water and change it to electricity. Others are powered by fossil fuels. Still others use nuclear power.

You will need: Reference materials on generating electricity
Pen or pencil

Do the following: Explain how each form of energy is changed to electrical energy in:

hydroelectric generating plants ______________________

fossil fuel generating plants ______________________

nuclear generating plants ______________________

Name ______________________ **Class** __________ **Date** __________

Activity 13. Earth and Space Science Word Search

You will need: Pen or pencil

Do the following: Find the word in the puzzle that matches each clue. Circle the word and write it on the line next to the clue.

1. A great cloud of particles and gases ______________
2. Solid rock which once may have covered the earth ______________
3. Scientist who studies the earth ______________
4. Outer surface of the earth ______________
5. Four-fifths of the earth's interior ______________
6. Center of the earth ______________
7. Instrument for recording trembling of the earth ______________
8. Shaking of the crust ______________
9. Small chunks of rocklike material ______________
10. One theory of how the earth began ______________
11. Possible explanation of things ______________
12. Earth is one ______________

E	K	A	U	Q	H	T	R	A	E	S	Q	G	S
C	D	S	T	E	N	A	L	P	I	H	U	E	E
E	O	A	N	S	E	U	Q	S	E	K	A	O	I
K	P	L	A	N	E	T	E	S	I	M	A	L	S
A	D	Y	L	A	N	H	I	U	Q	E	S	O	M
U	S	E	P	I	T	E	T	I	N	A	R	G	O
Q	U	A	N	O	S	L	S	O	M	E	N	I	G
H	U	E	P	H	T	I	X	I	V	R	L	S	R
T	B	Y	H	P	S	Z	O	D	A	O	A	T	A
R	H	M	A	L	U	B	E	N	D	C	R	Q	P
A	S	T	S	U	R	C	E	L	T	N	A	M	H

Name ______________________ **Class** __________ **Date** __________

Activity 14. Rocks and Minerals

You will need: Hand lens
Mineral and rock handbook
Strong bag

Do the following:

- Collect as many different looking rocks as you can.
- Examine each specimen with the hand lens. Can you tell what kind of rock (igneous, sedimentary, or metamorphic) each sample is? ______________________

  ____

- Separate rock samples with a layered appearance into one group. What type of rocks are these? ______________________

  ____

  ____

  Do you see tiny mineral crystals in any of the specimens?

  ____

  What type of rock do you think those are? ______________________

  Metamorphic rock tends to have larger crystals than igneous. It is also very hard. Do you have any metamorphic rocks in your collection? ______________________

  Notice the main rock formations near where you collected your rocks. Do all the rocks in your collection resemble the formations? ______________________
  If the rocks are different from the formations, where do you think they came from? ______________________

  ____

  How did they get there? ______________________

- Check the rocks you collected against a mineral handbook.
- Try to identify the crystals in the rocks.

Name ______________________ **Class** __________ **Date** __________

Activity 15. The Water Cycle

You will need: Your textbook
References about water
Pen or pencil
Your imagination

The *water cycle* is a name that scientists use. It describes how the same water is used over and over again in nature. It begins in the ocean, and may sometimes travel far away from the ocean. But it always returns to the ocean.

The water cycle begins with some ocean water evaporating. It may collect into clouds as water vapor, and fall as rain or snow over land many miles away from the ocean. It may water crops, be drunk by animals or soak into the ground. Eventually though, it will return to the ocean.

Below are two possible water cycles.

Do the following:

- Draw a water cycle where the water soaks into the ground.
- Draw a water cycle where an animal drinks the water.

Name ________________________ **Class** __________ **Date** __________

Activity 16. The Solar System

You will need: Your textbook
References on the solar system
Pen or pencil

Do the following:
- Draw a map of the solar system.
- Name each planet.
- Next to each planet list its moons (if any). Be sure to include the asteroid belt!

ANSWER KEY TO ADDITIONAL ACTIVITIES

Activity 1: Living Things Word Search

1. protoplasm
2. cellulose
3. cells
4. nucleus
5. vacuole
6. chlorophyll
7. amoeba
8. paramecium
9. cell wall
10. cilia
11. algae
12. carbon

N	V	A	C	C	E	L	S	T	U	A
O	P	R	O	T	O	P	L	A	S	M
B	V	C	E	L	L	U	L	O	S	E
R	A	C	U	O	E	S	E	M	P	B
A	C	E	S	U	E	L	C	U	N	A
C	H	L	O	R	O	P	H	Y	L	L
H	N	L	V	U	F	E	O	G	N	J
L	R	W	C	I	L	I	A	O	S	A
O	P	A	R	A	M	E	C	I	U	M
P	V	L	Y	D	R	S	S	E	L	A
N	A	L	E	N	A	M	O	E	B	A

Activity 2: Classifying Living Things

Answers will vary but should include: all are animals, all have backbones and all reproduce sexually.
Answers will vary.

Activity 3: Flowering Plants

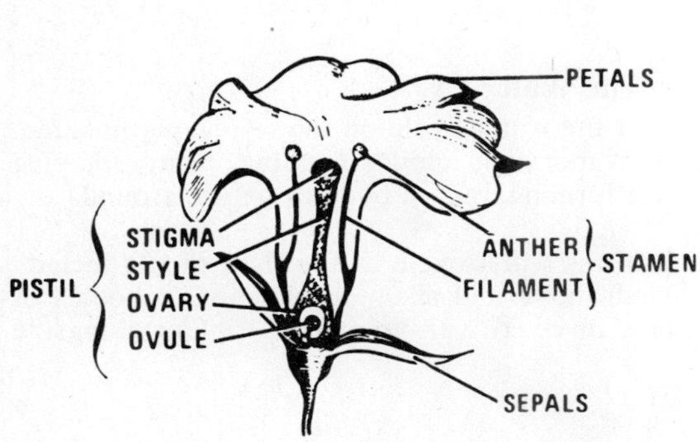

Activity 4: The Cave Dweller in Me

Answers will vary.

Activity 5: Digestive System Word Search

N	O	I	T	S	E	G	I	D	E	R	E
O	S	E	M	Y	Z	N	E	O	N	T	S
G	A	S	T	R	I	C	J	U	I	C	E
S	E	O	I	E	N	A	L	B	T	T	E
R	R	P	W	N	I	D	I	P	S	U	S
O	C	H	O	N	L	Y	V	Y	E	H	D
S	N	A	N	I	A	M	E	F	T	N	N
I	A	G	T	N	Y	L	R	U	N	E	A
C	P	U	T	E	T	L	O	O	I	G	L
N	I	S	P	E	P	M	R	M	T	A	G
I	T	R	A	V	E	S	R	A	L	O	M

Activity 6: The Circulatory System

Answers will vary but the following should be included: artery, capillary, vein, lungs, left ventricle, right ventricle, left atrium, right atrium.

Activity 7: Skeletal System Word Search

1. skeleton
2. patella
3. fibula
4. marrow
5. femur
6. clavicle
7. scapula
8. skull
9. cranium
10. hinge
11. humerus
12. ligament
13. tendon
14. biceps

T	S	B	R	F	H	U	M	E	R	U	S
E	C	L	A	V	I	C	L	E	L	P	S
N	A	W	E	A	N	B	G	L	E	K	C
D	P	O	N	N	C	N	U	C	E	A	R
O	U	R	A	Y	I	K	I	L	J	I	A
N	L	R	K	H	S	B	E	O	A	C	N
S	A	A	L	L	E	T	A	P	A	I	I
F	E	M	U	R	O	R	E	N	N	T	U
I	A	O	T	N	E	M	A	G	I	L	M

Activity 8: A Healthy Menu

Answers will vary.

Activity 9: Observe a Chemical Change

The color of the steel wool has turned reddish brown.
Yes
Yes
It is crumbly; it stains.
Rust
Rust is a compound rather than a mixture because rust cannot be separated into iron and oxygen by physical means.
Answers will vary.
A new substance is produced in a chemical change.

Activity 10: Series and Parallel Circuits

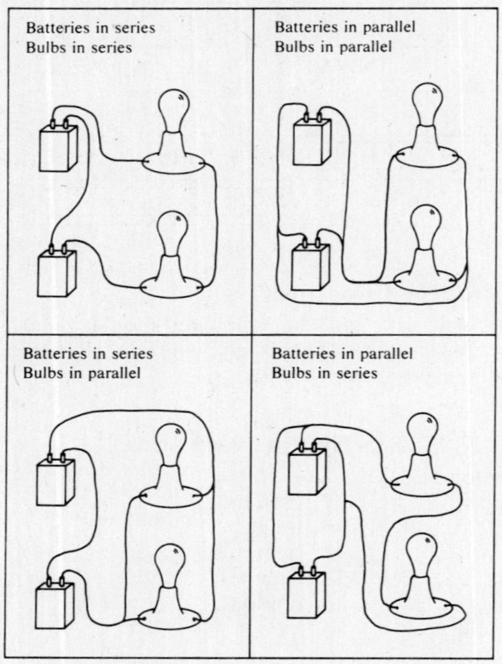

Activity 11: Make a Psaltery

No answers.

Activity 12: Changing Energy to Electrical Energy

Hydroelectric generating plants use falling water (such as waterfalls) to power the turbines which produce the electricity. This falling water is mechanical energy which is being converted to electricity.

Fossil fuel generating plants use fossil fuels instead of falling water to power the turbines in a generator. The fossil fuels are burned and this heat energy is converted to electricity.

Nuclear generating plants use the nuclear energy from the splitting of an atom to power the turbines. Nuclear energy is being converted to electricity.

Activity 13: Earth & Space Science Word Search

1. nebula
2. granite
3. geologist
4. crust
5. mantle
6. core
7. seismograph
8. earthquake
9. planetesimals
10. collision
11. hypothesis
12. planet

E	K	A	U	Q	H	T	R	A	E	S	Q	G	S
C	D	S	T	E	N	A	L	P	I	H	U	E	E
E	O	A	N	S	E	U	Q	S	E	K	A	O	I
K	P	L	A	N	E	T	E	S	I	M	A	L	S
A	D	Y	L	A	N	H	I	U	Q	E	S	O	M
U	S	E	P	I	T	E	T	I	N	A	R	G	O
Q	U	A	N	O	S	L	S	O	M	E	N	I	G
H	U	E	P	H	T	I	X	I	V	R	L	S	R
T	B	Y	H	P	S	Z	O	D	A	O	A	T	A
R	H	M	A	L	U	B	E	N	D	C	R	Q	P
A	S	T	S	U	R	C	E	L	T	N	A	M	H

Activity 14: Rocks and Minerals

Answers will vary.
Sedimentary
Answers will vary.
Igneous
Answers will vary.
Answers will vary.
Answers will vary.
Glaciers, rivers

Activity 15: The Water Cycle

The drawing of the water cycle on the left should include: ocean, water evaporating, clouds forming, clouds moving over land, rain forming, rain in puddles on the ground.

The drawing of the water cycle on the right should include the steps listed above and also: rain landing in bodies of water, i.e., lake or river, with an animal drinking alongside it.

Activity 16: The Solar System

The drawing should include: Mercury, Venus, Earth (1 moon), Mars (2 moons), Jupiter (16 moons), asteroid belt, Saturn (16 moons), Uranus (5 moons), Neptune (2 moons) and Pluto (1 moon).

Name ______________________________ Class __________ Date __________

UNIT TEST FOR UNIT 1

Matching Exercise Match each term in **Column A** with its description in **Column B**. Write the letter of the description in the space beside the correct term.

Column A	*Column B*
______ **1.** cells	**a.** thick, round structure usually found at the center of a cell
______ **2.** protoplasm	**b.** a plant that reproduces by seeds
______ **3.** nucleus	**c.** process of gradual change
______ **4.** vertebrate	**d.** earliest known human being
______ **5.** spermatophyte	**e.** building blocks of all living things
______ **6.** ovules	**f.** female sex cells in flowering plants
______ **7.** pollen grains	**g.** remains or imprints of early life
______ **8.** fossils	**h.** an animal with a backbone
______ **9.** evolution	**i.** male sex cells of flowering plants
______ **10.** Neanderthal	**j.** colorless, jellylike substance that is the basic living material

Multiple Choice Exercise Write the letter of the word or phrase that best completes each statement.

______ **1.** All of the following are found in both plant and animal cells except
(a) cell membranes **(b)** cytoplasm **(c)** vacuoles **(d)** chloroplasts

______ **2.** The waste materials that are made by a cell are stored in the
(a) nucleus **(b)** cytoplasm **(c)** vacuoles **(d)** cell wall

______ **3.** The response of living things to their environment is
(a) sensitivity **(b)** respiration **(c)** movement **(d)** circulation

______ **4.** The process in which digested food passes to different parts of the plant or animal is
(a) digestion **(b)** circulation **(c)** absorption **(d)** oxidation

______ **5.** The process by which oxygen is taken into a cell and combined with food to release energy is
(a) oxidation **(b)** respiration **(c)** digestion **(d)** absorption

______ **6.** The outermost living layer of all cells is the
(a) cell wall **(b)** cell membrane **(c)** nucleus **(d)** cytoplasm

______ **7.** The process by which green plants make their own food is
(a) digestion **(b)** photosynthesis **(c)** respiration **(d)** absorption

______ **8.** Cell reproduction is controlled by the
(a) nucleus **(b)** vacuoles **(c)** cell membrane **(d)** cytoplasm

______ **9.** A one-celled organism that does not have a well-defined nucleus belongs to the
(a) Plant Kingdom **(b)** Animal Kingdom
(c) Protist Kingdom **(d)** Moneran Kingdom

Name ______________________________ Class __________ Date __________

______ **10.** An animal whose body temperature does not change with the temperature of its environment is
(a) cold-blooded **(b)** warm-blooded
(c) an invertebrate **(d)** a bryophyte

______ **11.** Plants that have true roots, stems and leaves are classified as
(a) bryophytes **(b)** vertebrates **(c)** tracheophytes **(d)** algae

______ **12.** Mosses and liverworts are examples of
(a) spermatophytes **(b)** tracheophytes **(c)** bryophytes **(d)** algae

______ **13.** The smallest living organisms are
(a) lichens **(b)** viruses **(c)** protists **(d)** bacteria

______ **14.** An animal whose body temperature changes with the temperature of its environment is
(a) cold-blooded **(b)** warm-blooded
(c) an invertebrate **(d)** a bryophyte

______ **15.** The paramecium and ameba are examples of
(a) protists **(b)** monerans **(c)** bryophytes **(d)** algae

______ **16.** The spirogyra is an example of
(a) a protist **(b)** a moneran **(c)** a bryophyte **(d)** an alga

______ **17.** Petals play an important part in the reproductive process of flowers by
(a) producing seeds **(b)** producing leaves
(c) attracting insects to the leaves **(d)** producing fruit

______ **18.** The four layers of the woody stem are the
(a) bark, pith, cambium and palisade layer
(b) bark, pith, palisade layer and phloem
(c) phloem, xylem, palisade layer and phloem, xylem, palisade layer and pith
(d) pith, bark, cambium and wood

______ **19.** The transfer of pollen from one plant to another is
(a) germination **(b)** self-pollination
(c) cross-pollination **(d)** fertilization

______ **20.** The joining of the nucleus of a sperm cell with the nucleus of an egg cell is
(a) germination **(b)** fertilization **(c)** photosynthesis **(d)** pollination

______ **21.** Leaves get their green color from
(a) chloroplasts **(b)** chlorophyll **(c)** xylem **(d)** phloem

______ **22.** The stamens of a flowering plant are the
(a) female reproductive organs **(b)** female sex cells
(c) male reproductive organs **(d)** male sex cells

______ **23.** A plant embryo with stored food is a
(a) seed **(b)** fruit **(c)** pollen grain **(d)** cotyledon

______ **24.** The earth is now in the
(a) Precambrian Era **(b)** Cenozoic Era
(c) Mesozoic Era **(d)** Paleozoic Era

______ **25.** The most recent of the early human beings are the
(a) Java people **(b)** Peking people
(c) Neanderthals **(d)** Cro-Magnon people

Name ______________________________ Class __________ Date __________

UNIT TEST FOR UNIT 2 (CHAPTERS 1–4)

Matching Exercise Match each term in **Column A** with its description in **Column B**. Write the letter of the description in the space beside the correct term.

Column A	*Column B*
______ **1.** tissue	**a.** the largest gland in the human body
______ **2.** organ	**b.** tissues that work together to perform a special function
______ **3.** glands	**c.** sheet of muscle that lies between the chest and abdominal cavity
______ **4.** liver	**d.** carry blood away from the heart
______ **5.** veins	**e.** organs that give off enzymes
______ **6.** arteries	**f.** stops food from entering the trachea
______ **7.** capillaries	**g.** voice box
______ **8.** larynx	**h.** a group of similar cells
______ **9.** diaphragm	**i.** smallest kind of blood vessel
______ **10.** epiglottis	**j.** carry blood to the heart

Multiple Choice Exercise Write the letter of the word or phrase that best completes each statement.

______ **1.** All of the following are part of blood tissue except
(a) plasma **(b)** platelets **(c)** adipose **(d)** white blood cells

______ **2.** Messages are carried from one part of the body to other parts of the body by
(a) connective tissue **(b)** epithelial tissue
(c) blood tissue **(d)** nerve tissue

______ **3.** Protection and secretion are the two main functions of
(a) connective tissue **(b)** epithelial tissue
(c) blood tissue **(d)** cartilage and bone

______ **4.** Muscle that can be controlled is
(a) involuntary muscle **(b)** cardiac muscle
(c) epithelial muscle **(d)** voluntary muscle

______ **5.** Muscle is connected to bone by
(a) ligaments **(b)** tendons **(c)** cartilage **(d)** nerve tissue

______ **6.** Cartilage and bone are two types of
(a) fatty tissue **(b)** nerve tissue **(c)** muscle tissue **(d)** connective tissue

______ **7.** Cardiac muscle is found
(a) in the arms and legs **(b)** in the stomach
(c) only in the heart **(d)** in the heart and stomach

______ **8.** During digestion, enzymes
(a) change the chemical makeup of food
(b) change the chemical makeup of nerve tissue
(c) are secreted by the lungs
(d) control the action of the gall bladder

Name ______________________________ Class __________ Date __________

______ **9.** The life activity by which the body takes in oxygen and uses it to oxidize food and release energy is
(a) respiration **(b)** breathing **(c)** inhalation **(d)** exhalation

______ **10.** The outside of the crown of a tooth is the
(a) dentin **(b)** pulp **(c)** enamel **(c)** rennin

______ **11.** Small finger-like projections that cover the walls of the small intestine are called
(a) pulp **(b)** villi **(c)** gastric glands **(d)** dentin

______ **12.** Undigested wastes pass out of the body through the
(a) small intestine **(b)** large intestine **(c)** rectum **(d)** liver

______ **13.** The two organs that help in digestion, but are not part of the alimentary canal are the
(a) esophagus and liver **(b)** liver and pancreas
(c) esophagus and tongue **(d)** pancreas and esophagus

______ **14.** The enzyme that digests milk protein is
(a) pepsin **(b)** rennin **(c)** pytalin **(d)** chyme

______ **15.** The gall bladder is a small sac attached to the
(a) appendix **(b)** pancreas **(c)** liver **(d)** stomach

______ **16.** The mechanical process of taking air into the lungs is
(a) respiration **(b)** breathing **(c)** exhalation **(d)** digestion

______ **17.** Bleeding from a cut is stopped by
(a) red blood cells **(b)** plasma **(c)** platelets **(d)** white blood cells

______ **18.** The liquid part of the blood is
(a) hemoglobin **(b)** plasma **(c)** red blood cells **(d)** platelets

______ **19.** The red, iron-rich protein found in red blood cells is
(a) hemoglobin **(b)** platelet **(c)** oxygen **(d)** plasma

______ **20.** The function of white blood cells is to
(a) stop the bleeding from a cut
(b) fight infection and destroy bacteria
(c) carry oxygen to all the cells of the body
(d) carry wastes produced by the cells out of the body

______ **21.** Blood picks up oxygen and rids itself of waste material in the
(a) bone marrow **(b)** lungs **(c)** liver **(d)** pancreas

______ **22.** Blood is pumped to all parts of the body by the
(a) atria **(b)** ventricles **(c)** capillaries **(d)** veins

______ **23.** The valves between the atria and ventricles
(a) allow blood to flow in only one direction
(b) supply the heart with oxygen
(c) help to fight infection and destroy bacteria
(d) carry the blood out of the heart

______ **24.** The tiny air sacs in the lungs are the
(a) bronchial tubes **(b)** alveoli **(c)** bronchi **(d)** cilia

______ **25.** The nostrils and trachea are lined with tiny hair-like structures called
(a) cilia **(b)** flagella **(c)** alveoli **(d)** villi

Name ______________________________ Class __________ Date __________

UNIT TEST FOR UNIT 2 (CHAPTERS 5–8)

Matching Exercise Match each term in **Column A** with its description in **Column B**. Write the letter of the description in the space beside the correct term.

	Column A	*Column B*
______	**1.** stimuli	**a.** outer, dead layer of skin cells
______	**2.** central nervous system	**b.** used to examine the ears
______	**3.** impulses	**c.** largest bone in the body
______	**4.** epidermis	**d.** brain and spinal cord
______	**5.** dermis	**e.** used to measure blood pressure
______	**6.** vertebrae	**f.** messages that travel through the nervous system
______	**7.** femur	**g.** used to examine the eyes
______	**8.** otoscope	**h.** changes in the environment
______	**9.** opthalmoscope	**i.** inner layer of skin cells
______	**10.** pressure gauge	**j.** bones in the backbone

Multiple Choice Exercise Write the letter of the word or phrase that best completes each statement.

______ **1.** Reasoning, thought and memory take place in the
(a) medulla **(b)** spinal cord **(c)** cerebrum **(d)** cerebellum

______ **2.** Breathing and heartbeat are controlled by the
(a) medulla **(b)** spinal cord **(c)** cerebrum **(d)** cerebellum

______ **3.** Messages are carried to the nerve cell body by the
(a) axons **(b)** dendrites **(c)** cerebrum **(d)** spinal cord

______ **4.** A message that is carried along a nerve fiber is
(a) an impulse **(b)** a stimulus **(c)** an axis **(d)** a response

______ **5.** The part of the brain that is responsible for coordinating the muscles is the
(a) spinal cord **(b)** cerebellum **(c)** medulla **(d)** cerebrum

______ **6.** The opening in the eye that allows light to enter is the
(a) lens **(b)** pupil **(c)** iris **(d)** retina

______ **7.** Impulses are carried from the eye to the brain by the
(a) cochlea **(b)** Eustachian tube **(c)** auditory nerve **(d)** optic nerve

______ **8.** Nerves that stimulate the auditory nerve to carry messages to the brain are located in the
(a) eye **(b)** ear canal **(c)** Eustachian tube **(d)** cochlea

______ **9.** The life activity by which waste materials are removed from the body is
(a) perspiration **(b)** excretion **(c)** digestion **(d)** respiration

______ **10.** The liquid wastes of the body are stored in the
(a) large intestine **(b)** small intestine **(c)** ureters **(d)** urinary bladder

______ **11.** Liquid wastes are carried from the urinary bladder to the outside of the body by the
(a) pelvis **(b)** ureter **(c)** urethra **(d)** large intestine

Name ______ Class ______ Date ______

______ **12.** All of the following are organs of excretion except the
(a) small intestine **(b)** large intestine **(c)** lungs **(d)** kidneys

______ **13.** The sweat glands, oil glands, hair follicles and nerve endings are found in the
(a) epidermis **(b)** dermis **(c)** lungs **(d)** kidneys

______ **14.** The liquid waste that is excreted by the sweat glands is
(a) urine **(b)** carbon dioxide **(c)** perspiration **(d)** water

______ **15.** The tissue that makes it possible for bones to move is
(a) muscle **(b)** cartilage **(c)** blood **(d)** skin

______ **16.** Red blood cells are manufactured in the
(a) tendons **(b)** ligaments **(c)** heart **(d)** bone marrow

______ **17.** The connective tissue that connects bone to bone is called
(a) tendon **(b)** ligament **(c)** cardiac muscle **(d)** involuntary muscle

______ **18.** All of the muscles of the body that are connected to the skeleton are
(a) voluntary muscles **(b)** involuntary muscles
(c) cardiac muscles **(d)** ligaments

______ **19.** The tough, flexible connective tissue that gives shape to the ears and nose is
(a) muscle **(b)** adipose **(c)** vertebrae **(d)** cartilage

______ **20.** The type of joint found in the elbow, fingers and wrist is the
(a) patella **(b)** ball-and-socket joint **(c)** hinge joint **(d)** vertebrae

______ **21.** The right and left hip bones and the lower part of the spinal cord make up the
(a) vertebrae **(b)** shoulder girdle **(c)** pelvic girdle **(d)** humerus

______ **22.** The largest bone in the human body is the
(a) vertebrae **(b)** patella **(c)** humerus **(d)** femur

______ **23.** The disease caused by a lack of hemoglobin in the blood is
(a) anemia **(b)** leukemia **(c)** goiter **(d)** cardiac disease

______ **24.** A disease caused by a lack of vitamins is
(a) anemia **(b)** a deficiency disease
(c) dental plaque **(d)** high blood pressure

______ **25.** Scurvy develops from a lack of
(a) vitamin A **(b)** vitamin D **(c)** vitamin C **(d)** vitamin B

______ **26.** Two harmful materials that build up in the lungs of smokers are
(a) carbon monoxide and tar **(b)** carbon monoxide and nicotine
(c) tar and nicotine **(d)** carbon dioxide and tar

______ **27.** The body organ that is most likely to become damaged from drinking alcoholic beverages is the
(a) liver **(b)** brain **(c)** lung **(d)** heart

______ **28.** Fish, fish oils and dairy products are good sources of
(a) vitamin A **(b)** vitamin B_1 **(c)** vitamin D **(d)** vitamin C

______ **29.** Carbohydrates consist of the
(a) fats and oils **(b)** fats and sugars
(c) oils and starches **(d)** sugars and starches

______ **30.** Leafy green vegatables are a good source of
(a) vitamin A **(b)** vitamin B **(c)** vitamin C **(d)** vitamin D

Name ______________________ Class __________ Date __________

UNIT TEST FOR UNIT 3

Matching Exercise Match each term in **Column A** with its description in **Column B**. Write the letter of the description in the space beside the correct term.

	Column A		*Column B*
______	**1.** amu	**a.**	material through which electrons can flow
______	**2.** electron	**b.**	unit of heat
______	**3.** neutron	**c.**	the taking in of light
______	**4.** proton	**d.**	energy from the sun
______	**5.** conductor	**e.**	negatively charged subatomic particle
______	**6.** circuit	**f.**	irregular vibrations at irregular times
______	**7.** absorption	**g.**	generates electricity from solar energy
______	**8.** refraction	**h.**	neutral subatomic particle
______	**9.** opaque	**i.**	unit of temperature
______	**10.** calorie	**j.**	path of electricity
______	**11.** degree	**k.**	positively charged subatomic particle
______	**12.** echo	**l.**	the bending of light
______	**13.** noise	**m.**	unit for atomic mass
______	**14.** photovoltaic cell	**n.**	a reflected sound
______	**15.** solar energy	**o.**	material through which no light can pass

Multiple Choice Exercise Write the letter of the word or phrase that best completes each statement.

______ **1.** The smallest part of a compound that has all of the properties of the compound is
(a) an atom **(b)** an element **(c)** a proton **(d)** a molecule

______ **2.** The negatively charged particles that surround the nucleus of an atom are the
(a) protons **(b)** neutrons **(c)** electrons **(d)** amu's

______ **3.** An ordinary atom has the same number of protons
(a) and neutrons **(b)** and electrons
(c) neutrons and electrons **(d)** and amu's

______ **4.** The law of conservation of matter states that
(a) matter cannot be created or destroyed
(b) matter can be created and destroyed
(c) matter cannot be changed
(d) matter and energy are the same thing

______ **5.** The amount of space an object takes up is its
(a) mass **(b)** formula **(c)** volume **(d)** atomic number

Name ______________________ Class __________ Date __________

______ **6.** The atomic mass of an element is equal to the
(a) number of protons
(b) number of electrons
(c) number of protons and electrons
(d) number of protons and neutrons

______ **7.** The atomic number of an element is equal to the
(a) number of protons
(b) number of neutrons
(c) number of electrons and neutrons
(d) number of protons and neutrons

______ **8.** A negatively charged atom has
(a) fewer electrons than protons
(b) fewer electrons than neutrons
(c) more electrons than protons
(d) more electrons than neutrons

______ **9.** The force that pushes electricity through a circuit is
(a) amperage **(b)** resistance **(c)** voltage **(d)** ohm

______ **10.** A parallel circuit
(a) has only one path for electrons to follow
(b) has more than one path for electrons to follow
(c) has only one bulb lit at a time
(d) has no resistance

______ **11.** The unit for measuring resistance is the
(a) volt **(b)** amp **(c)** circuit **(d)** ohm

______ **12.** Materials that do not conduct electricity well are
(a) conductors **(b)** insulators **(c)** magnets **(d)** electrodes

______ **13.** A liquid through which electricity can flow is
(a) an electrode **(b)** a terminal **(c)** an electrolyte **(d)** an insulator

______ **14.** Magnetite and lodestone are examples of
(a) temporary magnets **(b)** permanent magnets
(c) electromagnets **(d)** natural magnets

______ **15.** A lens that is thicker in the center than it is at the edges is
(a) a convex lens **(b)** a concave lens **(c)** a reflector **(d)** a retina

______ **16.** Objects that give off their own light are
(a) luminous **(b)** illuminated **(c)** hyperopic **(d)** opaque

______ **17.** Materials that do not allow light to pass through them easily are
(a) transparent **(b)** opaque **(c)** hyperopic **(d)** incandescent

______ **18.** A person who is myopic is
(a) nearsighted **(b)** farsighted
(c) color blind **(d)** nearsighted and color blind

______ **19.** The type of light given off by electricity passing through gas in a tube is
(a) incandescent light **(b)** fluorescent light
(c) dispersed light **(d)** reflected light

______ **20.** A material through which some light can pass through is
(a) transparent **(b)** translucent **(c)** opaque **(d)** incandescent

Name ______________________ Class __________ Date __________

______ **21.** When white light passes through a prism, it
(a) separates into its component colors **(b)** is absorbed
(c) is reflected **(d)** is diffused

______ **22.** Heat moves through a solid by
(a) convection **(b)** conduction **(c)** radiation **(d)** an insulator

______ **23.** The heat content of food is measured in
(a) degrees **(b)** small calories **(c)** large Calories **(d)** thermostats

______ **24.** The basic unit of mass measurement is the
(a) liter **(b)** degree **(c)** gram **(d)** calorie

______ **25.** Heat moves through liquids and gases by
(a) convection **(b)** conduction **(c)** radiation **(d)** an insulator

______ **26.** Heat travels from the sun to Earth by
(a) convection **(b)** conduction **(c)** radiation **(d)** insulation

______ **27.** Most scientists measure temperature on the
(a) Fahrenheit scale **(b)** Celsius scale **(c)** Kelvin scale **(d)** Calorie scale

______ **28.** Of the following, the best conductor of heat is
(a) a solid **(b)** a liquid **(c)** a gas **(d)** a vacuum

______ **29.** Of the following, the worst conductor of sound is
(a) a solid **(b)** a liquid **(c)** a gas **(d)** a vacuum

______ **30.** The number of vibrations made by a vibrating object in one second is its
(a) pitch **(b)** loudness **(c)** amplitude **(d)** frequency

______ **31.** Radio waves are
(a) forms of nuclear energy **(b)** forms of electrical energy
(c) forms of heat energy **(d)** electromagnetic

______ **32.** The reflection of sound is
(a) an echo **(b)** an amplitude **(c)** a pitch **(d)** loudness

______ **33.** Sound is a form of energy caused by
(a) nuclear energy **(b)** vibrating objects
(c) radiant energy **(d)** electrical impulses

______ **34.** The splitting of an atom to produce two smaller atoms and a great amount of energy is
(a) fusion **(b)** fission **(c)** geothermal energy **(d)** a chemical reaction

______ **35.** The energy provided by the sun is
(a) nuclear energy **(b)** geothermal energy
(c) solar energy **(d)** heat energy

Name ______________________________ Class __________ Date __________

UNIT TEST FOR UNIT 4

Matching Exercise Match each term in **Column A** with its description in **Column B**. Write the letter of the description in the space beside the correct term.

	Column A		*Column B*
______	**1.** crust	**a.**	formed from the cementing together of rock particles
______	**2.** earthquake	**b.**	moon of Jupiter
______	**3.** seismograph	**c.**	molten rock inside the earth
______	**4.** sedimentary rock	**d.**	instrument that measures wind velocity
______	**5.** igneous rock	**e.**	large group of stars
______	**6.** bedrock	**f.**	tool that detects and records earthquakes
______	**7.** magma	**g.**	instrument that measures relative humidity
______	**8.** lava	**h.**	common fuel used in rockets
______	**9.** anemometer	**i.**	formed from the cooling and hardening of magmas
______	**10.** hygrometer	**j.**	group of stars that form a pattern in the sky
______	**11.** barometer	**k.**	molten rock on the earth's surface
______	**12.** galaxy	**l.**	layer of rock below the topsoil
______	**13.** constellation	**m.**	instrument that measures air pressure
______	**14.** Io	**n.**	outer layer of the earth
______	**15.** LOX	**o.**	shaking or trembling of the earth's crust

Multiple Choice Exercise Write the letter of the word or phrase that best completes each statement.

______ **1.** A scientist who studies the earth and what it is made of is
(a) an astronomer **(b)** a minerologist **(c)** a seismologist **(d)** a geologist

______ **2.** The tool that scientists use to detect and record earthquake waves is the
(a) seismograph **(b)** barometer **(c)** hypothesis **(d)** telescope

______ **3.** The outer layer of the earth is the
(a) core **(b)** planetisimal **(c)** crust **(d)** mantle

______ **4.** According to the latest theory of the beginning of the solar system, the sun was formed by
(a) nuclear fusion
(b) nuclear fission
(c) a near collision with a passing star
(d) a cloud of dust with extremely low temperatures

______ **5.** Possible answers to scientific questions are called
(a) facts **(b)** experiments **(c)** proofs **(d)** hypotheses

Name ______________________ Class ________ Date ________

______ **6.** A nebula is
(a) a cloud of dustlike particles and gas
(b) a large planet
(c) a small star
(d) a material made of molten iron and nickel

______ **7.** The outer core of the earth is thought to be made of
(a) granite **(b)** a solid layer of rock
(c) molten iron and nickel **(d)** solid iron and nickel

______ **8.** The color of a finely powdered mineral is the mineral's
(a) streak **(b)** luster **(c)** hardness **(d)** cleavage

______ **9.** The ability of a mineral to break apart along smooth, flat surfaces is
(a) streak **(b)** fracture **(c)** hardness **(d)** cleavage

______ **10.** The amount of time it takes for one-half of the atoms in a radioactive element to decay is the element's
(a) streak **(b)** cleavage **(c)** half-life **(d)** composition

______ **11.** The type of rock formed from the heating and cooling of magmas is the
(a) metamorphic rock **(b)** sedimentary rock **(c)** igneous rock **(d)** shale

______ **12.** The hardest known mineral is
(a) granite **(b)** talc **(c)** feldspar **(d)** diamond

______ **13.** The most abundant mineral in the earth's crust is
(a) granite **(b)** talc **(c)** oxygen **(d)** feldspar

______ **14.** Cement is an example of
(a) an igneous rock **(b)** a sedimentary rock
(c) an artificial rock **(d)** a metamorphic rock

______ **15.** Molten rock that reaches the earth's surface is
(a) metamorphic rock **(b)** coal **(c)** lava **(d)** soil

______ **16.** The process in which soil and rock are loosened and carried to another place is
(a) erosion **(b)** chemical weathering
(c) mechanical weathering **(d)** irrigation

______ **17.** The upper part of water-holding soil is
(a) groundwater **(b)** the water table **(c)** irrigation **(d)** bedrock

______ **18.** Scientists believe that the earth is between
(a) 3 and 4 million years old **(b)** 30 and 40 million years old
(c) 3 and 4 billion years old **(d)** 30 and 40 billion years old

______ **19.** Topsoil is made up of
(a) sand, clay and rock **(b)** sand, rock and humus
(c) sand, humus and lava **(d)** sand, clay and humus

______ **20.** The breaking and sliding of rock is
(a) folding **(b)** faulting **(c)** magma **(d)** erosion

______ **21.** The breaking down of rock into smaller pieces without the removal of the pieces is
(a) weathering **(b)** chemical erosion **(c)** faulting **(d)** irrigation

______ **22.** The layer of the atmosphere closest to the earth's surface is the
(a) ozone layer **(b)** troposphere **(c)** exosphere **(d)** ionosphere

Name ____________________ Class ________ Date ________

______ **23.** The instrument used to measure relative humidity is the
(a) anemometer **(b)** barometer **(c)** hygrometer **(d)** seismograph

______ **24.** An anemometer is used to measure
(a) relative humidity **(b)** wind velocity
(c) wind direction **(d)** air pressure

______ **25.** The scientific study of the weather is
(a) astronomy **(b)** minerology **(c)** geology **(d)** meteorology

______ **26.** A large amount of air that has the same temperature and humidity is
(a) a warm front **(b)** a cold front **(c)** an air mass **(d)** a stationary front

______ **27.** The ocean of air that surrounds the earth is the
(a) ionosphere **(b)** troposphere **(c)** stratosphere **(d)** atmosphere

______ **28.** Water vapor in the air is
(a) humidity **(b)** precipitation **(c)** an air mass **(d)** a sea breeze

______ **29.** The small masses of stone that orbit the sun are
(a) planetisimals **(b)** planetoids **(c)** meteorites **(d)** galaxies

______ **30.** A scientist who studies the solar system is
(a) a meteorologist **(b)** a constellation
(c) an astronomer **(d)** a seismologist

______ **31.** A group of stars that forms a pattern in the sky is a
(a) planetoid **(b)** satellite **(c)** constellation **(d)** nebula

______ **32.** Pieces of metal and rock that travel through space and sometimes land on Earth are
(a) planetoids **(b)** meteoroids **(c)** nebula **(d)** meteorites

______ **33.** A small body that revolves around a larger body is a
(a) constellation **(b)** nebula **(c)** satellite **(d)** galaxy

______ **34.** The farthest planet from Earth is
(a) Mars **(b)** Neptune **(c)** Mercury **(d)** Pluto

______ **35.** The speed a rocket must reach to leave the earth's atmosphere is its
(a) drag **(b)** thrust **(c)** escape velocity **(d)** propellant

______ **36.** The propellant of a space vehicle consists of the
(a) rocket fuel and oxidizer **(b)** combustion chamber
(c) passengers and cargo **(d)** lightweight covering of the rocket

______ **37.** The first satellite to orbit Earth was
(a) Sputnik **(b)** Viking I **(c)** Apollo 17 **(d)** Skylab

______ **38.** The Voyager spacecraft explored
(a) the moon **(b)** Earth **(c)** Jupiter and its moons **(d)** Mars

______ **39.** A rocket engine is different from a jet engine because it
(a) does not require an outside source of oxygen
(b) runs on helium
(c) runs on hydrogen
(d) requires an outside source of oxygen

______ **40.** The part of a spaceship that holds the passengers and instruments is the
(a) airframe **(b)** propellant **(c)** payload **(d)** combustion chamber

ANSWER KEY TO UNIT TESTS

Test	Answer	Chapter/ Objective	Page
UNIT 1: Matching	**1.** e	1/B1	6
	2. j	1/A3	6
	3. a	1/B4(1)	7
	4. h	2/C1	14
	5. b	2/D3	19
	6. f	3/C3	26
	7. i	3/C3	27
	8. g	4/A2	31
	9. c	4/A1	31
	10. d	4/B7	36
Unit 1: Multiple-Choice	**1.** d	1/B6	7
	2. c	1/B4(4)	7
	3. a	1/A1	5
	4. c	1/C3	9
	5. b	1/C2	8
	6. b	1/B4(3)	7
	7. b	1/C1	8
	8. a	1/B4(1)	7
	9. d	2/B	14
	10. b	2/C3	17
	11. c	2/D2	19
	12. c	2/D2	19
	13. b	2/B	14
	14. a	2/C3	17
	15. a	2/A	14
	16. d	2/A	14
	17. c	3/C1	26
	18. d	3/A3	23
	19. c	3/C4	27
	20. b	3/C5(1)	27
	21. b	3/A5	24
	22. c	3/C2	26
	23. a	3/C5(2)	27
	24. d	4/B6	36
	25. d	4/B9	36
UNIT 2 (1–4): Matching	**1.** f	1/A	45
	2. b	1/A	45
	3. e	2/A3	51
	4. a	2/D1	56
	5. j	3/C1	63
	6. d	3/C1	62
	7. i	3/C1	64
	8. g	4/B3	69
	9. c	4/A4	67
	10. f	4/B3	69
Multiple-Choice	**1.** c	1/A8	46
	2. d	1/A4	46
	3. b	1/A2	45
	4. d	1/A4	46
	5. b	1/A6	46
	6. d	1/A5	46
	7. c	1/A4	46
	8. a	2/A2	51
	9. a	4/A1	67
	10. c	2/B4	53
	11. b	2/C1	55
	12. c	2/B11	54
	13. b	2/D1	56
	14. b	2/B8	53
	15. c	2/D2	56
	16. b	4/A2	67
	17. c	3/B1	61
	18. b	3/B1	61
	19. a	3/B2	61
	20. b	3/B1	61
	21. b	3/A5	60
	22. b	3/A4	60
	23. a	3/A4	60
	24. b	4/B5	70
	25. a	4/B2	69
UNIT 2 (5–8): Matching	**1.** h	5/B5	74
	2. d	5/B6	74
	3. f	5/B3	85
	4. a	6/B2	85
	5. i	6/B1	85
	6. j	7/B3	91
	7. c	7/B7	91
	8. b	8/A1(4)	98
	9. g	8/A1(5)	98
	10. e	8/A1(3)	98
Multiple-Choice	**1.** c	5/C4	75
	2. a	5/C4	75
	3. b	5/B2	74
	4. a	5/B3	74
	5. b	5/C4	75
	6. b	5/D2	77
	7. d	5/D2	77
	8. d	5/D6	78
	9. b	6/A1	83
	10. d	6/A5	84
	11. c	6/A5	84
	12. a	6/A2	83
	13. b	6/B1	85
	14. c	6/B5	85
	15. a	7/E1	94
	16. d	7/A	89
	17. b	7/z zD1	93
	18. a	7/E2	94
	19. d	7/B3	91
	20. c	7/C2	92
	21. c	7/B6	91
	22. d	7/B7	91
	23. a	8/B2	100
	24. b	8/B2	100
	25. c	8/B2(5)	100
	26. c	8/C2	102
	27. a	8/D4	102
	28. c	8/B3	100
	29. d	8/B1	99
	30. c	8/B2(5)	100

Test	Answer	Chapter/ Objective	Page
UNIT 3: Matching	**1.** m	1/B2	111
	2. e	1/A2	110
	3. h	1/A2	110
	4. k	2/A1	124
	5. a	2/B5	126
	6. j	2/B1	124
	7. c	3/C3	138
	8. l	3/C3	139
	9. o	3/C2	137
	10. b	4/A3	146
	11. i	4/B1	146
	12. n	5/C4	158
	13. f	5/D1	158
	14. g	6/C4	169
	15. d	6/C3	169
Multiple-Choice	**1.** d	1/D3	113
	2. c	1/A2	110
	3. b	1/A2	110
	4. a	1/D5	114
	5. c	1/A	109
	6. d	1/B2	111
	7. a	1/B2	111
	8. c	2/A2	122
	9. c	2/B4	124
	10. b	2/B7	127
	11. d	2/B6	126
	12. b	2/B5	126
	13. c	2/B3	124
	14. d	2/C2	129
	15. a	3/E3	141
	16. a	3/C3	138
	17. b	3/C2	137
	18. a	3/E6	141
	19. b	3/B	136
	20. b	3/C2	137
	21. a	3/D2	140
	22. b	4/D2	149
	23. c	4/A4	146
	24. c	4/A3	146
	25. a	4/D4	150
	26. c	4/D5	150
	27. b	4/B6	147
	28. a	4/D3	150
	29. d	5/B2	156
	30. d	5/A3	156
	31. d	5/E3	161
	32. a	5/C4	158
	33. b	5/A1	155
	34. b	6/C1	169
	35. c	6/C3	169

Test	Answer	Chapter/ Objective	Page
UNIT 4: Matching	**1.** n	1/B4	179
	2. o	1/C1	180
	3. f	1/C2	180
	4. a	2/B2	187
	5. i	2/B2	187
	6. l	3/D1	203
	7. c	3/C6	202
	8. k	3/C6	202
	9. d	4/C2	213
	10. g	4/C2(6)	214
	11. m	4/C2(1)	212
	12. e	5/A4	220
	13. j	5/A7	220
	14. b	6/C7	238
	15. h	6/A3	234
Multiple-Choice	**1.** d	1/B3	179
	2. a	1/C1	180
	3. c	1/B4	179
	4. a	1/A2	178
	5. d	1/A1	177
	6. a	1/A2	177
	7. c	1/B6	179
	8. a	2/A3	184
	9. d	2/A3	185
	10. c	2/C2	194
	11. c	2/B1	187
	12. d	2/A4	184
	13. d	2/A5	185
	14. c	2/B6	190
	15. c	3/C6	202
	16. a	3/B5	199
	17. b	3/D3	204
	18. c	3/A	197
	19. d	3/D1	203
	20. b	3/C3	202
	21. a	3/B1	198
	22. b	4/A	207
	23. c	4/C2(6)	214
	24. b	4/C2(3)	213
	25. d	4/C1	212
	26. c	4/B5	210
	27. d	4/A	207
	28. a	4/B5	210
	29. b	5/D1	228
	30. c	5/A	219
	31. c	5/A7	220
	32. d	5/D2	228
	33. c	5/B5	223
	34. d	5/B4	223
	35. c	6/A3	234
	36. a	6/A2	234
	37. a	6/C1	236
	38. c	6/C7	238
	39. a	6/A5	234
	40. c	6/A2	234